Carbon Nanotubes and Nanosheets for Sustainable Solutions

Carbon Nanotubes and Nanosheets for Sustainable Solutions

Guest Editor

Muralidharan Paramsothy

MDPI

Basel • Beijing • Wuhan • Barcelona • Belgrade • Novi Sad • Cluj • Manchester

Guest Editor
Muralidharan Paramsothy
NanoWorld Innovations
(NWI), Singapore
Singapore

Editorial Office
MDPI AG
Grosspeteranlage 5
4052 Basel, Switzerland

This is a reprint of the Special Issue, published open access by the journal *Nanomaterials* (ISSN 2079-4991), freely accessible at: www.mdpi.com/journal/nanomaterials/special_issues/GXDA8MFOB0.

For citation purposes, cite each article independently as indicated on the article page online and using the guide below:

Lastname, A.A.; Lastname, B.B. Article Title. *Journal Name* **Year**, *Volume Number*, Page Range.

ISBN 978-3-7258-3118-0 (Hbk)
ISBN 978-3-7258-3117-3 (PDF)
https://doi.org/10.3390/books978-3-7258-3117-3

Cover image courtesy of Muralidharan Paramsothy

Contents

About the Editor

Muralidharan Paramsothy

Dr. Muralidharan Paramsothy is an early-retired scientist based in Singapore. Before retiring by choice, he was active for almost 5 years in supporting condition monitoring as well as intelligent energy projects in the Agency for Science, Technology and Research (A*STAR). Prior to that, he was active for almost 10 years in contributing to nanomaterials and nano(science)technology endeavors in Singapore. Overall, he has extensively managed academic as well as industrial stakeholders. He has numerous internationally published peer-reviewed articles, the Nanoscale Electro Negative Interface Density or NENID theory, and intellectual property pertaining to optical fiber sensor utilization to his name. He has organized a fair number of academic symposia in the USA and Canada. He has also served more flexibly as a Guest Editor in themed topics in journals, as well as held less-flexible, traditional roles on journal editorial boards. He is driven by work that benefits everybody, not that which merely promotes the individual. Turning 50 in 2025, the Guest Editor lovingly dedicates the current as well as immediately preceding Special Issues to his late father, Paramsothy Subramaniam, who taught Physiology at the National University of Singapore for about three decades before passing on peacefully in 1994.

Preface

This Special Issue is a continuation of the preceding Special Issue titled "70th Year Anniversary of Carbon Nanotube Discovery - Focus on Real World Solutions". For a complete reference, the Editorial of the current Special Issue is listed as Paramsothy, M. Carbon Nanotubes and Nanosheets for Sustainable Solutions. *Nanomaterials* 2024, 14, 1649. Turning 50 in 2025, the Guest Editor lovingly dedicates both Special Issues to his late father, Paramsothy Subramaniam, who taught Physiology at the National University of Singapore for about three decades before passing on peacefully in 1994. Twelve representative articles have been published in this Special Issue. The first eight articles are on currently impactful applications utilizing CNTs and/or graphene:

Biomedical Application: Seratonin Nanosensor;

Agriculture Application: Seed Priming;

Energy Applications: Superconductivity & Supercapacitor;

Thermal Application: Nanocomposite for Road De-Icing;

Structural Application: Vibration and Damping.

The last four articles are on the synthesis and fundamental nanoscale aspects of CNTs and/or graphene:

Synthesis and Structure: Nanofiber Growth on Nanowire;

Electronic Structure: Pseudopotential Method;

Crystallographic Structure: Defect Study & Engineering.

Including mixed country authorship, this Special Issue continuation is also of a truly global nature. In addition, the Guest Editor began publishing critical work on the CNT–polymer interface towards the mechanical properties of the nanocomposite up to just over two decades ago.

Muralidharan Paramsothy
Guest Editor

nanomaterials

Communication

Directed Evolution of Near-Infrared Serotonin Nanosensors with Machine Learning-Based Screening

Seonghyeon An [1,†], Yeongjoo Suh [1,†], Payam Kelich [2], Dakyeon Lee [1,3], Lela Vukovic [2] and Sanghwa Jeong [1,*]

[1] Department of Biomedical Convergence Engineering, Pusan National University, Yangsan 50612, Republic of Korea
[2] Department of Chemistry and Biochemistry, University of Texas at El Paso, El Paso, TX 79968, USA
[3] Department of Chemistry, Pohang University of Science and Technology (POSTECH), Pohang 37673, Republic of Korea
* Correspondence: sanghwa.jeong@pusan.ac.kr; Tel.: +82-51-510-8539
† These authors contributed equally to this work.

Abstract: In this study, we employed a novel approach to improve the serotonin-responsive ssDNA-wrapped single-walled carbon nanotube (ssDNA-SWCNT) nanosensors, combining directed evolution and machine learning-based prediction. Our iterative optimization process is aimed at the sensitivity and selectivity of ssDNA-SWCNT nanosensors. In the three rounds for higher serotonin sensitivity, we substantially improved sensitivity, achieving a remarkable 2.5-fold enhancement in fluorescence response compared to the original sequence. Following this, we directed our efforts towards selectivity for serotonin over dopamine in the two rounds. Despite the structural similarity between these neurotransmitters, we achieved a 1.6-fold increase in selectivity. This innovative methodology, offering high-throughput screening of mutated sequences, marks a significant advancement in biosensor development. The top-performing nanosensors, N2-1 (sensitivity) and L1-14 (selectivity) present promising reference sequences for future studies involving serotonin detection.

Keywords: directed evolution; carbon nanotube; serotonin; nanosensor; machine learning; fluorescence; near-infrared

Citation: An, S.; Suh, Y.; Kelich, P.; Lee, D.; Vukovic, L.; Jeong, S. Directed Evolution of Near-Infrared Serotonin Nanosensors with Machine Learning-Based Screening. *Nanomaterials* **2024**, *14*, 247. https://doi.org/10.3390/nano14030247

Academic Editor: Domenica Tonelli

Received: 6 January 2024
Revised: 20 January 2024
Accepted: 22 January 2024
Published: 23 January 2024

1. Introduction

Optical biosensors are techniques that enable the detection of biological and chemical substances [1] and have many advantages over conventional analytical methods, such as being highly sensitive, specific, and cost-effective [2,3]. Carbon nanotubes, a novel nanomaterial used in these biosensors, are promising building blocks for biosensor applications with unique electronic and optical properties [4]. Among them, single-walled carbon nanotubes (SWCNTs) are 1D nanomaterials with a graphene layer rolled into a tube [5], which has a high surface-to-volume ratio, and is advantageous for surface functionalization [6]. SWCNTs are functionalized with single-stranded DNA (ssDNA), a promising strategy among various functionalizations in biotechnology, and can be dispersed in an aqueous phase and emit fluorescence in the near-infrared region. Especially, the second near-infrared (nIR-II) fluorescence of SWCNT, whose wavelength is around 1000–1700 nm, is useful for detection and imaging as it exhibits very low background and high penetration depths into biological tissues [7]. Most organic fluorophores suffer from photobleaching under high and long illumination. However, even at high fluences, SWCNTs show continuous photostability in fluorescence emission, which is adequate for long-life sensors [8]. The diverse chirality of SWCNTs facilitates high throughput screening and hyperspectral imaging and can be utilized as optical sensors for human disease biomarkers as their different chiral properties in the same dispersion solution allow them to react differently to target analytes [9].

We sought to utilize these SWCNT biosensors to examine their fluorescence response to neurotransmitters, specifically serotonin (5-hydroxytryptamine, 5HT). 5HT is a biomolecule that acts as a neurotransmitter and is known to be associated with depression, mania, and anxiety disorders [10–12]. To efficiently sense 5HT, SWCNTs can be functionalized with a specific ssDNA sequence. The ssDNA-wrapped SWCNT complexes formed should ideally recognize only the specific molecules and report their detection optically, via fluorescence changes in response to the target substance [13,14].

To develop the novel 5HT optical nanosensors with high sensitivity and selectivity, we applied the directed evolution protocol to improve the sensor performance of certain ssDNA-SWCNT complexes. Directed evolution is a method mainly utilized in protein engineering that mimics the process of natural selection to manipulate proteins or nucleic acids toward a user-defined goal [15–17]. We started with the 5HT-responsive sequence found in our previous report as the original sequence [13], then applied the directed evolution techniques to develop higher-performance ssDNA sequences for 5HT nanosensors [18] (Scheme 1). Using a similar method, Lambert et al. also applied directed evolution to find the dopamine (DA)-sensitive SWCNT nanosensor [18]; the team improved the sensor properties of the DNA-wrapped SWCNT through DNA mutations. They used tens of ssDNAs randomly mutated from the original sequence to screen the fluorescence response of the ssDNA-SWCNT complexes upon the addition of DA. Several mutants in each cycle exhibited fluorescence enhancement compared to the initial ssDNA-SWCNT complex and the best mutated sequence was selected.

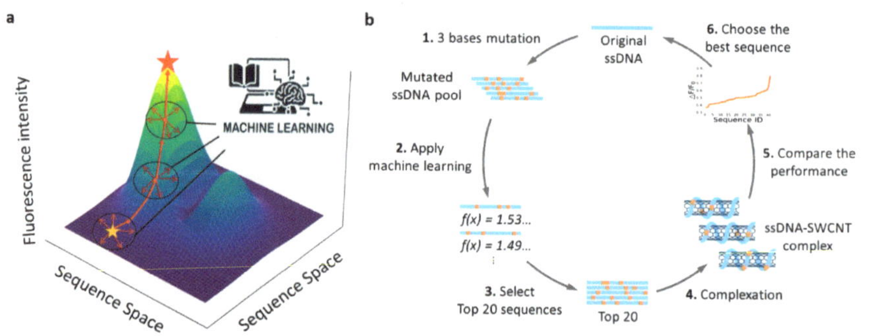

Scheme 1. (**a**) Directed evolution with the help of ML model. Application of ML model during directed evolution to predict the sequences that could show better sensor response. (**b**) Protocol of directed evolution of serotonin (5-hydroxytryptamine, 5HT)-responsive ssDNA-SWCNT nanosensors protocol with the help of ML-based screening.

Following the previous research, we did the modified directed evolution with 3-base mutation for a total of 3 rounds via machine learning (ML) prediction. The total number of 3-base mutated ssDNA sequences is very large, up to 22,032 sequences per round; it is impractical to test all of them. Therefore, we applied an ML model that predicts 5HT-specific sequences from a previous paper to determine the $\Delta F/F_0$ (increase in fluorescence intensity of 5HT over DI water) results at 1195 nm for each sequence and performed experiments on the top 20 sequences [19]. In this paper, we repeated the evolution process of finding sequences that are responsive to 5HT in terms of sensitivity and selectivity. In this approach, the combination strategy of directed evolution and ML model prediction for fluorescence measurements using ssDNA-SWCNTs offers great potential in terms of time and cost-effectiveness and high throughput in finding the most sensitive sequences for target substances [20].

2. Materials and Methods

2.1. Materials

Super-purified HiPCo SWCNTs (Batch #HS37-027) were purchased from NanoIntegris (Boisbriand, QC, Canada). DI water, serotonin hydrochloride, dopamine hydrochloride, acetylcholine chloride, γ-aminobutyric acid (GABA), glutamate (Glu), uric acid (UA), and ascorbic acid (AA) were purchased from Sigma-Aldrich(St. Louis, MO, USA). All the DNA molecules were purchased and used without further purification from Integrated DNA Technologies (IDT, Clarville, IA, USA).

2.2. Fabrication of ssDNA-SWCNT Nanosensor

The nanosensor dispersion was fabricated as reported previously in the literature [13]. The mixture solution was prepared by adding 1.0 mg ssDNA and 1.0 mg SWCNT to a 2-mL e-tube and filling the remaining volume with 1× PBS to a total volume of 1 mL. The mixture was bath-sonicated for 3 min to ensure that the SWCNTs were well dispersed in the e-tube. The microtube was then tip-sonicated in an ice bath at 50% amplitude for 20 min (VCX-130, SONICS & Materials, Inc., Newtown, CT, USA). After sonication, the microtubes were centrifuged at 21,000× g for 1 h, and 800 μL of the supernatant was collected out of a total volume of 1 mL to remove sediment. The final ssDNA-SWCNT suspension was diluted 100 times (10 μL of the sensor to 990 μL of DI water) to measure the absorbance. The concentration of the ssDNA-SWCNT complex was calculated by measuring its absorbance at 632 nm (extinction coefficient = 0.036 mg·L^{-1} cm^{-1}) [21].

2.3. Fluorescence Imaging of ssDNA-SWCNT Nanosensors Immobilized on a Glass Substrate

A glass coverslip (thickness = 0.13 mm, Ø18) is soaked in 10% APTES in anhydrous ethanol for 10 min. Then the coverslip was rinsed with DI water and blown with N2 gas. The APTES-coated glass coverslip was immersed in 50 mg/L of SWCNT and incubated at 4 °C for 30 min. After incubation, the coverslip was softly rinsed with DI water. The coverslip was then fixed in the imaging chamber (Chamlide-AC, Live Cell Instrument, Seoul, Republic of Korea). The surface-immobilized nanosensors were imaged on an inverted fluorescence microscope with a 50× objective lens by 721 nm laser excitation [13]. The fluorescence image was captured with a NINOX 640 camera (Raptor Photonics, Larne, UK) through a 950-nm long pass filter. Before the 5HT addition, the imaging chamber was filled with 990 μL of 1× PBS. After 30 s, 10 μL of 10 mM 5HT was injected into the imaging chamber for a final 5HT concentration of 100 μM. Fluorescence images were captured at 0.5 s intervals with exposure time = 500 ms.

2.4. Fluorescence Measurement

The ssDNA-SWCNT complex suspension was diluted to 10 mg/L by 1× PBS before the fluorescence measurement. 5HT is the main substance used in terms of sensitivity, and DA, acetylcholine (Ach), GABA, Glu, UA, and AA were used to compare the fluorescence intensity to 5HT in terms of selectivity. Analytes were dissolved in DI water to make a stock solution, with concentrations of 1 mM DA, 5HT, Ach, GABA, Glu, AA, and 100 μM UA. Because of the low solubility of UA, 100 μM UA stock solution was used. 50 μL of each stock solution was added to 450 μL of sensor solution for a final analyte concentration of 100 μM except 10 μM for UA. 50 μL of DI water was added to the sensor solution for measuring baseline fluorescence. The fluorescence was measured after 20 min incubation and the change in the fluorescence spectrum was measured. A 721-nm laser (PSU-H-LED laser, CNI laser, Changchun, China) was used as an excitation light source, and detection was performed using an InGaAs photodiode array detector (NIRQuest, Ocean Insight, Orlando, FL, USA). Fluorescence measurements were performed in triplicate, and the average value was calculated and used for data analysis.

2.5. Directed Evolution with Machine Learning Screening

To find sensors that are highly sensitive and selective to 5HT, we used a directed evolution approach. First, the initial 30-mer ssDNA sequence 5′-CCCCCCAGCCCTTCACCACC-AACTCCCCCCCC-3′, which is known to have a fluorescent response to 5HT [13], was used as the original sequence. Three bases in the center 18 bases were randomly mutated to the one of A, G, C, or T except for two fixed 6-mer $(C)_6$ on each flank to obtain a diverse DNA pool. This yielded a total of 22,032 sequences. For ML, we applied multiple ML regression models to obtain the ensemble-predicted $\Delta F/F_0$ values, as reported in Ref [19]. This ensemble model method predicts $\Delta F/F_0$ values for DNA-SWCNT complexes with the defined DNA sequences. We then ranked the sequences according to the responsiveness to 5HT and ensured that an ensemble of ML classification models based on convolutional neural networks (developed and validated in Ref [19]) also classified the highest-ranked sequences as high response sequences. The ensemble model produced the multiple predicted $\Delta F/F_0$ for 5HT, and we calculated the mean value as the indicator (mean_r). The top 20 sequences with the highest predicted $\Delta F/F_0$ were chosen for experimental characterization [19]. We experimentally measured the change in fluorescence in response to 5HT and DA. Among the 20 sensors, the best-performing sequence was selected and used as the starting sequence for the next round. The sensor performance was compared in terms of sensitivity and selectivity, and each performance parameter was calculated based on the fluorescence peak at 1202 nm. Depending on the performance, we aimed for a total of three rounds.

3. Results

3.1. Directed Evolution with Machine Learning Prediction

To fabricate a sensor with better performance for 5HT, we applied directed evolution technology for the optimal ssDNA-SWCNT complex sensor. Directed evolution is a technique used in protein engineering that mimics the process of natural selection to guide the evolution of high-affinity proteins or nucleic acids to the target molecules [21]. Based on a previous paper that applied this technology to ssDNA, we concurrently mutate and test the ssDNA sequences with SWCNT to find a sequence that is more sensitive in response to 5HT (Scheme 1). Using the original sequence 5′-CCCCCCAGCCCTTCACCACCAACTCCCCCC-3′ [13], which is known to respond to 5HT in the previous paper [19], we mutated 3 bases from the original sequence to produce 22,032 sequences, as the first mutated DNA library (Figure 1a). With this mutated library, we applied the ensemble ML model from our previous work [19] to predict the fluorescence response of each ssDNA-SWCNT complex to 5HT. This ensemble ML model consists of 2 regression models based on support vector machine approach using radial basis function (RBF) and sigmoid kernels and 9 convolutional neural network classification models. This ensemble ML model (regression) produces the min, max, and mean values of predicted fluorescence change of ssDNA-SWCNT complex to 5HT as $\Delta F/F_0$ ($\Delta F/F_0 = (F - F_0)/F_0$, where F is the fluorescence intensity for the 5HT and F_0 is the baseline fluorescence intensity before 5HT addition). Among the predicted $\Delta F/F_0$ and corresponding sequence, the top 20 sequences with the highest $\Delta F/F_0$ were selected by using the mean value as an indicator (Figure 1b,c). The same sequences were also checked to be high response to 5HT using the classification models. We experimentally fabricated ssDNA-SWCNT complexes with 20 new ssDNA sequences and measured their $\Delta F/F_0$ to find out the best-performing sequence. This best sequence is used as the starting sequence for the next round of directed evolution.

a

Sequence ID	DNA sequence (from 5′ to 3′)
Original	CCCCCCAGCCCTTCACCACCAACTCCCCCC
1	CCCCCCGGCCCTTGACCGCCAACTCCCCCC
2	CCCCCCAACCCTGCACCACCATCTCCCCCC
3	CCCCCCAGCCCTTCACCACCATACCCCCCC
4	CCCCCCAGCCCGTAACCACCAATTCCCCCC
5	CCCCCCTGCCCTACTCCACCAACTCCCCCC
⋮	⋮
22,032	CCCCCCAGTCCTTCGCAACCAACTCCCCCC

b

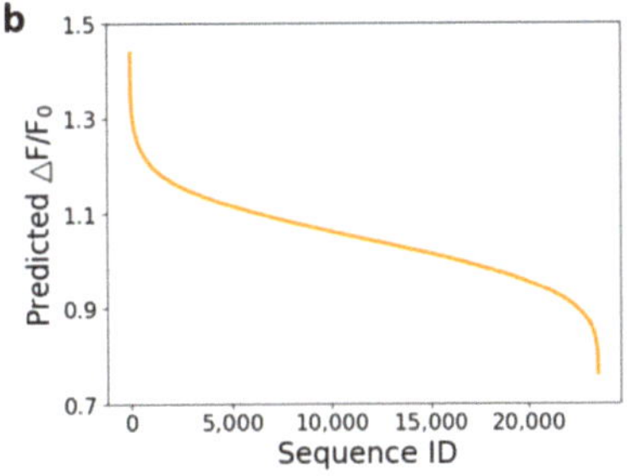

c

Sequence ID	DNA sequence (from 5′ to 3′)	Predicted $\Delta F/F_0$	Sequence ID	DNA sequence (from 5′ to 3′)	Predicted $\Delta F/F_0$
1	CCCCCCAGCCCAACACAACCAACTCCCCCC	1.533	11	CCCCCCAGCCCAACACCACCAGCTCCCCCC	1.456
2	CCCCCCAGCCCTACACAACCAGCTCCCCCC	1.491	12	CCCCCCAGCCCGACACAACCAACTCCCCCC	1.451
3	CCCCCCAGCCCTACACAACCACCTCCCCCC	1.480	13	CCCCCCAGCCCTACACAACCAACTCCCCCC	1.449
4	CCCCCCAGCCCTACACAACCAACGCCCCCC	1.479	14	CCCCCCAGCACTACACAACCAACTCCCCCC	1.448
5	CCCCCCAGCCCTACACATCCAACTCCCCCC	1.469	15	CCCCCCAGCCCATCACAACCAACGCCCCCC	1.447
6	CCCCCCAGCCCTACCCAACCAACTCCCCCC	1.468	16	CCCCCCAGCCCTAAACAACCAACTCCCCCC	1.444
7	CCCCCCAACCCTACACAACCAACTCCCCCC	1.467	17	CCCCCCAGCCCATCACAACCAGCTCCCCCC	1.442
8	CCCCCCAGGCCTACACAACCAACTCCCCCC	1.466	18	CCCCCCAGCCCTACACAACAAACTCCCCCC	1.441
9	CCCCCCAGCCCAACACCACCACCTCCCCCC	1.463	19	CCCCCCAGCCCAACACGACCAACTCCCCCC	1.440
10	CCCCCCAGCCCTAGACAACCAACTCCCCCC	1.458	20	CCCCCCAGCCCTACACAACCAACCCCCCCC	1.439

Figure 1. ssDNA sequences and predicted scores for directed evolution. (**a**) Examples of 3-bases mutated sequences from an original sequence. We tested 22,032 mutated sequences per round by ML model. The mutated bases were highlighted in red color. (**b**) Ensemble the mean value plot of $\Delta F/F_0$ predicted by the ML model for all 22,032 mutated sequences in round 1, sorted in descending order. $\Delta F/F_0$ is defined as the ratio of increased fluorescence intensity upon 5HT addition. (**c**) Top 20 sequences from the ML model with the highest predicted $\Delta F/F_0$.

3.2. *Towards 5HT Nanosensors with Higher Sensitivity*

At first, we applied this directed evolution strategy to find the better 5HT nanosensors in terms of sensitivity. We measured the 5HT responsive fluorescence change of ssDNA-SWCNT complexes from the top 20 sequences of 22,032 mutated sequences. Sensitivity to 5HT in those sensors could be compared by $\Delta F/F_0$ values at 1202 nm (Figure 2a). ($\Delta F/F_0 = (F - F_0)/F_0$, where F is the fluorescence intensity after 5HT addition, and F_0 is the fluorescence intensity after DI water addition.) Among the 20 sequences, the N1-12 sequence showed the highest $\Delta F/F_0$ of 2.366, which is 2.5 times larger than $\Delta F/F_0$ of 0.949 from the original sequence. The N1-12 sequence was used as the starting sequence for round 2 of ML (Figure S1), and the top 20 of the 22,032 mutated sequences from N1-12 were tested in the same protocol. The best sequence in round 2, N2-1, showed an improved $\Delta F/F_0$ of 3.705 (Figure S2). Using N2-1 as the starting sequence for round 3, we found another mutated sequence of N3-5 with a slightly improved $\Delta F/F_0$ of 3.765. $\Delta F/F_0$ values from all tested sequences were plotted in Figure 2b. Sensitivity improvement by directed evolution seems to be saturated at round 2 because the $\Delta F/F_0$ of N2-1 and N3-5 do not show a statistically significant difference (Figure 2c). Our directed evolution protocol improves the $\Delta F/F_0$ of the ssDNA-SWCNT nanosensor from $\Delta F/F_0 = 0.949$ to $\Delta F/F_0 = 3.765$, and this evolution is almost saturated at round 2. Experiments were conducted to compare the performance of the sensors made from the best-performing ssDNA in each round. The response of the sensors to 5HT in 1× PBS was compared. The normalized $\Delta F/F_0$ values for 5HT concentrations of 0.1 μM, 1 μM, 5 μM, 10 μM, 30 μM, 70 μM, and 100 μM are shown experimentally (Figure 2d). K_d values were obtained via the Hill equation [22].

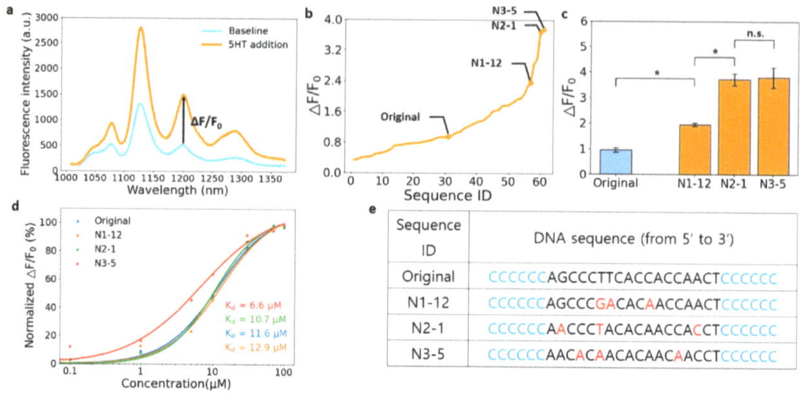

Figure 2. Sensitivity-targeted directed evolution up to 3 rounds. (**a**) Fluorescence spectra of original sequence ssDNA-SWCNT complex solution with and without 5HT addition. $\Delta F/F_0$ was calculated from the fluorescence intensity at 1202 nm and used as a criterion for sensitivity. (**b**) Experimental $\Delta F/F_0$ value plot for the tested sequences. N1-12, N2-1, and N3-5 mark the best-performing values in each round 1, 2, and 3, respectively. Note that the best performing $\Delta F/F_0$ value increases with each passing round. (**c**) $\Delta F/F_0$ values of the best-performing sequences and the original sequences for each round. * indicates statistically significant results at $p < 0.05$, n.s. indicates not statistically significant results at $p > 0.05$. N3-5 is not statistically significant when *t*-tested with N2-1, so the evolution is saturated at round 2. (**d**) Calibration curves for normalized $\Delta F/F_0$ as a function of 5HT concentration. Each calibration curve represents ssDNA-SWCNT from original (blue), N1-12 (orange), N2-1 (green), and N3-5 (red) sequences. Experiments were performed in triplicate for each concentration from 0.1 to 100 μM. (**e**) The best-performing sequences in each round of the sensitivity experiment.

Hill equation:

$$\Delta F/F_0 = \frac{(\Delta F/F_0)_{max} \times [5HT]^n}{K_d + [5HT]^n}, \qquad (1)$$

K_d = dissociation constant, n = cooperativity.

The dissociation constant (K_d) indicates the strength of binding between ssDNA-SWCNT and the analyte of 5HT. The smaller the K_d value, the higher the binding force between the two, and the larger the K_d value, the lower the binding force between the two. According to the hill equation, the K_d values of the original, N1-12, N2-1, and N3-5 are 11.6 μM, 12.9 μM, 10.7 μM, and 6.6 μM, respectively. Those nanosensors showed similar K_d around ~10 μM. Also, the K_d value is decreasing as the number of rounds increases.

3.3. Towards 5HT Nanosensors with Higher Selectivity

Herein, 'selectivity' refers to the ratio, which represents the fluorescence increase for DA and 5HT calculated as 5HT/DA (DA to 5HT fluorescence increase ratio). A higher value indicates that the nanosensor is more sensitive to 5HT than DA, i.e., having a good performance for 5HT.

Following the sensitivity test, we compared the fluorescence changes with DA to round-find sequences that respond specifically to 5HT. Since 5HT and DA have similar chemical structures, there is cross-sensitivity in fluorescence intensity. To avoid this cross-sensitivity, we wanted to check the performance of the sensor from the selectivity point of view and find a sequence with high selectivity for 5HT. We analyzed the data by measuring the reactivity to 5HT and the reactivity to DA and calculating the ratios. We measured the $\Delta F_{5HT}/F_0$ and $\Delta F_{DA}/F_0$ values at 1202 nm (Figure 3a,b). and divided them to obtain the 5HT/DA value. Selectivity was compared by the fluorescence intensity value of 5HT/DA, which is the ratio of the fluorescence change measured at the 1202 nm wavelength. The selectivity was defined as 5HT/DA = $\{(F_{5HT} - F_0)/F_0\}/\{(F_{DA} - F_0)/F_0\}$ where F_{5HT} is the fluorescence intensity after 5HT addition, F_{DA} is the fluorescence intensity after DA addition, and F_0 is the baseline fluorescence intensity at 1202 nm. Among the 20 sequences, the L1-14 sequence showed the highest 5HT/DA of 0.799, which is 1.6 times larger than the 5HT/DA of 0.508 from the original sequence. L1-14 sequence was used as the starting sequence for round 2, and the top 20 of the 22,032 mutated sequences from L1-14 were tested in the same protocol (Figure S3). The best sequence in round 2, N2-6, showed an improved 5HT/DA of 0.592. 5HT/DA values from all tested sequences were plotted in Figure 3c. Selectivity improvement by directed evolution seems to be saturated at round 1 because the $\Delta F_{5HT}/F_{DA}$ of N2-6 shows a statistically significant difference but does not show an increase in performance over the original in round 1 (Figure 3d). We measured the selectivity of our fluorescent nanosensors to different neurochemicals and biomolecules (Figure 3e). When checking the selectivity for original and L1-14, both sensors show minimal response to Ach, GABA, Glu, and UA. AA shows a larger response to the sensor compared to DA, but the effect is lower at L1-14. AA is already known to increase the fluorescence of ssDNA functionalized SWCNT [23], naturally, our experiments showed a higher $\Delta F/F_0$ for AA than for DA. Our directed evolution protocol improves the $\Delta F_{5HT}/F_{DA}$ of ssDNA-SWCNT nanosensor from $\Delta F_{5HT}/F_{DA} = 0.508$ to $\Delta F_{5HT}/F_{DA} = 0.799$, and this evolution is almost saturated at round 1. Compared to the original sequence, the L1-14 sequence has a 3 bp mutation at positions 4, 7, and 11, all of which are A mutations (Figure 3f).

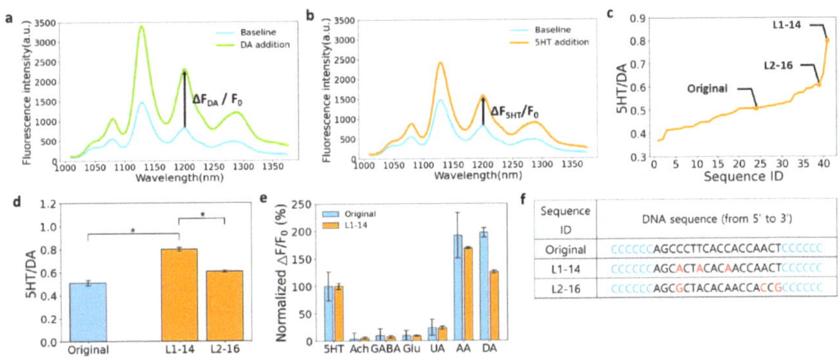

Sequence ID	DNA sequence (from 5' to 3')
Original	CCCCCCAGCCCTTCACCACCAACTCCCCCC
L1-14	CCCCCCAGCACTACACAACCAACTCCCCCC
L2-16	CCCCCCAGCGCTACACAACCACCGCCCCCC

Figure 3. Selectivity-targeted directed evolution up to 2 rounds. (**a**) Fluorescence spectra of original sequence ssDNA-SWCNT complex solution with and without DA addition. (**b**) Fluorescence spectra of original sequence ssDNA-SWCNT complex solution with and without 5HT addition. (**c**) Selectivity value plot for the entire sequences, listed in ascending order. Selectivity was calculated as the ratio of the $\Delta F/F_0$ value of 5HT to DA (5HT/DA), which is $(\Delta F_{5HT}/F_0)/(\Delta F_{DA}/F_0)$. L1-14 and L2-16 indicate the best-performing values in rounds 1 and 2, respectively. It can be seen that the performance decreased in L2 rather than in L1. (**d**) 5HT/DA values of the best-performing sequences and the original sequences for each round. * indicates statistically significant results at $p < 0.05$, n.s. indicates not statistically significant results at $p > 0.05$. L2-16 is statistically significant when *t*-tested with L1-14. However, since the performance is lower than L-14, it can be seen that it is saturated in L1. (**e**) $\Delta F/F_0$ measured for 100 μM Ach, GABA, Glu, AA and 10 μM UA regarding $\Delta F/F_0$ of 100 μM 5HT. Experiments were performed on the original and L1-14 sequence, with three replicates for each analyte. (**f**) The best-performing sequences in each round of the selectivity experiment.

3.4. Fluorescence Image Analysis of Surface-Immobilized Nanosensors after 5HT Treatment

Fluorescence images were obtained for the N2-1 ssDNA-SWCNT nanosensor, which showed the best performance in sensitivity. SWCNTs were adsorbed to an APTES-coated glass coverslip, filled with 1× PBS, and spiked with 5HT to observe the fluorescence response. 5HT was injected at 30 s for a final concentration of 100 μM and the change in $\Delta F/F_0$ was observed for 1 min. Overall, we can observe the fluorescence increase upon treatment with 5HT (Figure 4a,b). On average, the nanosensor fluorescence increased up to 40% $\Delta F/F_0$ after 5HT treatment (Figure 4c). Since the sensors showed good responsiveness to 5HT while anchored to the surface, it is expected that the sensors will also show good response to neurotransmitters when coated around the neural cells.

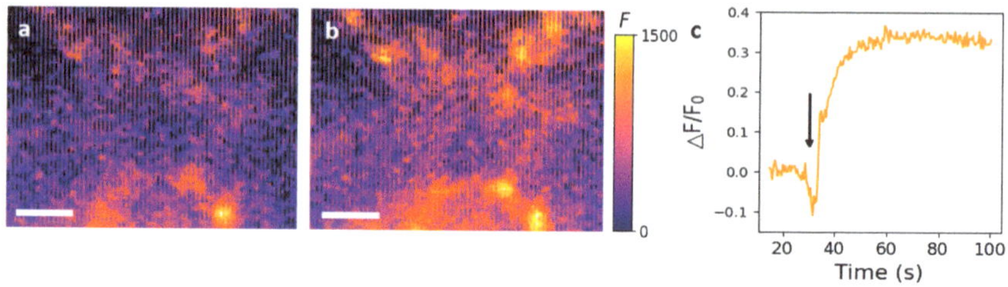

Figure 4. Fluorescence images and analysis of fluorescence changes over time for the N2-1 sensor. (**a**,**b**) Fluorescence images of surface-immobilized N2-1 sensors before and after 5HT treatment on a glass coverslip (scale bar, 10 μm). (**c**) Plot showing fluorescence change over time for multiple sensor regions. 5HT was injected at the time point indicated by the arrow.

4. Discussion

In this study, we tested 100 mutated sequences in multiple rounds of evolution (total sensitivity 3 rounds + selectivity 2 rounds = 5 rounds) from the original ssDNA sequence. By overcoming the physical limitations of sequence-by-sequence testing using conventional directed evolution, ML allowed us to scale up our experimental approach. The high throughput 5HT sensors we found are expected to serve as real-time sensing for neurodegenerative disorders including Parkinson's disease, Huntington's disease, and schizophrenia.

In the ssDNA functionalized SWCNT sensor, the sensitivity and selectivity have been improved in various methods. Specifically, the introduction of XNA, a chemically modified form of DNA, has resulted in an improved turn-on response and increased stability [24]. However, the chemical modification of DNA, such as XNA preparation, incurs higher costs and longer testing times for each sample. Additionally, Lambert et al. reported the application of directed evolution to the dopamine nanosensor for higher sensitivity in ssDNA-SWCNT [18]. However, it is noteworthy that the conventional approach involved screening the mutated DNA library through random selection.

In contrast, our methodology integrates directed evolution with machine learning screening, enabling a cost-effective and time-efficient exploration of diverse DNA-assisted binding moieties on SWCNT. In comparison to previous directed evolution methods for ssDNA-SWCNT, our approach utilizes a machine learning model to predict more sensitive mutated sequences for 5HT. This enhances our ability to discover superior sequences, offering a more targeted approach compared to the random selection of mutated sequences.

The experimental results showed that there were commonly mutated bases, such as the seventh base of the sequences that were predicted to have high performance, and that the sixth base sequence was mutated in the order of G $\rightarrow$ T $\rightarrow$ A during the three rounds of sensitivity. In addition, the mutated bases were mostly changed to adenine, and in the final selected sequences, N2-1 at sensitivity and L1-14 at selectivity, three bases in the mutation site were changed to A. In addition to the original sequence, which was used as a reference in this study, we expect to find a sequence that performs better by conducting similar experiments on other sequences known to bind well to 5HT, such as 5HT aptamer [14,25]. Finally, analyzing these selected sequences will enable the discovery of 5HT sensors based on performance, aiding in more accurate predictions from an ML perspective. The biomedical application of SWCNT faces challenges, including concerns about toxicity, the necessity for improved biocompatibility, and ensuring long-term stability. Addressing issues related to biodistribution, immunogenicity, production scalability, and ethical considerations is crucial for advancing the responsible use of SWCNTs in biomedical engineering. Ongoing research efforts are essential to overcome these limitations and unlock the full potential of SWCNTs in clinical applications.

5. Conclusions

In summary, we used a combination of measuring the neurotransmitter fluorescence with ssDNA-SWCNT complexes and directed evolution technology to evaluate the ssDNA-SWCNT sensor performance by comparing the fluorescence response to randomly mutated sequences. The final selected sequences, N2-1 (5′-CCCCCCAACCCTACACAACCACCTCC-CCCC-3′) and L1-14 (5′-CCCCCCAGCACTACACAACCAACTCCCCCC-3′) can serve as reference sequences for future studies using ssDNA-SWCNT sensors to detect 5HT. In this study, we did not find any sequences with a selectivity greater than 1.0, meaning that the fluorescence intensity for 5HT exceeded that for DA. In this regard, we believe that conducting the same study for other neurotransmitters (acetylcholine, norepinephrine, GABA, etc.) in terms of selectivity will enhance our understanding of 5HT and allow us to discover sensors with superior performance.

Supplementary Materials: The following supporting information can be downloaded at: https://www.mdpi.com/article/10.3390/nano14030247/s1, Figure S1. Top 20 sequences in each round of machine learning and the predicted $\Delta F/F_0$ results. (a) N1, L1 (b) N2 (c) N3 (d) L2; Figure S2. The $\Delta F/F_0$ values of the top 20 sequences for each round in sensitivity, sorted in ascending order. (a) N1 (b) N2 (c) N3; Figure S3. The 5HT/DA values of the top 20 sequences for each round in selectivity, sorted in ascending order. (a) L1 (b) L2

Author Contributions: Conceptualization, methodology, software, validation, formal analysis, investigation, visualization, S.J., S.A., Y.S., P.K., D.L. and L.V.; writing—original draft preparation, S.A. and Y.S.; writing—reviewing and editing, S.J., S.A., Y.S., P.K. and L.V.; project administration, resources, supervision, S.J. All authors have read and agreed to the published version of the manuscript.

Funding: This work was supported by a 2-Year Research Grant from Pusan National University.

Data Availability Statement: Data can be shared upon request to Sanghwa Jeong.

Acknowledgments: We acknowledge the computer time provided by the Texas Advanced Computing Center (TACC).

Conflicts of Interest: The authors declare no conflicts of interest.

References

1. Singh, P. *Surface Plasmon Resonance: A Boon for Viral Diagnostics*; Elsevier Inc.: Amsterdam, The Netherlands, 2017; ISBN 978-0-12-809633-8.
2. Ligler, F.S. Perspective on optical biosensors and integrated sensor systems. *Anal. Chem.* **2009**, *81*, 519–526. [CrossRef] [PubMed]
3. Chen, C.; Wang, J. Optical biosensors: An exhaustive and comprehensive review. *Analyst* **2020**, *145*, 1605–1628. [CrossRef] [PubMed]
4. Kruss, S.; Hilmer, A.J.; Zhang, J.; Reuel, N.F.; Mu, B.; Strano, M.S. Carbon nanotubes as optical biomedical sensors. *Adv. Drug Deliv. Rev.* **2013**, *65*, 1933–1950. [CrossRef] [PubMed]
5. Dubey, R.; Dutta, D.; Sarkar, A.; Chattopadhyay, P. Functionalized carbon nanotubes: Synthesis, properties and applications in water purification, drug delivery, and material and biomedical sciences. *Nanoscale Adv.* **2021**, *3*, 5722–5744. [CrossRef] [PubMed]
6. Prieto-Simón, B.; Bandaru, N.M.; Saint, C.; Voelcker, N.H. Tailored carbon nanotube immunosensors for the detection of microbial contamination. *Biosens. Bioelectron.* **2015**, *67*, 642–648. [CrossRef]
7. Ackermann, J.; Metternich, J.T.; Herbertz, S.; Kruss, S. Biosensing with fluorescent carbon nanotubes. *Angew. Chem. Int. Ed.* **2022**, *61*, e202112372. [CrossRef] [PubMed]
8. Boghossian, A.A.; Zhang, J.; Barone, P.W.; Reuel, N.F.; Kim, J.-H.; Heller, D.A.; Ahn, J.H.; Hilmer, A.J.; Rwei, A.; Arkalgud, J.R.; et al. Near-Infrared Fluorescent Sensors based on Single-Walled Carbon Nanotubes for Life Sciences Applications. *ChemSusChem* **2011**, *4*, 848–863. [CrossRef]
9. Hendler-Neumark, A.; Bisker, G. Fluorescent single-walled carbon nanotubes for protein detection. *Sensors* **2019**, *19*, 5403. [CrossRef]
10. Mohammad-Zadeh, L.F.; Moses, L.; Gwaltney-Brant, S.M. Serotonin: A review. *J. Vet. Pharmacol. Ther.* **2008**, *31*, 187–199. [CrossRef]
11. Liu, L.; Zhong, M.; Dong, J.; Chen, M.; Shang, J.; Yue, Y. 5-hydroxytryptamine (5-HT) positively regulates pigmentation via inducing melanoblast specification and melanin synthesis in zebrafish embryos. *Biomolecules* **2020**, *10*, 1344. [CrossRef]

12. Graeff, F.G.; Guimarães, F.S.; De Andrade, T.G.; Deakin, J.F. Role of 5-HT in stress, anxiety, and depression. *Pharmacol. Biochem. Behav.* **1996**, *54*, 129–141. [CrossRef] [PubMed]
13. Jeong, S.; Yang, D.; Beyene, A.G.; Bonis-O'Donnell, J.T.D.; Gest, A.M.M.; Navarro, N.; Sun, X.; Landry, M.P. High-throughput evolution of near-infrared serotonin nanosensors. *Sci. Adv.* **2019**, *5*, eaay3771. [CrossRef] [PubMed]
14. Dinarvand, M.; Neubert, E.; Meyer, D.; Selvaggio, G.; Mann, F.A.; Erpenbeck, L.; Kruss, S. Near-infrared imaging of serotonin release from cells with fluorescent nanosensors. *Nano Lett.* **2019**, *19*, 6604–6611. [CrossRef] [PubMed]
15. Yang, K.K.; Wu, Z.; Arnold, F.H. Machine-learning-guided directed evolution for protein engineering. *Nat. Methods* **2019**, *16*, 687–694. [CrossRef] [PubMed]
16. Arnold, F.H. Design by directed evolution. *Acc. Chem. Res.* **1998**, *31*, 125–131. [CrossRef]
17. Packer, M.S.; Liu, D.R. Methods for the directed evolution of proteins. *Nat. Rev. Genet.* **2015**, *16*, 379–394. [CrossRef] [PubMed]
18. Lambert, B.; Gillen, A.J.; Schuergers, N.; Wu, S.; Boghossian, A.A. Directed evolution of the optoelectronic properties of synthetic nanomaterials. *Chem. Commun.* **2019**, *55*, 3239–3242. [CrossRef]
19. Kelich, P.; Jeong, S.; Navarro, N.; Adams, J.; Sun, X.; Zhao, H.; Landry, M.P.; Vuković, L. Discovery of DNA–Carbon nanotube sensors for serotonin with machine learning and near-infrared fluorescence spectroscopy. *ACS Nano* **2021**, *16*, 736–745. [CrossRef]
20. Hie, B.L.; Shanker, V.R.; Xu, D.; Bruun, T.U.J.; Weidenbacher, P.A.; Tang, S.; Wu, W.; Pak, J.E.; Kim, P.S. Efficient evolution of human antibodies from general protein language models. *Nat. Biotechnol.* **2023**. [CrossRef]
21. Mayerhöfer, T.G.; Mutschke, H.; Popp, J. Employing Theories Far beyond Their Limits—The Case of the (Boguer-) Beer–Lambert Law. *ChemPhysChem* **2016**, *17*, 1948–1955. [CrossRef]
22. Kruss, S.; Landry, M.P.; Ende, E.V.; Lima, B.M.; Reuel, N.F.; Zhang, J.; Nelson, J.; Mu, B.; Hilmer, A.; Strano, M. Neurotransmitter Detection Using Corona Phase Molecular Recognition on Fluorescent Single-Walled Carbon Nanotube Sensors. *J. Am. Chem. Soc.* **2014**, *136*, 713–724. [CrossRef] [PubMed]
23. Polo, E.; Kruss, S. Impact of Redox-Active Molecules on the Fluorescence of Polymer-Wrapped Carbon Nanotubes. *J. Phys. Chem. C* **2016**, *120*, 3061–3070. [CrossRef]
24. Gillen, A.J.; Kupis-Rozmyslowicz, J.; Gigli, C.; Schuergers, N.; Boghossian, A.A. Xeno Nucleic Acid Nanosensors for Enhanced Stability Against Ion-Induced Perturbations. *J. Phys. Chem. Lett.* **2018**, *9*, 4336–4343. [CrossRef] [PubMed]
25. Nakatsuka, N.; Heard, K.J.; Faillétaz, A.; Momotenko, D.; Vörös, J.; Gage, F.H.; Vadodaria, K.C. Sensing serotonin secreted from human serotonergic neurons using aptamer-modified nanopipettes. *Mol. Psychiatry* **2021**, *26*, 2753–2763. [CrossRef]

nanomaterials

MDPI

Article

Seed Priming with Single-Walled Carbon Nanotubes Grafted with Pluronic P85 Preserves the Functional and Structural Characteristics of Pea Plants

Sashka Krumova [1], Asya Petrova [2], Nia Petrova [1,3], Svetozar Stoichev [1], Daniel Ilkov [2], Tsonko Tsonev [1], Petar Petrov [4], Dimitrina Koleva [5] and Violeta Velikova [1,2,*]

1 Institute of Biophysics and Biomedical Engineering, Bulgarian Academy of Sciences, "Acad. G. Bonchev" Str., Bl. 21, 1113 Sofia, Bulgaria; sashka.b.krumova@gmail.com (S.K.); zlatkova.nia@gmail.com (N.P.); svetlio85@abv.bg (S.S.); tsonev@gmail.com (T.T.)
2 Institute of Plant Physiology and Genetics, Bulgarian Academy of Sciences, "Acad. G. Bonchev" Str., Bl. 21, 1113 Sofia, Bulgaria; petrova_assya@abv.bg (A.P.); daniel_ilkov@yahoo.com (D.I.)
3 Institute of Plant Biology, Biological Research Centre, Temesváry krt. 62, 6726 Szeged, Hungary
4 Institute of Polymers, Bulgarian Academy of Sciences, "Acad. G. Bonchev" Str., Bl. 103, 1113 Sofia, Bulgaria; ppetrov@polymer.bas.bg
5 Faculty of Biology, Sofia University, "St. Kliment Ohridsky", 1000 Sofia, Bulgaria; koleva@biofac.uni-sofia.bg
* Correspondence: violet@bio21.bas.bg

Abstract: The engineering of carbon nanotubes in the last decades resulted in a variety of applications in electronics, electrochemistry, and biomedicine. A number of reports also evidenced their valuable application in agriculture as plant growth regulators and nanocarriers. In this work, we explored the effect of seed priming with single-walled carbon nanotubes grafted with Pluronic P85 polymer (denoted P85-SWCNT) on *Pisum sativum* (var. RAN-1) seed germination, early stages of plant development, leaf anatomy, and photosynthetic efficiency. We evaluated the observed effects in relation to hydro- (control) and P85-primed seeds. Our data clearly revealed that seed priming with P85-SWCNT is safe for the plant since it does not impair the seed germination, plant development, leaf anatomy, biomass, and photosynthetic activity, and even increases the amount of photochemically active photosystem II centers in a concentration-dependent manner. Only 300 mg/L concentration exerts an adverse effect on those parameters. The P85 polymer, however, was found to exhibit a number of negative effects on plant growth (i.e., root length, leaf anatomy, biomass accumulation and photoprotection capability), most probably related to the unfavorable interaction of P85 unimers with plant membranes. Our findings substantiate the future exploration and exploitation of P85-SWCNT as nanocarriers of specific substances promoting not only plant growth at optimal conditions but also better plant performance under a variety of environmental stresses.

Keywords: agronanotechnology; carbon nanotubes; chlorophyll fluorescence; leaf anatomy; nanoparticles; photoprotection; photosynthesis; plant biomass; seed germination; seed priming

Citation: Krumova, S.; Petrova, A.; Petrova, N.; Stoichev, S.; Ilkov, D.; Tsonev, T.; Petrov, P.; Koleva, D.; Velikova, V. Seed Priming with Single-Walled Carbon Nanotubes Grafted with Pluronic P85 Preserves the Functional and Structural Characteristics of Pea Plants. *Nanomaterials* **2023**, *13*, 1332. https://doi.org/10.3390/nano13081332

Academic Editor: Muralidharan Paramsothy

Received: 14 March 2023
Revised: 7 April 2023
Accepted: 10 April 2023
Published: 11 April 2023

1. Introduction

The process of seed germination is essential for agricultural practices—it needs to be uniform and rapid to achieve the best field performance characteristics and crop yields under different and constantly fluctuating environmental conditions (reviewed in [1]). A number of seed coating and priming techniques are being investigated and exploited to achieve beneficial effects, including recent advancements in agronanotechnology. The exploration of nano-sized agrochemicals for seed treatment is extensively reviewed in [2–5]. Both positive (plant growth enhancement, environmental safety, improved plant stress resistance) and negative (toxicity, environmental pollution) effects of carbon-based nanopesticides are thoroughly discussed in [5]. The molecular mechanism of nanoparticles action on seeds is not yet clarified; however, it appears that it is dependent on the utilized type of nanoagent, plant species, and mode of interaction between them.

Despite some clues on the effects of nanoparticles, in particular multi-(MWCNT) and single-(SWCNT) walled carbon nanotubes, on plant growth (reviewed in [4,6]), this emerging topic remains largely unexplored. Both MWSCNT and SWCNT are shown to exert positive or negative effects on plants' physiological states depending on their characteristics (size, surface functionalization, charge, concentration), type of application, plant species, and growth conditions. Bare SWCNT, in particular, have been found to affect *Zea mays'* root morphology via increased expression of genes involved in seminal root development and decreased expression of genes controlling root hair formation [5]. Root length was also shown to increase upon seed priming with 40 mg/L SWCNT for *Raphanus sativus* and *Brassica rapa* [7], and *Salvia macrosiphon*, *Capsicum annuum*, *Festuca arundinaces* [8]. A wide SWCNT concentration range was tested on *Hyoscyamus niger* seeds by [9], which clearly demonstrated the positive effect on root length and germination percentage up to a concentration of 200 mg/L and an inhibiting effect of higher SWCNT doses.

The study of Cañas et al. [10] demonstrated the strongly individual response of different crops to SWCNT treatments—*Allium cepa* and *Cucumis sativus* seed exposure to non-functionalized SWCNT resulted in enhancement of root length; however, the *Lycopersicon esculentum* root elongation was inhibited under the same treatment. Functionalization of SWCNT with poly-3-aminobenzenesulfonic acid that was utilized to increase the nanoparticles dispersibility resulted in a negative effect on the root length for *Lactuca sativa*, while *Brassica oleracea* and *Daucus carota* were found to be insensitive to the applied treatments [10].

In 2009, Khodakovskaya et al. [11] for the first time showed that MWCNT are able to penetrate the seed coat of tomato seeds and accumulate more water as a result of incubation in Murashige and Skoog medium. The authors suggested that MWCNT enhance the water uptake either by generation of more pores in the seed coat or by regulation of the gating of seed coat water channels (aquaporins). Besides the higher water uptake, the authors also recorded faster germination and higher biomass production over 27 days of seedlings growth. Later on, additional evidence for MWCNT-induced enhancement of plant development are presented by Lahiani et al. [12] and Khodakovskaya et al. [13]. A general trend for increased germination percentage of salvia (*Salvia macrosiphoti* Boiss.), pepper (*Capsicum anntium* L.), and tall fescue (*Festuca arundinacea* Sch.) upon increase in SWCNT concentration was observed by [8]. The authors further revealed that SWCNT are also able to penetrate seed coats and roots. This process depended on the treatment conditions and the cell wall composition in different plant species. SWCNT were also proposed to be able to physically interact with (and penetrate) the cell membranes, including the outer and inner chloroplast membranes, by different mechanisms, such as fluid-phase endocytosis, lipid exchange envelope penetration, and passive diffusion [14–16]. They were also suggested to interact with thylakoid membrane lipids [15], pigments [17], and proteins involved in the photosynthetic electron transport [16,18–24]. Recent studies also demonstrate the possibility for electron leakage and transfer between thylakoids and SWCNT without tight physical interaction with the photosystems [25]. However, the SWCNT–seed interaction is still not thoroughly investigated.

Generally, carbon nanotubes are prone to aggregation, and different strategies to overcome this problem are exploited—polymer-grafting being one of them. To the best of our knowledge, so far only chitosan-wrapped (both covalently and non-covalently bound chitosan) SWCNT with or without additional PEGylation were utilized for successful chloroplast-targeted gene delivery by entering through stomata pores by leaf infiltration [26]. In our previous work, we explored the effect of foliar application of SWCNT grafted by Pluronic P85 (P85) triblock co-polymer (hereon denoted P85-SWCNT) on intact *Pisum sativum* plants [27,28]. We utilized this polymer as a grafting agent for SWCNT since it ensures stable aqueous dispersion of these nanoparticles. We found that P85-SWCNT exerted strong effects on the leaf surface modification and photosynthetic functioning of both photosystem II (PSII) and photosystem I (PSI), which clearly indicated penetration and transportation of this nanomaterial within plant tissues and organelles [27,28]. Interestingly, P85 behaved differently from P85-SWCNT—it had no effect on the examined morphological characteristics; however, when applied in certain concentrations, it led to the increase in some photosynthesis-related parameters, such as maximum efficiency of PSII photochemistry, actual photochemical efficiency of PSII, and fraction of open PSII centers [27]. This prompted us to further explore the application of both P85-SWCNT and P85 in the field of nanoagronomy, this time in regard to seed priming. As far as we are aware, no studies have been performed on pea seed interactions with either SWCNT or MWCNT, although this crop is among the major food resources in Europe [29], and therefore, improvements in its growth technology will have a strong economic impact.

The aim of the present study was to investigate how the initial stages of plant development are affected by seed pre-sawing treatment with P85-SWCNT. We hypothesized that the effect of this priming agent on important agronomical traits, such as seeds germination, plant biomass accumulation, and structural and functional leaf characteristics, is concentration dependent and therefore tested a wide concentration range.

The presented results shed light on the seed/P85-SWCNT interactions and their implications in nanoagriculture. In particular, we demonstrate that P85-SWCNT in concentrations up to 100 mg/L are safe priming agents, i.e., they largely preserve pea leaf physiological characteristics.

2. Materials and Methods

2.1. Chemicals and Reagents

Poly(ethylene oxide)$_{26}$-block-poly(propylene oxide)$_{40}$-block-poly(ethylene oxide)$_{26}$ triblock copolymer "Pluronic" P85 (used as obtained) was purchased from BASF (Ludwigshafen/Rhein, Germany). SWCNT (>77% carbon) were purchased from Sigma-Aldridge (Darmstadt, Germany) and used without purification. Glutaraldehyde (electron microscopy grade, 25% solution in water) and sodium phosphate buffer were obtained from Thermo Fisher Scientific, Waltham, MA, USA).

2.2. Pluronic P85 and P85-SWCNT Dispersions

Polymer suspensions with concentrations of 0.04, 0.2, 1, 5, 10, and 30 g/L were prepared in distilled water, and aliquots were further used for dispersion of 0.4, 2, 10, 50, 100, and 300 mg/L SWCNT (i.e., 100:1 P85:SWCNT *w*/*w* ratio), respectively, upon sonication for 15 min [30]. The polymer solution and P85-SWCNT dispersions were additionally sonicated for 30 min before the start of the seed priming procedure.

2.3. Seed Priming

For each individual biological experiment, 50 *Pisum sativum* (var. RAN-1) seeds were soaked in P85 or P85-SWCNT dispersions of the above defined concentrations under continuous shaking at 25 °C for a period of 6 h. Control (hydro-primed) seeds were soaked in distilled water under the same experimental conditions. The seed weights were determined before the start of the imbibition and every 2 h after initial seed soakings. Seed imbibitions (extent of priming mixture components absorption) were determined in relation to the seed dry weights. After the priming procedure, the seeds were re-dried to the original moisture content (which took 10–12 days) and later stored in the dark for an additional 2 days until further use. Before subsequent germination, the primed seeds were soaked in distilled water for 2 h. Scheme 1 presents the experimental design steps. Each priming treatment was repeated three times.

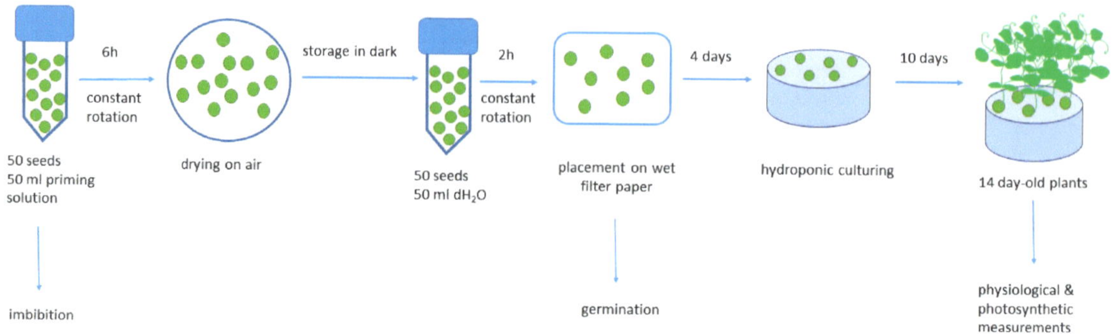

Scheme 1. Experimental design. For further details see text in Section 2.3.

2.4. Seed Germination and Plant Development

For characterization of the germination process, the primed seeds were placed on wet filter paper in the dark. Root length (threshold of 2 mm) was checked every 24 h for a period of 4 days. The following seed germination parameters as defined in Ranal et al. [31] were determined: germinability (G), i.e., the germination percentage, mean germination time (MT), mean germination rate (MR), and synchrony of the germination process (Z).

On the fourth day of germination, the seeds were placed in containers for hydroponic growth filled with tap water. Plants were grown for an additional 10 days in climatic chamber providing 12 h photoperiod, 400 μmol/m^{2}s photon flux density (SMD LED 6500K lamps), 22 °C ambient temperature, and 65% air humidity. At the end of this period, the plants were either used for fresh and dry biomass determination or subjected to anatomical and physiological measurements. The mean % developed plants was estimated by the percentage of developed plants relative to the total number of sown seeds for each variant. The survival index (SVI) was determined by multiplication of the dry weight/plant by the percentage of developed plants at the 14th day of growth for each treatment. Specific leaf area (SLA) and leaf dry mass per unit area (LMA) were determined on fully developed leaves from the second and third leaf pairs. Ten leaf pairs for each individual set of experiments were analyzed. Individual leaf areas were measured using the Image-Pro Plus 6.0 software (Media Cybernetics, Silver Spring, MD, USA). Leaves were dried at 105 °C for 30 min and then at 60 °C to constant weight. LMA was calculated from the ratio of dry leaf biomass and leaf area (g/m^2).

2.5. Leaf Anatomy

Anatomical studies were performed as described in [32]. Briefly, leaf sections (1–2 mm^2) were collected from the middle part of the second and third well developed leaves and fixed in 3% (*m*/*v*) glutaraldehyde in 0.1 M sodium phosphate buffer, pH 7.4. At least 30 transversal sections per treatment were mounted on slides in glycerol and studied with a light microscope. Images were taken by a digital camera Nikon Eclipse 50i (Nikon Solutions Co., Ltd. Tokyo, Japan). Leaf anatomy was characterized by measuring the thickness of the leaf, palisade and spongy parenchyma, and adaxial and abaxial epidermis.

2.6. Leaf Pigments and Chlorophyll Fluorescence

Adaxial surface leaf pigments were measured by the Dualex instrument manufactured by ForceA (Orsay, France). The ratio of chlorophyll (Chl) and flavonoids levels were used to determine the nitrogen balance index (NBI).

Fluorescence imaging was performed by means of IMAGING-PAM MAXI version produced by H. Walz GmbH (Effeltrich, Germany) supplemented with blue excitation light unit (IMAG-MAX/L LED) and IMAG-K7 CCD camera. Prior to the measurements, the plants were dark adapted for 30 min. The experimental setup and the detailed parameter definitions are as in [27]. The efficiency of the photosynthetic apparatus was assessed on the basis of the following parameters: F_v/F_m—maximum quantum yield of photosystem II (PSII) (determined in dark-adapted state), Φ_{PSII}—quantum efficiency of PSII photochemistry (determined in light-adapted state as in [33]), NPQ—non-photochemical quenching of Chl *a* fluorescence [34]; qL—measure of the fraction of open PSII reaction centers [35]. After the dark adaptation period, leaves were exposed to continuous actinic illumination for 15 min and saturation pulses were applied every 1 min to follow NPQ kinetics. All Chl fluorescence measurements were performed on intact plants.

2.7. Statistical Evaluation

Experimental data were subjected to Student's *t*-test, and three different levels of statistical significance were defined, i.e., $p < 0.1$ (*), $p < 0.05$ (**), $p < 0.01$ (***), as shown in the figure and table captions.

3. Results

3.1. Seed Germination

As a first step in our study, we checked different periods of hydro-seed priming in the range of 4–24 h. Optimal results for seed imbibition and germination were obtained for 6 h treatment, and all further experiments presented in this work were based on this priming protocol.

Next, we examined the imbibition process in seeds primed with P85-SWCNT and compared the results with those obtained for hydro- (control) and P85-primed ones. For all tested treatments, the imbibition level was similar to the control, and for a period of 6 h it reached ca. 120% from the initial value determined for dry seeds (Table 1).

Table 1. Imbibition (%), of hydro-, P85-SWCNT-, and P85-primed pea seeds for a period of 2–6 h. Mean values (±SE, n = 150). The data are not statistically different from the hydro-primed sample according to Student's *t*-test at $p < 0.01$.

		Imbibition (%)	
Concentration	**2 h**	**4 h**	**6 h**
		H_2O (control)	
	70 ± 5	105 ± 8	126 ± 6
(mg/L)		**P85-SWCNT**	
0.4	60 ± 8	94 ± 7	123 ± 10
2	67 ± 5	105 ± 4	129 ± 3
10	63 ± 3	100 ± 1	125 ± 1
50	59 ± 8	101 ± 11	124 ± 7
100	59 ± 8	103 ± 14	123 ± 11
300	54 ± 9	96 ± 12	120 ± 10
(g/L)		**P85**	
0.04	64 ± 11	96 ± 11	121 ± 12
0.2	75 ± 11	106 ± 10	127 ± 10
1	68 ± 13	95 ± 13	118 ± 9
5	72 ± 14	104 ± 18	122 ± 16
10	72 ± 12	106 ± 16	123 ± 15
30	71 ± 14	104 ± 19	121 ± 18

The germination process was further characterized on the basis of the parameters defined in [31]. For all tested P85-SWCNT concentrations, the mean germinability (G) value was similar to the hydro-primed control. However, this parameter was lower in P85-treated seeds than the hydro-primed control and the respective P85-SWCNT variants, though the differences were not statistically significant (Table 2). The mean time of germination (MT) and mean rate of germination (MR) remained close to the control values for all treatments.

The synchrony of germination (Z) was significantly higher than the control only for the 100 mg/L P85-SWCNT treatment (Table 2).

Table 2. Germination of hydro-, P85-SWCNT, and P85-primed pea seeds for a period of 4 days. Germination parameters (germinability (G, %), mean germination time (MT, day), mean germination rate (MR, seeds per day), and synchrony of the germination process (Z)) were estimated as in [31]. Mean ± SE. Statistically significant differences from the control sample at $p < 0.05$ (**) according to Student's *t*-test are indicated.

Concentration	G (%)	MT (Day)	Z	MR (Seeds per Day)
		H_2O (control)		
	55 ± 5	2.15 ± 0.1	0.42 ± 0.04	0.47 ± 0.02
(mg/L)		P85-SWCNT		
0.4	52 ± 7	2.26 ± 0.06	0.38 ± 0.03	0.44 ± 0.01
2	58 ± 5	2.11 ± 0.10	0.42 ± 0.02	0.48 ± 0.02
10	54 ± 4	2.2 ± 0.05	0.46 ± 0.04	0.45 ± 0.01
50	53 ± 4	2.33 ± 0.17	0.55 ± 0.07	0.44 ± 0.03
100	42 ± 18	2.43 ± 0.66	0.73 ± 0.10 **	0.45 ± 0.12
300	60 ± 6	2.31 ± 0.62	0.44 ± 0.12	0.47 ± 0.13
(g/L)		P85		
0.04	35 ± 13	2.43 ± 0.39	0.39 ± 0.08	0.42 ± 0.07
0.2	35 ± 7	2.61 ± 0.46	0.34 ± 0.06	0.40 ± 0.07
1	42 ± 14	2.66 ± 0.41	0.35 ± 0.06	0.39 ± 0.06
5	32 ± 22	2.09 ± 0.31	0.44 ± 0.04	0.49 ± 0.07
10	34 ± 18	2.06 ± 0.56	0.45 ± 0.01	0.52 ± 0.14
30	27 ± 23	2.40 ± 0.60	0.74 ± 0.26	0.44 ± 0.11

The root development of primed seeds was followed for a period of 4 days. The P85-SWCNT treatment did not lead to changes in root growth relative to the hydro-primed control, while significantly shorter root length was observed in P85 primed seeds (Figure 1).

3.2. *Plant Development*

Plants developed from primed seeds were characterized on the 14th day of their growth at the stage of three well-developed leaf pairs. P85-SWCNT priming did not induce any negative changes in the number of developed plants, LMA, TDB, and SVI as compared to the hydro-primed control, with the sole exception of 100 mg/L concentration that also resulted in lower SVI (Figure 2). For all tested P85 concentrations, the mean % developed plants was lower than the hydro-primed control, but the differences were not statistically significant for the 1 and 5 g/L P85 treatments. The LMA reduced significantly for the 1, 5, and 30 g/L P85 treatments, while all other tested concentrations did not considerably affect this parameter. P85 application (with exception of 10 g/L) led also to largely reduced TBD and SVI values (Figure 2).

3.3. *Leaf Anatomical Traits*

The leaves of control *Pisum sativum* plants were typical bifacial, amphistomatic with an average thickness of the lamina of 268 ± 18 μm (Table 3). The photosynthesizing parenchyma (determined by the distance between adaxial and abaxial epidermis) represented about 80% of the thickness of the lamina. It consisted of a single-row palisade parenchyma and 7–8 rows of spongy parenchyma with a clearly differentiated collecting layer (Figure S1a,b). Epidermal tissues were relatively uniformly structured by tangentially flattened basal epidermal cells and had almost evenly distributed stomata on both surfaces.

Histological examination of the P85-SWCNT and P85 variants showed no anatomical changes for samples primed with low concentrations of P85-SWCNT (0.4, 2 and 10 mg/L) and P85 (0.04, 0.2 and 1 g/L); however, differences in leaf anatomical traits were observed in variants treated with higher concentrations of P85-SWCNT (50, 100 and 300 mg/L) and P85 (5, 10 and 30 g/L). Palisade cells in P85-SWCNT were not typical in shape and had strongly reduced contact, while those in P85 variants were with typically elongated cylindrical shape and with well-defined structural contact ensuring the normal symplast transport (Figure S1). The higher P85-SWCNT concentrations also resulted in significantly decreased leaf thickness (Table 3). The concentration range of 50–300 mg/L P85-SWCNT resulted in lamina thickness decrease of 22–26%, while application of 5–30 g/L P85 resulted in 28–41% reduction. The average thickness of the leaf lamina was reduced by 24% in the P85-SWCNT and 41% in P85-treated variants compared to the hydro-primed control. This was due to the substantial decrease of the spongy parenchyma by 28–30% for P85-SWCNT and 54–64% for P85 variants. No significant changes in palisade parenchyma were detected in either of those variants. As a consequence, the mesophyll thickness decreased by 25–31% and 33–40% for P85-SWCNT and P85 samples, respectively, compared to the control (Table 3).

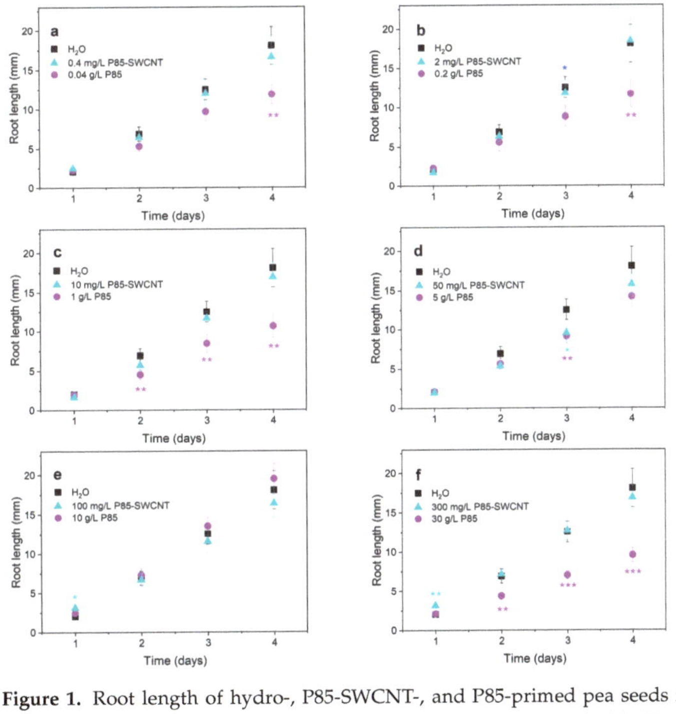

Figure 1. Root length of hydro-, P85-SWCNT-, and P85-primed pea seeds for the initial 4 days after seeds sowing. Data for 0.4 mg/L (**a**), 2 mg/L (**b**), 10 mg/L (**c**), 50 mg/L (**d**), 100 mg/L (**e**) and 300 mg/L (**f**) P85-SWCNT (triangles), as well as the corresponding P85 treatments (circles), are presented in different panels. For clarity each panel contains data for hydro-primed samples (squares) as well. Mean ± SE. Statistically significant differences from the control sample at $p < 0.1$ (*), $p < 0.05$ (**) and $p < 0.01$ (***) according to Student's *t*-test are indicated.

3.4. *Leaf Physiological Characteristics*

The total Chl content, assessed on the adaxial leaf surface, remained close to the control one for P85-SWCNT treatments. The Chl level exhibited non-linear P85 concentration dependence—it was significantly higher for 0.04 and 0.2 g/L and lower for 30 g/L P85 variants (Figure 3a). The flavonoid content was lower for 100 mg/L P85-SWCNT and 1 and 30 g/L P85 but higher for 0.4–1 mg/L P85-SWCNT and 10 g/L P85 treatments (Figure 3b). The NBI was significantly higher than the control only for 100 mg/L P85-SWCNT and 1 g/L P85 but was slightly inhibited for 10 g/L P85-primed plants (Figure 3c).

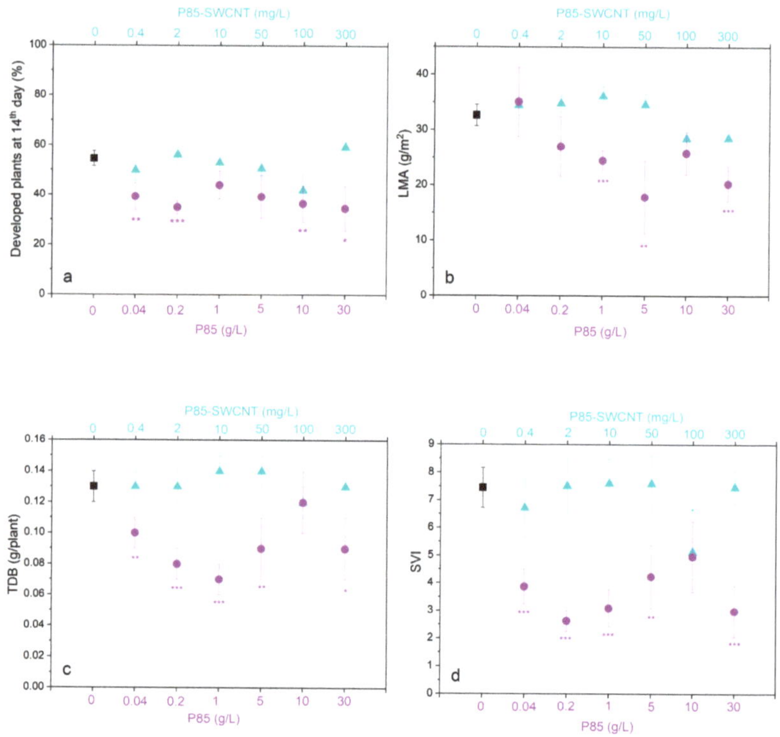

Figure 2. Plant growth parameters determined for hydro-, P85-SWCNT-, and P85- primed seeds at the 14th day after seeds sowing. Percentage developed plants (**a**), leaf dry mass per unit leaf area (LMA, (**b**)), total dry biomass (TDB, (**c**)) and survival index (SVI, (**d**)). Mean ± SE. Statistically significant differences from the control sample at $p < 0.1$ (*), $p < 0.05$ (**), $p < 0.01$ (***) according to Student's *t*-test are indicated with asterisk in the respective colors (violet for P85 and cyan for P85-SWCNT treatments).

Table 3. Anatomical characteristics of *Pisum sativum* leaves developed from hydro-primed control seeds and seeds primed in dispersions with different concentrations of P85-SWCNT and P85. Mean ± SE. Statistically significant differences from the control sample at $p < 0.05$ (**), $p < 0.01$ (***) according to Student's *t*-test are indicated.

Concentration	Leaf Thickness (μm)	Adaxial and Abaxial Epidermis Thickness (μm)	Mesophyll Thickness (μm)	Palisade Parenchyma Thickness (μm)	Spongy Parenchyma Thickness (μm)
			H_2O (control)		
	268 ± 18	52 ± 7	215 ± 16	41 ± 7	173 ± 10
(mg/L)			P85-SWCNT		
50	201 ± 14 **	49 ± 6	149 ± 14 **	38 ± 8	111 ± 15 ***
100	209 ± 15 **	42 ± 5	161 ± 10 **	45 ± 9	122 ± 16 **
300	198 ± 12 ***	37 ± 5	154 ± 11 ***	44 ± 8	107 ± 18 **
(g/L)			P85		
5	159 ± 11 ***	26 ± 4 ***	129 ± 9 ***	46 ± 9	79 ± 16 ***
10	165 ± 12 ***	25 ± 3 ***	145 ± 12 ***	49 ± 5	98 ± 16 ***
30	158 ± 11 ***	26 ± 4 ***	129 ± 9 ***	38 ± 9	93 ± 17 ***

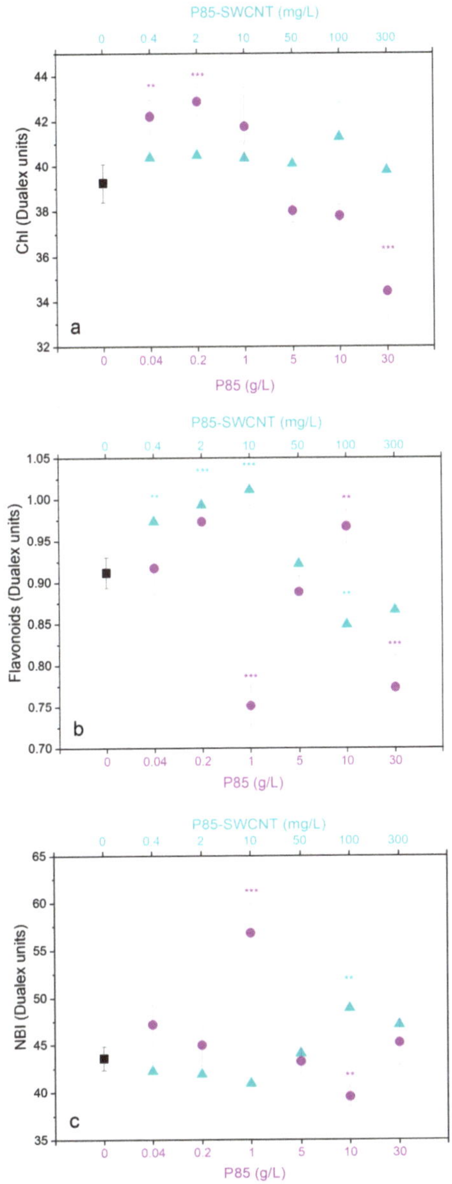

Figure 3. Adaxial leaf surface characterization for plants developed from hydro-, P85-SWCNT, and P85-primed seeds. Data are obtained on intact leaves at the 14th day after seeds sowing. Total chlorophyll (Chl) content (**a**), total flavonoids content (**b**), nitrogen balance index (NBI, (**c**)). Mean ± SE. Statistically significant differences from the control sample at $p < 0.05$ (**), $p < 0.01$ (***) according to Student's *t*-test are indicated with the respective colors (violet for P85 and cyan for P85-SWCNT treatments).

Among all P85-SWCNT treated variants, only 100 mg/L exhibited slightly but significantly higher (at $p < 0.01$) maximum quantum yield of PSII than the control, while this parameter was higher for all P85 treatments, with the exception of 10 g/L (Figure 4a).

The Φ_{PSII} parameter increased proportionally to the applied P85-SWCNT amount, but for P85 treatments it was 14–19% higher than the control already at the lowest applied concentration (Figure 4b). A concentration dependence in qL was observed for both P85-SWCNT and P85 treatments (Figure 4c). The steady-state NPQ values reached after 15 min of illumination gradually decreased with the increase in P85-SWCNT concentration, while for all P85 treatments they were largely reduced (by ca. 25%), as compared to hydro-primed seeds (Figure 4d).

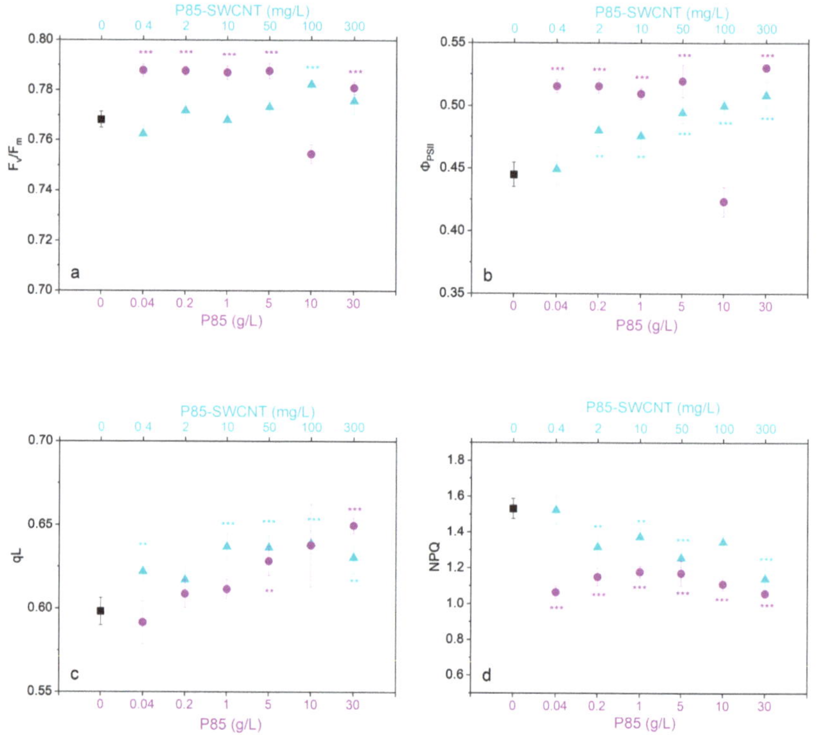

Figure 4. Chlorophyll fluorescence parameters determined on intact leaves of 14-day-old plants developed from hydro-, P85-SWCNT-, and P85-primed seeds. Maximum quantum yield of photosystem II in the dark (F_v/F_m, (**a**)), quantum efficiency of PSII photochemistry in the light (Φ_{PSII}, (**b**)), fraction of open PSII reaction centers (qL, (**c**)), and non-photochemical quenching of Chl *a* fluorescence determined after 15 min of illumination (NPQ, (**d**)). Mean ± SE. Statistically significant differences from the control sample at $p < 0.05$ (**), $p < 0.01$ (***) according to Student's *t*-test are indicated with the respective colors (violet for P85 and cyan for P85-SWCNT treatments).

The kinetics of NPQ development followed a similar trend for hydro- and P85-SWCNT-primed samples in the concentration range of 0.4–100 mg/L, i.e., a first phase associated with the initial increase in NPQ values within the first 5 min after illumination and a second one related to attainment of (quasi) steady-state NPQ value after 15 min of illumination. Only for the 300 mg/L P85-SWCNT, the first phase was essentially missing, and the NPQ values for the whole period of 15 min of illumination were lower than in hydro-primed variants (Figure 5). Interestingly, the P85 treatments followed a different course since no clear maximum was visible, and the NPQ values reached steady state already in the first 5 min of illumination; they also remained dramatically lower than the control ones for the tested light period.

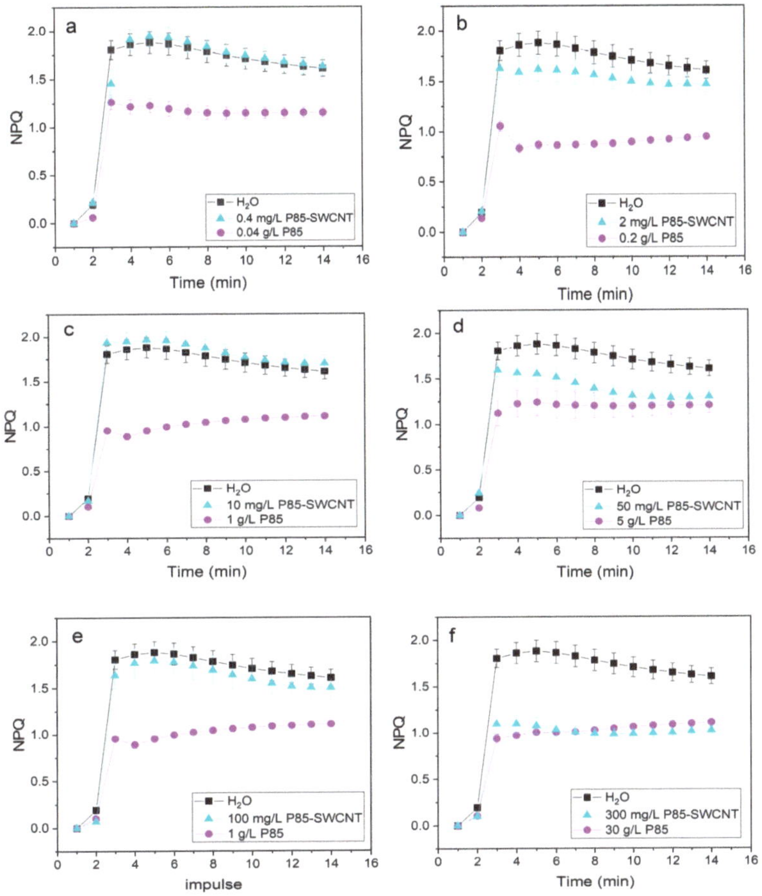

Figure 5. Traces of NPQ kinetics recorded for 14-day old plants developed from hydro-, P85-SWCNT-, and P85-primed seeds. Data for 0.4 mg/L (**a**), 2 mg/L (**b**), 10 mg/L (**c**), 50 mg/L (**d**), 100 mg/L (**e**) and 300 mg/L (**f**) P85-SWCNT (triangles), as well as the corresponding P85 treatments (circles), are presented in different panels. For clarity each panel contains data for hydro-primed samples (squares) as well. Mean ± SE. For further experimental details, see Materials and Methods.

4. Discussion

4.1. Advancements in Nanopriming as New Tool for Germination Improvement

Among the different types of interactions between nanomaterials and plants, the seed priming technique emerges as a widely used approach for improvement of the germination process of a variety of crops [1,3]; however, its further elaboration continues to be of high research interest. The advancements in nanotechnology prompted the researchers to study novel nanomaterials as priming agents; however, this field is still in its infancy. To the best of our knowledge, there are only two reports in the literature regarding the exploitation of carbon nanotubes for this purpose. Yousefi et al. [36] utilized 10–100 mg/L MWCNT as the priming agent for hopbush seeds, which resulted in enhanced germination and biomass accumulation, including in drought stress conditions. However, Lopez-Vargas et al. [37] did not observe any effect on germination after tomato seed priming with graphene and MWCNT (applied in the range 10–1000 mg/L). On the contrary, the authors reported negative impacts on root length and hypocotyl biomass. While 1000 mg/L concentration had a multitude of negative effects, the lower concentrations of the two nanomaterials exhibited distinct and sometimes contrasting effects on a large number of growth-related parameters and enzymatic activities, with strong concentration dependence [37]. The hormesis effect of carbonaceous nanomaterials (effectiveness vs. nanotoxicity) is documented in the literature for different nanomaterials, plant species, and treatment protocols and is recently reviewed in [38].

To gain further fundamental knowledge on the effect of seed priming with carbon nanotubes, in this work, for the first time, we explored the impact of P85-SWCNT on the early stages of pea plant development. It should be noted that in the above-mentioned reports, bare MWCNT were utilized, although this poses a problem with preserving those nanomaterials stably in a non-aggregated state, especially within the plant tissues and organelles. To overcome this obstacle, in our work, we utilized P85-SWCNT (i.e., polymer grafted) that are known to form stable dispersions for a prolonged period. However, since P85 molecules (unimers) are still present in the utilized dispersions, it was also needed to probe their individual action on seeds priming. Although pluronics in general are widely used for biomedical purposes [39], their interactions with seeds and plants are scarcely investigated. To shine some light on this problem, we also characterized the effect of Pluronic P85 as the priming agent. This study is also justified by a number of reports identifying different polymers as beneficial seed coating agents—for example, the study of Samal et al. [40] showed that the coating of cowpea seeds with methyl cellulose, ethyl cellulose, polyvinyl pyrrolidone, hydroxy propyl cellulose, and methyl vinyl acetate polymers results in significantly higher seed germination and better field performance and seed yield per hectare after 6 months of storage in dry conditions. Seeds coating with water absorbent materials is regarded as a novel technique for seeds preservation from drought stress in arid areas [41]. A recent development of polymeric nanofibrous coatings composed of cellulose diacetate is shown to be promising for slow active ingredients delivery to seeds [42].

Our data clearly demonstrate that P85-SWCNT generally appear safe for the plant, at least below 300 mg/L, while P85 exerts strong inhibiting effects, especially on plant biomass accumulation and photoprotection ability. These contrasting effects are further discussed in detail below.

4.2. Optimal Conditions for Pea Seed Priming with P85-SWCNT

For the initial 4 days after seeds sowing, we did not detect significant effects of either P85-SWCNT- or P85-priming on seeds imbibition during the 6 h priming procedure and on the germination process. Exploring the plant growth parameters (% developed plants, LMA, TBD, and SVI) of 14 day-old plants developed from P85-SWCNT primed seeds revealed that they are very similar to the hydro-primed ones. The leaf anatomical characteristics (leaf and spongy parenchyma thickness) were significantly affected only for 50–300 mg/L P85-SWCNT treatments; however, the adaxial and abaxial epidermis thickness was preserved as in the control samples. In functional terms, P85-SWCNT primed variants also did not exhibit large deviations as compared to the hydro-primed ones. The total adaxial surface Chl content and Chl fluorescence parameters determined on dark-adapted intact leaves did not vary from the control plants. In light-adapted conditions, however, those variants exhibited higher PSII yield at concentrations above 2 mg/L P85-SWCNT and increased qL values for all tested concentrations, strongly suggesting the presence of more photochemically functional (open) PSII reactions centers with the increase in P85-SWCNT concentration. The NPQ traces recorded for P85-SWCNT-primed variants resembled closely the one of hydro-primed variants, with the exception of 300 mg/L P85-SWCNT.

The fact that P85-SWCNT do not affect seed characteristics but do affect photosynthetic parameters of developed plants strongly suggests that they penetrate the seed without affecting its metabolism and translocate to the photosynthetic organelles where they exert their action. Indeed, numerous authors state that CNT can travel through phloem and xylem and accumulate in different plant organs/organelles, including the chloroplast [4]. The mechanism by which CNT interact with seeds and plant tissues is still unclear; however, two possibilities are proposed—perforation of membranes (i.e., pores formation) and transport via aquaporins [11].

4.3. Harmful Effects of Priming with P85 Polymer

P85 treatments exhibited a number of harmful effects at different growth stages—starting from the seed germination (i.e., the root formation stage), followed by thinner lamina and decreased LMA (at concentration ≥ 1 g/L P85), associated with lower biomass accumulation and plants survival and strong inhibition of NPQ (both the initial phases of its development and in its (quasi)steady-state value after 15 min of illumination), which strongly influences the capability of plants to cope with light stress. The lower LMA and leaf lamina thickness is believed to be related to lower physical strength and shorter leaf lifespans [43–45], but it also relates to photosynthesis efficiency via allocation of nitrogen within photosynthetic tissue, CO_2 diffusion regulation, and light penetration within photosynthetic tissues [45]. Based on our results, we assume that the lower LMA in P85-treated variants results in higher photon flux reaching the photosynthetic apparatus, which leads to higher PSII quantum yield. This statement is supported by the fact that the arrangement of palisade cells in P85 plants should in principle facilitate light channeling deeper into the leaf [46–49]. However, the decrease in mesophyll thickness (due to lower spongy parenchyma) in P85-primed samples must affect the light scattering and focusing within the leaf as well as the gas conductivity processes and is probably associated with lower tolerance for high light intensity [48,50]. In a number of species, lower mesophyll thickness is also correlated with larger capacity for chloroplast movement—another tool to cope with short-term light intensity fluctuations [51]. Furthermore, the thinner epidermis in P85 variants is likely to reduce the leaf protection from the harmful UV-B rays [48]. All those features have long-term consequences for the plant growth of P85 variants as well as for their photosynthetic and photoprotection function since thinner mesophyll was also shown to be related to incomplete development of the NPQ process [52,53]. The data clearly show that in P85-primed variants, a large portion of the PSII excitation energy is not safely dissipated as heat (a major mechanism for photoprotection) but emitted as fluorescence (hence the high PSII quantum yield). Furthermore, those variants appear to be trapped in a partially photoinhibited state that does not allow for full development of the NPQ process. Thus, it could be expected that P85 priming will enhance the plant susceptibility to high light exposure and consequently photooxidative damages to the photosynthetic apparatus, which however requires further experimental verification. Recent studies showed that the time course of NPQ is related to biomass accumulation [54–56]. Indeed, in our study we found a similar correlation—a significant reduction in biomass accumulation is accompanied with substantially lower NPQ values in P85 variants.

The variety of negative effects of P85 priming can be explained with the low size of P85 unimers that can easily transverse the seed coat and cell wall, and enter the plant cells. It is well known that due to their amphiphilic nature pluronic unimers and micelles do interact with biological membranes and modify their properties (reviewed in [56]).

5. Conclusions

Achieving sustainable agriculture is a major goal for the present and future generations since there is an urgent need for reduction of environmental pollution, sustainable utilization of the natural resources, development of crops that would resist climate change challenges, and optimization of food production and price [57].

Here, we explored a number of growth- and physiology-related parameters as markers for the effect of pea seed priming with Pluronic P85 tri-block copolymer grafted single-walled carbon nanotubes (P85-SWCNT). Detailed exploration of the seed germination process, plant biometry characteristics, leaf anatomy, physiology, and photosynthetic activity revealed that P85-SWCNT seed priming in the concentration range of 0.4–100 mg/L is safe for pea plant development and even stimulates photosystem II photochemistry. Therefore, P85-SWCNT nanoparticles appear as suitable object for further development as nanocarriers of specific substances acting as plant growth regulators. Importantly, anatomical and physiological changes occurring in P85-SWCNT- and P85-treated plants revealed that those two agents have distinct modes of action. On one hand, it appears that P85-SWCNT application supports plants to largely overcome the negative impact of the P85 polymer. On the other hand, however, the fact that a large amount of P85 unimers is in a bound form in P85-SWCNT particles might be the reason for the much smaller negative effect exerted by the free P85 unimers in P85-SWCNT dispersion. Nevertheless, the evaluation of P85-SWCNT toxicity, in particular, requires a further dedicated study.

Supplementary Materials: The following supporting information can be downloaded at: https://www.mdpi.com/article/10.3390/nano13081332/s1, Figure S1: Leaf sections obtained for 14-day-old plants.

Author Contributions: Conceptualization, V.V. and S.K.; methodology, V.V., T.T., P.P., A.P., D.I., N.P., S.S. and D.K.; software, T.T.; formal analysis, V.V., T.T., A.P. and S.K.; data curation, V.V.; writing—original draft preparation, S.K. and V.V.; writing—review and editing, T.T., P.P., A.P., N.P., S.S. and D.K.; visualization, S.K.; supervision, V.V. and S.K.; project administration, S.K. and V.V.; funding acquisition, V.V. and S.K. All authors have read and agreed to the published version of the manuscript.

Funding: This research was funded by Bulgarian Science Fund, grant number KΠ-06-H36/8/13.12.2019.

Data Availability Statement: The data are contained within the article and the Supplementary Material.

Conflicts of Interest: The authors declare no conflict of interest.

References

1. Mondal, S.; Bose, B. Seed Priming: An Interlinking technology between seeds, seed germination and seedling establishment. In *Plant Reproductive Ecology—Recent Adwances*; Rustagi, A., Chaudhry, B., Eds.; IntechOpen: London, UK, 2021; pp. 1–16.
2. Chetyrkina, M.R.; Fedorov, F.S.; Nasibulin, A.G. In vitro toxicity of carbon nanotubes: A systematic Review. *RSC Adv.* **2022**, *12*, 16235–16256. [CrossRef] [PubMed]
3. Nile, S.H.; Thiruvengadam, M.; Wang, Y.; Samynathan, R.; Shariati, M.A.; Rebezov, M.; Nile, A.; Sun, M.; Venkidasamy, B.; Xiao, J.; et al. Nano-priming as emerging seed priming technology for sustainable agriculture–recent developments and futureperspectives. *J. Nanobiotechnol.* **2022**, *20*, 254. [CrossRef] [PubMed]
4. Safdar, M.; Kim, W.; Park, S.; Gwon, Y.; Kim, Y.O.; Kim, J. Engineering plants with carbon nanotubes: A sustainable agriculture approach. *J. Nanobiotechnol.* **2022**, *20*, 275. [CrossRef] [PubMed]
5. Shelar, A.; Nile, S.H.; Singh, A.V.; Rothenstein, D.; Bill, J.; Xiao, J.; Chaskar, M.; Kai, G.; Patil, R. Recent advances in nano-enabled seed treatment strategies for sustainable agriculture: Challenges, risk assessment, and future perspectives. *Nano-Micro Lett.* **2023**, *15*, 54. [CrossRef]
6. Yan, S.; Zhao, L.; Li, H.; Zhang, Q.; Tan, J.; Huang, M.; He, S.; Li, L. Single-walled carbon nanotubes selectively influence maize root tissue development accompanied by the change in the related gene expression. *J. Hazard. Mater.* **2013**, *246-247*, 110–118. [CrossRef]
7. Haghighi, M.; da Silva, J.A.T. The effect of carbon nanotubes on the seed germination and seedling growth of four vegetable species. *J. Crop Sci. Biotechnol.* **2014**, *17*, 201–208. [CrossRef]
8. Pourkhaloee, A.; Haghigh, M.; Saharkhiz, M.J.; Jouzi, H.; Doroodmand, M.M. Carbon nanotubes can promote seed germination via seed coat penetration. *Seed Technol.* **2011**, *33*, 155–169.
9. Hatami, M.; Hadian, J.; Ghorbanpour, M. Mechanisms underlying toxicity and stimulatory role of single-walled carbon nanotubes in Hyoscyamus niger during drought stress simulated by polyethylene glycol. *J. Hazard. Mater.* **2017**, *324*, 306–320. [CrossRef]
10. Cañas, J.E.; Long, M.; Nations, S.; Vadan, R.; Dai, L.; Luo, M.; Ambikapathi, R.; Lee, E.H.; Olszyk, D. Effects of functionalized and nonfunctionalized single-walled carbon nanotubes on root elongation of select crop species. *Environ. Toxicol. Chem.* **2008**, *27*, 1922–1931. [CrossRef]
11. Khodakovskaya, M.V.; Dervishi, E.; Mahmood, M.; Xu, Y.; Li, Z.; Watanabe, F.; Biris, A.S. Carbon nanotubes are able to penetrate plant seed Coat and dramatically affect seed germination and plant growth. *ACS Nano* **2009**, *3*, 3221–3227. [CrossRef]
12. Lahiani, M.H.; Dervishi, E.; Chen, J.; Nima, Z.; Gaume, A.; Biris, A.S.; Khodakovskaya, M.V. Impact of carbon nanotube exposure to seeds of valuable crops. *ACS Appl. Mater. Interfaces* **2013**, *5*, 7965–7973. [CrossRef]
13. Khodakovskaya, M.V.; Kim, B.S.; Kim, J.N.; Alimohammadi, M.; Dervishi, E.; Mustafa, T.; Cernigla, C.E. Carbon nanotubes as plant growth regulators: Effects on tomato growth, reproductive system, and soil microbial community. *Small* **2013**, *9*, 115–123. [CrossRef]
14. Liu, Q.; Chen, B.; Wang, Q.; Shi, X.; Xiao, Z.; Lin, J.; Fang, X. Carbon nanotubes as molecular transporters for walled plant cells. *Nano Lett.* **2009**, *9*, 1007–1010. [CrossRef]

15. Lew, T.T.S.; Wong, M.H.; Kwak, S.-Y.; Sinclair, R.; Koman, V.B.; Strano, M.S. Rational design principles for the transport and subcellular distribution of nanomaterials into plant protoplasts. *Small* **2018**, *14*, 1802086. [CrossRef] [PubMed]
16. Giraldo, J.P.; Landry, M.P.; Faltermeier, S.M.; McNicholas, T.P.; Iverson, N.M.; Boghossian, A.A.; Reuel, N.F.; Hilmer, A.J.; Sen, F.; Brew, J.A.; et al. Plant nanobionics approach to augment photosynthesis and biochemical sensing. *Nat. Mater.* **2014**, *13*, 400–408. [CrossRef]
17. Ghasemi-Kooch, M.; Dehestani, M. Interaction of photosynthetic pigments with single-walled carbon nanotube (15, 15): A molecular dynamics study. *Adsorption* **2018**, *24*, 43–51. [CrossRef]
18. Dorogi, M.; Balint, Z.; Mikó, C.; Vileno, B.; Milas, M.; Hernadi, K.; Forró, L.; Varó, G.; Nagy, L. Stabilization effect of single-walled carbon nanotubes on the functioning of photosynthetic reaction centers. *J. Phys. Chem. B* **2006**, *110*, 21473–21479. [CrossRef]
19. Mackowski, S. Hybrid nanostructures for efficient light harvesting. *J. Phys. Condens. Matter* **2010**, *22*, 193102. [CrossRef] [PubMed]
20. Hajdu, K.; Szabó, T.; Magyar, M.; Bencsik, G.; Németh, Z.; Nagy, K.; Magrez, A.; Forró, L.; Váró, G.; Hernádi, K.; et al. Photosynthetic reaction center protein in nanostructures. *Phys. Stat. Sol. B* **2011**, *248*, 2700–2703. [CrossRef]
21. Nagy, L.; Magyar, M.; Szabó, T.; Hajdu, K.; Giotta, L.; Dorogi, M.; Milano, F. Photosynthetic machineries in nano-systems. *Curr. Protein Pept. Sci.* **2014**, *15*, 363–373. [CrossRef]
22. Wiwatowski, K.; Duzynska, A.; Swiniarski, M.; Szalkowski, M.; Zdrojek, M.; Judek, J.; Mackowski, S.; Kaminska, I. Energy transfer from natural photosynthetic complexes to single-wall carbon nanotubes. *J. Lumin.* **2016**, *170*, 855–859. [CrossRef]
23. Orlanducci, S.; Fulgenzi, G.; Margonelli, A.; Rea, G.; Antal, T.K.; Lambreva, M.D. Mapping single walled carbon nanotubes in photosynthetic algae by single-cell confocal raman microscopy. *Materials* **2020**, *13*, 5121. [CrossRef] [PubMed]
24. Antal, T.K.; Volgusheva, A.A.; Kukarskikh, G.G.; Lukashev, E.P.; Bulychev, A.A.; Margonelli, A.; Orlanducci, S.; Leo, G.; Cerri, L.; Tyystjärvi, E.; et al. Single-walled carbon nanotubes protect photosynthetic reactions in *Chlamydomonas reinhardtii* against photoinhibition. *Plant Physiol. Biochem.* **2022**, *192*, 298–307. [CrossRef] [PubMed]
25. Lambreva, M.D.; Akhtar, P.; Sipka, G.; Margonelli, A.; Lambrev, P.H. Fluorescence quenching in thylakoid membranes induced by single-walled carbon nanotubes. *Photochem. Photobiol. Sci* **2023**. *Epub ahead of print*. [CrossRef] [PubMed]
26. Kwak, S.Y.; Lew, T.T.S.; Sweeney, C.J.; Koman, V.B.; Wong, M.H.; Bohmert-Tatarev, K.; Snell, K.D.; Seo, J.S.; Chua, N.H.; Strano, M.S. Chloroplast-selective gene delivery and expression in planta using chitosan-complexed singlewalled carbon nanotube carriers. *Nat. Nanotechnol.* **2019**, *14*, 447–455. [CrossRef]
27. Velikova, V.; Petrova, N.; Kovács, L.; Petrova, A.; Koleva, D.; Tsonev, T.; Taneva, S.; Petrov, P.; Krumova, S. Single-walled carbon nanotubes modify leaf micromorphology, chloroplast ultrastructure and photosynthetic activity of pea plants. *Int. J. Mol. Sci.* **2021**, *22*, 4878. [CrossRef]
28. Petrova, N.; Paunov, M.; Petrov, P.; Velikova, V.; Goltsev, V.; Krumova, S. Polymer-modified single-walled carbon nanotubes affect photosystem II photochemistry, intersystem electron transport carriers and photosystem I end acceptors in pea plants. *Molecules* **2021**, *26*, 5958. [CrossRef]
29. Karkanis, A.; Ntatsi, G.; Kontopoulou, C.K.; Pristeri, A.; Bilalis, D.; Savvas, D. Field pea in European cropping systems: Adaptability, biological nitrogen fixation and cultivation practices. *Not. Bot. Horti. Agrobo Cluj Napoca* **2016**, *44*, 325–336. [CrossRef]
30. Petrov, P.D.; Georgiev, G.L. Fabrication of super-macroporous nanocomposites by deposition of carbon nanotubes onto polymer cryogels. *Eur. Polym. J.* **2012**, *48*, 1366–1373. [CrossRef]
31. Ranal, M.A.; de Santana, D.G.; Ferreira, W.R.; Mendes-Rodrigues, C. Calculating germination measurements and organizing spreadsheets. *Rev. Bras. Botânica* **2009**, *32*, 849–855. [CrossRef]
32. Velikova, V.; Arena, C.; Izzo, L.G.; Tsonev, T.; Koleva, D.; Tattini, M.; Roeva, O.; De Maio, A.; Loreto, F. Functional and Structural Leaf Plasticity Determine Photosynthetic Performances during Drought Stress and Recovery in Two Platanus orientalis Populations from Contrasting Habitats. *Int. J. Mol. Sci.* **2020**, *21*, 3912. [CrossRef]
33. Genty, B.; Briantais, J.M.; Baker, N.R. The relationship between the quantum yield of photosynthetic electron transport and quenching of chlorophyll fluorescence. *Biochim. Biophys. Acta* **1989**, *990*, 87–92. [CrossRef]
34. Bilger, W.; Björkman, O. Temperature dependence of violaxanthin de-epoxidation and non-photochemical fluorescence quenching in intact leaves of *Gossypium hirsutum* L. and *Malva parviflora* L. *Planta* **1991**, *184*, 226–234. [CrossRef] [PubMed]
35. Baker, N.R. Chlorophyll fluorescence: A probe of photosynthesis in vivo. *Ann. Rev. Plant Biol.* **2008**, *59*, 89–113. [CrossRef] [PubMed]
36. Yousefi, S.; Kartoolinejad, D.; Naghdi, R. Effects of priming with multi-walled carbon nanotubes on seed physiological characteristics of hopbush (Dodonaea viscosa L.) under drought stress. *Int. J. Environ. Res.* **2017**, *74*, 528–539. [CrossRef]
37. López-Vargas, E.R.; González-García, Y.; Pérez-Álvarez, M.; Cadenas-Pliego, G.; González-Morales, S.; Benavides-Mendoza, A.; Cabrera, R.I.; Juárez-Maldonado, A. Seed priming with carbon nanomaterials to modify the germination, growth, and antioxidant status of tomato seedlings. *Agronomy* **2020**, *10*, 639. [CrossRef]
38. Wu, Q.; Fan, C.; Wang, H.; Han, Y.; Tai, F.; Wu, J.; Li, H.; He, R. Biphasic impacts of graphite-derived engineering carbon-based nanomaterials on plant performance: Effectiveness vs. nanotoxicity. *Adv. Agrochem.* **2023**, *in press*. [CrossRef]
39. Batrakova, E.V.; Kabanov, A.V. Pluronic block copolymers: Evolution of drug delivery concept from inert nanocarriers to biological response modifiers. *J. Control. Release* **2008**, *130*, 98–106. [CrossRef]
40. Samal, A.; Mohanty, S.; Das, S.; Das, B.C.; Beura, J.K. Polymer coating of cowpea seeds (Vigna unguiculata) for improving its storability and performance. *J. Pharm. Innov.* **2021**, *10*, 592–597.

41. Su, L.Q.; Li, J.G.; Xue, H.; Wang, X.F. Super absorbent polymer seed coatings promote seed germination and seedling growth of Caragana korshinskii in drought. *J. Zhejiang Univ. Sci B.* **2017**, *18*, 696–706. [CrossRef]
42. Farias, B.V.; Pirzada, T.; Mathew, R.; Sit, T.L.; Opperman, C.; Khan, S.A. Electrospun polymer nanofibers as seed coatings forcrop protection. *ACS Sustain. Chem. Eng.* **2019**, *7*, 19848–19856. [CrossRef]
43. Choong, M.F.; Lucas, P.W.; Ong, J.S.Y.; Pereira, B.; Tan, H.T.W.; Turner, I.M. Leaf fracture toughness and sclerophylly: Their correlations and ecological implications. *New Phytol.* **1992**, *121*, 597–610. [CrossRef]
44. Wright, I.; Reich, P.; Westoby, M.; Ackerly, D.D.; Baruch, Z.; Bongers, F.; Cavender-Bares, J.; Chapin, T.; Cornelissen, J.H.; Diemer, M.; et al. The worldwide leaf economics spectrum. *Nature* **2004**, *428*, 821–827. [CrossRef] [PubMed]
45. Onoda, Y.; Wright, I.J.; Evans, J.R.; Hikosaka, K.; Kitajima, K.; Niinemets, Ü.; Poorter, H.; Tosens, T.; Westoby, M. Physiological and structural tradeoffs underlying the leaf economics spectrum. *New Phytol.* **2017**, *214*, 1447–1463. [CrossRef]
46. Vogelmann, T.C.; Martin, G. The functional significance of palisade tissue: Penetration of directional versus diffuse light. *Plant Cell Environ.* **1993**, *16*, 65–72. [CrossRef]
47. Terashima, I.; Saeki, T. Light environment within a leaf. I. Optical properties of paradermal sections of Camellia leaves with special reference to differences in the optical properties of palisade and spongy tissues. *Plant Cell Physiol.* **1984**, *24*, 1493–1501. [CrossRef]
48. Karabourniotis, G.; Liakopoulos, G.; Bresta, P.; Nikolopoulos, D. The Optical Properties of Leaf Structural Elements and Their Contribution to Photosynthetic Performance and Photoprotection. *Plants* **2021**, *10*, 1455. [CrossRef] [PubMed]
49. Cui, M.; Vogelmann, T.C.; Smith, W.K. Chlorophyll and light gradients in sun and shade leaves of Spinacia oleracea. *Plant Cell Environ.* **1991**, *14*, 493–500. [CrossRef]
50. Théroux-Rancourt, G.; Roddy, A.B.; Earles, J.M.; Gilbert, M.E.; Zwieniecki, M.A.; Boyce, C.K.; Tholen, D.; McElrone, A.J.; Simonin, K.A.; Brodersen, C.R. Maximum CO_2 diffusion inside leaves is limited by the scaling of cell size and genome size. *Proc. Biol. Sci. B.* **2021**, *288*, 20203145. [CrossRef]
51. Davis, P.A.; Caylor, S.; Whippo, C.W.; Hangarter, R.P. Changes in leaf optical properties associated with light-dependent chloroplast movements. *Plant Cell Environ.* **2011**, *34*, 2047–2059. [CrossRef]
52. Matos, F.S.; Wolfgramm, R.; Gonc alves, F.V.; Cavatte, P.C.; Ventrella, M.C.; DaMatta, F.M. Phenotypic plasticity in response to light in the coffee tree. *Environ. Exp. Bot.* **2009**, *67*, 421–427. [CrossRef]
53. Kalmatskaya, O.A.; Karavaev, V.A.; Tikhonov, A.N. Slow induction of chlorophyll a fluorescence excited by blue and red light in Tradescantia leaves acclimated to high and low light. *Photosynth. Res.* **2019**, *142*, 265–282. [CrossRef] [PubMed]
54. Kromdijk, J.; Głowacka, K.; Leonelli, L.; Gabilly, S.T.; Iwai, M.; Niyogi, K.K.; Long, S.P. Improving photosynthesis and crop productivity by accelerating recovery from photoprotection. *Science* **2016**, *354*, 857–861. [CrossRef] [PubMed]
55. De Souza, A.P.; Burgess, S.J.; Doran, L.; Hansen, J.; Manukyan, L.; Maryn, N.; Gotarkar, D.; Leonelli, L.; Niyogi, K.K.; Long, S.P. Soybean photosynthesis and crop yield are improved by accelerating recovery from photoprotection. *Science* **2022**, *377*, 851–854. [CrossRef] [PubMed]
56. Ghosh, D.; Mohapatra, S.; Dogra, V. Improving photosynthetic efficiency by modulating non-photochemical quenching. *Trends Plant Sci.* **2023**, *28*, 264–266. [CrossRef] [PubMed]
57. Dziergowska, K.; Michalak, I. The role of nanoparticles in sustainable agriculture. In *Smart Agrochemicals for Sustainable Agriculture*; Chojnacka, K., Saeid, A., Eds.; Elsevier Inc.: Amsterdam, The Netherlands, 2022; pp. 225–278. [CrossRef]

nanomaterials

MDPI

Article

Microstructure and Superconducting Properties of Bi-2223 Synthesized via Co-Precipitation Method: Effects of Graphene Nanoparticle Addition

Siti Nabilah Abdullah [1], Mohd Mustafa Awang Kechik [1,*], Aliah Nursyahirah Kamarudin [1], Zainal Abidin Talib [2], Hussein Baqiah [3], Chen Soo Kien [1], Lim Kean Pah [1], Muhammad Khalis Abdul Karim [1], Muhammad Kashfi Shabdin [1], Abdul Halim Shaari [1], Azhan Hashim [4], Nurbaisyatul Ermiza Suhaimi [4] and Muralidhar Miryala [5]

1 Laboratory of Superconductor and Thin Films, Department of Physics, Faculty of Science, Universiti Putra Malaysia, Serdang 43400, Malaysia
2 Department of Physics, College of Natural Sciences, Jeonbuk National University 567, Baekje-daero, Deokjin-gu, Jeonju-si 54896, Republic of Korea
3 Shandong Key Laboratory of Biophysics, Institute of Biophysics, Dezhou University, No. 566 University Rd. West, Dezhou 253023, China
4 Faculty of Applied Sciences, Universiti Teknologi MARA Pahang, Jengka 26400, Malaysia
5 Materials for Energy and Environmental Laboratory, Superconducting Materials, Shibaura Institute of Technology, 3 Chome-7-5 Toyosu, Koto, Tokyo 135-8548, Japan
* Correspondence: mmak@upm.edu.my

Abstract: The effects of graphene addition on the phase formation and superconducting properties of $(Bi_{1.6}Pb_{0.4})Sr_2Ca_2Cu_3O_{10}$ (Bi-2223) ceramics synthesized using the co-precipitation method were systematically investigated. Series samples of Bi-2223 were added with different weight percentages (x = 0.0, 0.3, 0.5 and 1.0 wt.%) of graphene nanoparticles. The samples' phase formations and crystal structures were characterized via X-ray diffraction (XRD), while the superconducting critical temperatures, T_c, were investigated using alternating current susceptibility (ACS). The XRD showed that a high-T_c phase, Bi-2223, and a small low-T_c phase, Bi-2212, dominated the samples. The volume fraction of the Bi-2223 phase increased for the sample with x = 0.3 wt.% and 0.5 wt.% of graphene and slightly reduced at x = 1.0 wt.%. The ACS showed that the onset critical temperature, $T_{c\text{-onset}}$, phase lock-in temperature, T_{cj}, and coupling peak temperature, T_P, decreased when graphene was added to the samples. The susceptibility–temperature (χ'-T) and (χ''-T) curves of each sample, where χ' and χ'' are the real and imaginary parts of the susceptibility, respectively, were obtained. The critical temperature of the pure sample was also measured.

Keywords: Bi-2223; co-precipitation method; graphene nanoparticles; critical temperature; critical current

check for updates

Citation: Abdullah, S.N.; Kechik, M.M.A.; Kamarudin, A.N.; Talib, Z.A.; Baqiah, H.; Kien, C.S.; Pah, L.K.; Abdul Karim, M.K.; Shabdin, M.K.; Shaari, A.H.; et al. Microstructure and Superconducting Properties of Bi-2223 Synthesized via Co-Precipitation Method: Effects of Graphene Nanoparticle Addition. *Nanomaterials* **2023**, *13*, 2197. https://doi.org/10.3390/nano13152197

Academic Editor: Oleksandr V. Dobrovolskiy

Received: 28 February 2023
Revised: 6 June 2023
Accepted: 15 June 2023
Published: 28 July 2023

1. Introduction

Bismuth strontium calcium copper oxide, BSCCO, has the general chemical formula $Bi_2Sr_2Ca_{n-2}Cu_nO_x$, and its critical temperatures, T_{cj}, for the first few members with Bi-2201, Bi-2212, and Bi-2223, are 20 K, 85 K, and 110 K, respectively [1–3]. Since Bi-2223 has the highest superconducting transition temperature, T_c, in their series, it is the most attractive for potential application.

BSCCO is a class of high-temperature superconductors (HTS) that has been widely investigated and applied in engineering, medical equipment, mining, and transport systems. In 2004, high-performance and long-length Bi-2223 HTS wires were successfully commercialized and well-received in the market [4–10]. These commercialized Bi-2223 wires or DI-BSCCO (Dynamically Innovative BSCCO) have been used in several cable projects, such as the Albany project in the United States [6]; the Yokohama project in Japan [7]; the

AmpaCity project in Germany [8]; the DC cable project in Saint Petersburg, Russia [9]; and possibly as superconducting feeder cables to railway lines in Japan and France, which is still being appraised [10]. Bulk superconducting materials have various applications, including current leads, electrical fault current limiters, and electromagnetic levitation systems. However, the limitations of the Bi-2223 system, such as weak critical current density, J_c, due to intergrain weak links, and weak flux pinning ability in bulk samples, have hindered progress. The current fabrication technique has succeeded in producing Bi-2223 wires with J_c as high as 7.4×10^4 A/cm^2 at 77.3 K and a self-magnetic field [11] for Bi-2223 thin films of 1.3×10^6 A/cm^2 at 70 K [12]. Magneto-optical analyses have revealed that for Bi-2223 wires, the local J_c had reached 2.5×10^5 A/cm^2 [13]. To extend the applications of HTS wires, their basic properties, particularly the performance of I_c, must be improved to enhance the J_c. One approach that has been broadly accepted is the addition or substitution of various elements into the BSCCO system. Numerous studies have shown that nanoparticles can be easily integrated and spread widely among the grains in the BSCCO system because of their small size. Promising outcomes on the properties of the Bi-2223 system, such as high T_c phase formation, improved intergrain connectivity, flux pinning ability, and critical current density, J_c, were achieved when nanosized dopants were integrated [14–20].

Graphene, a two-dimensional crystalline carbon material, has been used in many applications, especially electronics, due to its outstanding electrical, mechanical, and chemical properties [21,22]. Graphene is an important carbon-based nanomaterial with outstanding electrical, thermal, and mechanical properties [23–25]. Both graphene and Bi-2223 superconductors grow in platelet (sheet)-like microstructures. Therefore, using graphene nanoparticles as an additive in a Bi-2223 system is appealing. Several studies have been conducted on the incorporation of graphene into a superconducting system, such as graphene with yttrium barium copper oxide, YBCO [26–33], magnesium diboride, MgB_2 [34–36], and thallium-based high-temperature superconductors [37]. However, few investigations have been reported for BSCCO systems [38]. Undoubtedly, adding graphene to a superconductor material can have several beneficial effects, including acting as an inter-grain weak link, enhancing the critical current density (J_c), and reducing the average grain diameter.

Recent studies have found that single-phase Bi-2223 superconductors are usually challenging to synthesize because of the complexity of the reaction during phase formation. The co-precipitation method, a wet chemical technique, can often be considered a viable solution for addressing this issue. This is because powders generated through this method possess a reduced grain size, increased purity, and greater homogeneity than those produced using the solid-state method [39–46]. To the best of our knowledge, no work has been carried out on the addition of graphene nanoparticles into Bi-2223 using the wet chemical co-precipitation method. The outcome of this study has the potential to enhance the critical current density (J_c) of Bi-2223 superconductors in future studies. In this paper, we report on the effect of the critical temperature, T_c; the critical current density, J_c; and the structural and morphological properties of bulk $Bi_{1.6}Pb_{0.4}Sr_2Ca_2Cu_3O_y$ when graphene nanoparticles are integrated into the superconductors using the co-precipitation method.

2. Materials and Methods

The preparation of $Bi_{1.6}Pb_{0.4}Sr_2Ca_2Cu_3O_y$ was carried out via the co-precipitation method using bismuth(III) acetate $(CH_3COO)_3Bi$ (99%), lead(II) acetate $(CH_3COO)_2Pb{\cdot}3H_2O$ (99%), strontium acetate $(CH_3COO)_2Sr{\cdot}0.5H_2O$ (98%), calcium acetate $(CH_3COO)_2Ca{\cdot}XH_2O$ (97%), and copper(II) acetate $(CH_3COO)_2Cu{\cdot}H_2O$ (98–102%). All chemicals were purchased from Alfa Aesar and used without further purification. The powders were weighed and mixed in a 2:2:2:3 stoichiometric ratio and were dissolved in 500 mL acetate acid and labeled as solution A. Adding lead, Pb, into the BSCCO composition is able to effectively increase the volume fraction of the Bi-2223 phase [47]. Solution B was later prepared by dissolving 25 g of oxalic acid $C_2H_2O_4{\cdot}2H_2O$ (99.5–102.5%) in a solution of distilled water:propane-2-ol

(1:1.5) to obtain a concentration of 0.5 M. Solution A was stirred at 400 rpm at a temperature of 80 °C, while solution B was stirred at 400 rpm without heating. They were then soaked in an ice bath to achieve a temperature range of 0 °C to 2 °C. Solutions A and B were mixed until a uniform milky blue suspension was obtained. The blue suspension slurry was filtered, dried on a hot plate at 80 °C for 12 h, and ground into powder. The powder was initially calcined in an alumina boat at 730 °C for 12 h to obtain a good texture.

The calcined powders were ground and subjected to another calcination process at 850 °C for 24 h. The initial powder underwent a regrinding process and was combined with graphene sheets obtained from Nanostructured and Amorphous Materials, Inc. in Houston, TX, USA, with an average diameter ranging from 0.5 to 3 μm and a thickness between 0.55 to 1.2 nm. Different weight percentages of graphene sheets were added: 0.0 wt.%, 0.3 wt.%, 0.5 wt.%, and 1.0 wt.%. Subsequently, the mixture was pressed into pellets weighing 1.0 g each using hydraulic press under 5 tons of pressure. The pellets were sintered in a tube furnace at 850 °C for 48 h at a heating rate of 2 °C/min and cooling rate of 1 °C/min. X-ray diffraction (XRD) patterns of the samples were recorded using an X-ray diffraction diffractometer (Xpert Pro Panalytical Philips DY 1861 diffractometer) with CuKα radiation. The thermal material's stability was monitored using a thermogravimetric analyzer (Mettler Toledo, TGA/SDTA851e), and the measurement of AC magnetic susceptibility as a function of temperature was obtained using an AC Susceptometer (CryoBIND). The morphology and elemental analysis were analyzed using field emission scanning electron microscopy (FESEM, JEOL JSM-7100F). The J_c of the samples was determined using a four-point probe (4PP) with a 12 K closed-cycle He Cryostat system.

3. Results

3.1. Thermogravimetric Analysis

Thermogravimetric analysis (TGA) and differential thermogravimetry (DTG) curves versus temperature for pure Bi-2223 samples were obtained immediately after the co-precipitation process had been performed and before the sintering process was carried out. The TGA curve of Bi-2223 displayed in Figure 1 shows three different distinctive transitions at various temperatures and the same trend as that in the DTG curves. As the heating rate increased, the peak of the derivative thermogravimetric (DTG) curve shifted toward higher temperatures. The initial phase of the process lead to a 7.46% reduction in weight, which took place within the temperature range of 58.39 to 199.05 °C. This weight loss can be attributed to the evaporation of water and moisture present in the powder. Consequently, the water molecules derived from $Bi_2(C_2O_4)_3$, $Pb(C_2O_4)$, $Sr(C_2O_4)$, $Ca(C_2O_4)$, and $Cu(C_2O_4)$ contribute to the formation of dehydrated oxalate. The second phase occured between 200.10 °C and 508.82 °C, resulting in a weight loss of 33.54%. This weight reduction is primarily caused by the decomposition of $Bi_2(C_2O_4)_3$ into Bi_2O_3, $Pb(C_2O_4)$ into PbO, $Sr(C_2O_4)$ into $SrCO_3$, $Ca(C_2O_4)$ into $CaCO_3$, and $Cu(C_2O_4)$ into CuO. Previous studies have also reported similar findings regarding the thermal decomposition of individual oxalates of Bi, Sr, Ca, and Cu. However, for rare earth acetate, the decomposition typically occurs within the temperature range of approximately 200 to 300 °C [45]. The third transition in Figure 1 comprises a weight loss of 6.43% between 536.68 to 932.66 °C, representing the decomposition of $CaCO_3$ to CaO and $SrCO_3$ to SrO. These results are consistent with that of similar investigations [48]. Figure 1 also shows that the weight losses become almost constant after 850 °C, establishing the optimal temperature for calcination and sintering of BSSCO. This has been confirmed by several researchers reporting that the optimal temperature for the calcination and sintering process for BSSCO is in the range of 840 °C to 850 °C [49,50]. At this temperature, there is also a tendency to crystallize the (Bi,Pb)-2223 from Bi-2212 and other impurities [51].

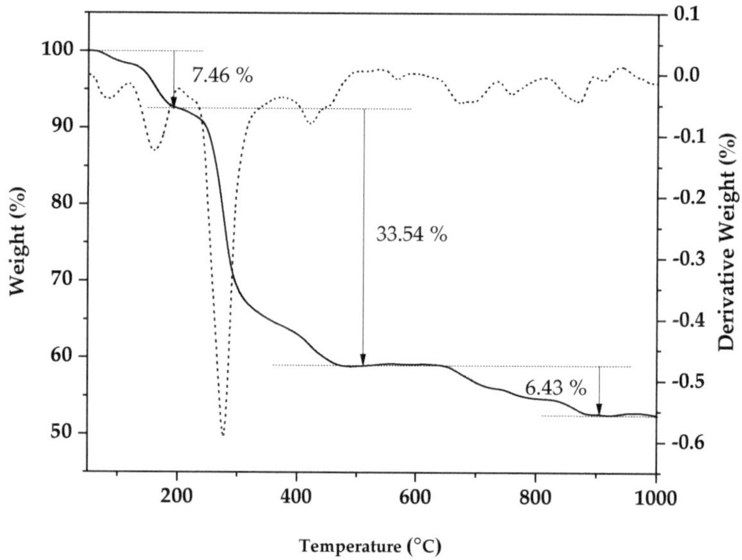

Figure 1. TGA (solid curve) and DTG (dashed curve) for pure Bi-2223 samples before the sintering process.

3.2. X-ray Diffraction (XRD) Analysis

The patterns of XRD for Bi-2223 samples integrated with the graphene nanoparticles of 0, 0.3, 0.5, and 1.0 wt.% are shown in Figure 2. The XRD patterns show that all the secondary phase peaks decreased for the sintered samples. This demonstrates that sintering has successfully reduced the secondary phases in Bi-2223 samples. The XRD peaks for all the samples indexed to Bi-2223 (ICSD: 98-008-0522). The highest intensity of Bi-2223 peaks was observed at 2θ ≈ 28.73°, which matches the Miller indices of the (0 1 7) plane. It can be seen from Figure 2 that the intensity of the peaks decreases when 0.3 wt.% addition of graphene nanoparticles to the samples and increases when 0.5 wt.% and 1.0 wt.% was added. At the same time, the peaks also slightly shifted towards the smaller angle of 2θ when the amount of graphene nanoparticles increased in the sample. However, some minor peaks belonging to Bi-2212 (ICSD: 98-003-0405) were also observed in all XRD patterns at $2\theta \approx 31.16°$, 35.45°, 41.24°, and 47.46°, which indicate the presence of a secondary phase. Unfortunately, no graphene-related peak appears in the XRD results due to the small amount of graphene added. Previous studies have found that the main graphene peak appears around 2θ equal to 9.86° [36], 10° [27,33], 12.85° [29], 26.5°, and 54.5° [37]. The number of phases present for Bi-2223 and Bi-2212 was calculated using the following expressions [52,53]:

$$\text{Bi-2223}(\%) = \frac{\sum I_{2223}}{\sum I_{2223} + \sum I_{2212}} \times 100\% \tag{1}$$

$$\text{Bi-2212}(\%) = \frac{\sum I_{2212}}{\sum I_{2212} + \sum I_{2223}} \times 100\% \tag{2}$$

where I is the peak intensity of the phases. Table 1 shows that the phase of Bi-2223 increases with graphene but declines when 1.0 wt.% graphene nanoparticles are added. A reverse trend was observed for the phase of Bi-2212. The Bi-2223 (0.0 wt.%) XRD pattern shows a tetragonal crystal structure with lattice parameters $a = b = 3.826$ Å and $c = 37.104$ Å. All the graphene-added samples given in Table 2 show almost identical lattice parameters, indicating that adding graphene nanoparticles did not distort the Bi-2223 crystal system or become part of the Bi-2223 crystal structure. This indicates that the nanoparticles are located in the middle of superconducting grains. Previous research has indicated that the presence of impurities in nanoparticles can serve in two ways: they can act as pinning centers that immobilize vortices or enhance the interconnection among the grains. This, in turn, can potentially enhance the critical current density J_c [54,55]. The crystallite size listed in Table 2 was calculated by choosing the highest peak and using the Scherrer equation [56]:

$$L = \frac{K\lambda}{B_{size}\ cos\theta} \tag{3}$$

where L is the crystallite size, K equal to 0.9 is a dimensionless shape factor, B_{size} is the line broadening at half the maximum intensity (FWHM), λ equal to 1.5406 Å is the X-ray wavelength for Cu K_α radiation, and θ is the Bragg angle. Table 2 shows that all the samples possess comparable crystallite sizes. The XRD patterns were also examined using the Williamson–Hall plot method, which also determined the crystallite size and the lattice strain of the sample. Previous studies have established that lattice strain and crystallite size are the primary reasons for the broadening of the diffraction peaks. The Williamson–Hall plot was obtained by utilizing the equation below [57]:

$$\beta_{hkl}\cos\theta = \frac{K\lambda}{L} + 4C_\varepsilon \sin\theta \tag{4}$$

where C_ε is the lattice strain. By plotting β_{hkl} cosθ against 4sinθ, the intercept ($K\lambda/L$) of the plot represents the crystallite size and the gradient represents the lattice strain (C_ε) of the samples [41]. The results shown in Table 2 demonstrate a marked difference between using the Williamson–Hall and Scherrer equations. The finding from the Williamson–Hall plot shows that adding graphene to the samples decreases the crystallite size. Both calculations show that the sample with 0.3 wt.% has the smallest crystallite size. The lattice strains in the sample do not show any noticeable changes after introducing graphene.

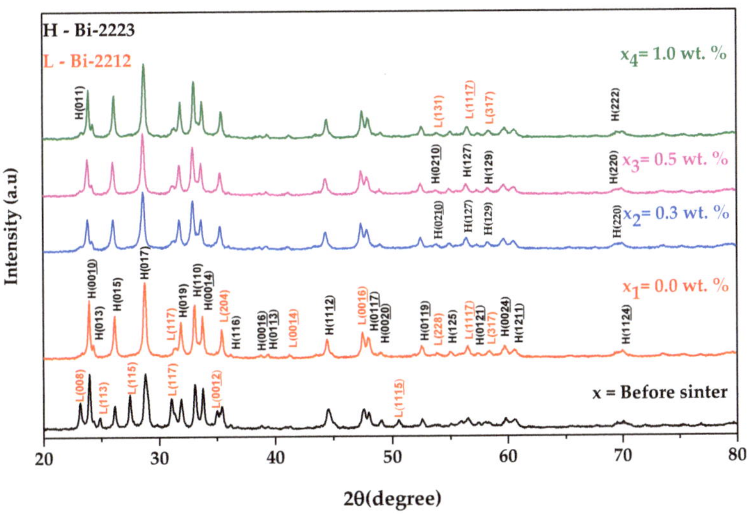

Figure 2. XRD analysis of Bi-2223 samples before and after the sintering process for samples with 0.0 wt.%, 0.3 wt.%, 0.5 wt.%, and 1.0 wt.% addition of graphene. The peak indicated by L and H represents the Bi-2212 and Bi-2223 phases, respectively.

Table 1. The percentage of Bi-2212 phase formed in Bi-2223 with 0.0 wt.%, 0.3 wt.%, 0.5 wt.%, and 1.0 wt.% addition of graphene.

Bi-2223 + x wt.% of Graphene	Intensity Fraction (%)	
	Bi-2223	Bi-2212
0.0	85.17	14.83
0.3	87.43	12.57
0.5	88.25	11.75
1.0	84.00	16.00

Table 2. The percentage of Bi-2212 phase formed in Bi-2223 with 0.0 wt.%, 0.3 wt.%, 0.5 wt.%, and 1.0 wt.% addition of graphene.

Bi-2223 + x wt.% of Graphene	Lattice Parameter (Å)			Crystallite Size (nm)		Unit Cell Volume (Å^3)	Lattice Strain (%)
	a-Axis	*b*-Axis	*c*-Axis	Scherrer	Williamson–Hall		
0.0	3.826 ± 0.000659	3.826 ± 0.000659	37.104 ± 0.008048	46.26	397	543.083	0.175
0.3	3.824 ± 0.000760	3.824 ± 0.000760	37.064 ± 0.009404	39.09	74.5	541.859	0.070
0.5	3.823 ± 0.000703	3.823± 0.000703	37.062 ± 0.008636	46.25	182	541.768	0.179
1.0	3.823 ± 0.000635	3.823 ± 0.000635	37.066 ± 0.007766	46.26	109	541.791	0.073

3.3. Critical Temperature, T_c Analysis

Figure 3 illustrates the temperature dependencies of the real component (χ') and imaginary component (χ'') of the samples on an applied field of 39.885 A/m and a frequency of 219 Hz, with the field applied parallel to the long dimension of the samples. As seen in the figure, there are two discernible decreases in the real part of AC susceptibility, χ'. The initial decline observed at the onset temperature, $T_{c\text{-onset}}$, can be attributed to the transition occurring within individual grains (intra-grain). On the other hand, the second decrease observed at the phase-locking temperature, T_{cj}, results from the superconducting coupling between the grains (inter-grain) [58]. In the temperature range between the onset of superconductivity ($T_{c\text{-onset}}$) and the critical temperature (T_{cj}), the superconducting grains are disconnected, making the system resistive. However, below T_{cj}, the grains become connected and are phase-locked, resulting in zero phase difference across the intragranular junctions. There is a definite transition for $T_{c\text{-onset}}$ (χ') to the lower temperature with increasing graphene addition. The decrease in $T_{c\text{-onset}}$ can be attributed to the reduction in hole concentration resulting from oxygen deficiency in the CuO chain [41]. The findings also show that the intragranular coupling between grains decreased by increasing the wt.% of graphene added to the samples based on the phase-locking temperature, T_{cj}. Prior work has indicated that the Bi-2212 low-T_c phase can affect the phase-locking, T_{cj} [58].

The peak observed in the imaginary part χ'' represents the measure of dissipation in the samples. The imaginary part of the AC susceptibility (ACS) in all the samples exhibited a single peak at the coupling peak temperature, T_p, corresponding to intergrain coupling. Notably, no secondary peak was observed near the critical temperature, T_c, indicating the absence of an intra-granular peak in any of the samples. This could indicate an excellent quality for the ceramic superconductor in all the samples where the coupling phase occurs immediately after the occurrence of superconducting inside the grain [59]. There could be various reasons for this phenomenon, including inadequate total grain volume and grain size relative to the penetration depth (λ). However, as shown in the table, shifting the T_p to lower temperatures with the introduction of graphene nanoparticles indicates the weakening of the grains and subsequently reducing the pinning force, while the $T_p/T_{c\text{-onset}}$

ratio indicates a weak coupling between the grains. The value of Josephson's current, I_o, was calculated by using the Ambegaokar–Baratoff theory [60,61]:

$$I_O = 1.57 \times 10^{-8} \left(\frac{T_{c\text{-onset}}^2}{T_{c\text{-onset}} - T_{cj}} \right) \tag{5}$$

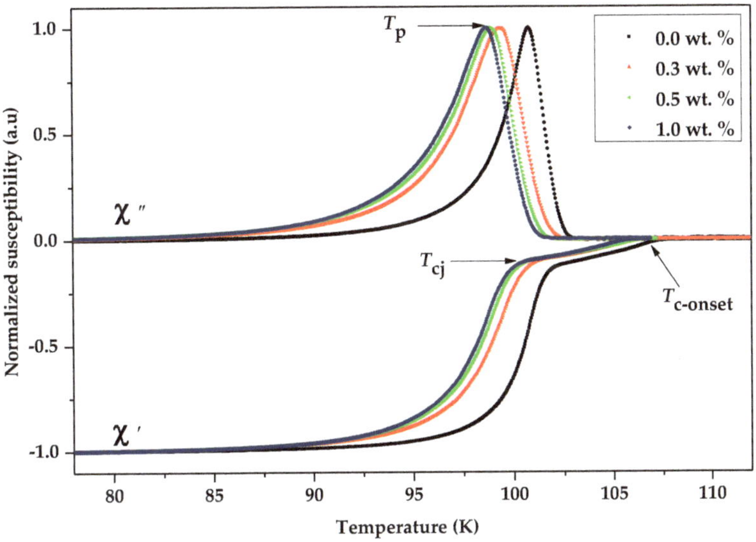

Figure 3. Graph of the normalized susceptibility of the real (χ') and imaginary part (χ'') against the temperature for samples Bi-2223 with 0.0 wt.%, 0.3 wt.%, 0.5 wt.%, and 1.0 wt.% addition of graphene.

Table 3 indicates that I_o decreases with the increasing amount of graphene nanoparticles in the samples. This may be attributed to an increased presence of unreacted graphene nanoparticles at the boundaries between grains. As a consequence, the tunneling of Josephson's current across the grains is decreased. In addition, the decreasing I_o also implies that excess graphene nanoparticles may degrade the grain connectivity.

Table 3. Coupling peak temperature, T_p, onset critical temperature, $T_{c\text{-onset}}$, phase lock-in temperature, T_{cj}, and Josephson current, I_o for Bi-2223 with 0.0 wt.%, 0.3 wt.%, 0.5 wt.%, and 1.0 wt.% addition of graphene.

Bi-2223 + x wt.% of Graphene	T_p (K)	$T_{c\text{-onset}}$ (K)	$T_p/T_{c\text{-onset}}$	T_{cj} (K)	I_o (μA)
0.0	100.78	107.69	0.936	101.91	31.50
0.3	99.10	107.21	0.924	100.94	28.80
0.5	98.92	106.60	0.928	100.23	28.01
1.0	98.78	106.54	0.927	99.90	26.84

3.4. Critical Current Density, J_c Analysis

The graph of voltage, V versus current, I, shown in Figure 4, displays the critical current, I_c, of the samples measured at 40 K in a zero magnetic field. The transport critical current density, J_c, was calculated using the equation $J_c = I_c/A$ [62], where A is the cross-sectional area of the bar-shaped samples. The critical current density increase is observed for samples with 0.3 wt.% and 1.0 wt.% addition of graphene but not for the 0.5 wt.%, as shown in Figure 5. The improvement in J_c is attributable to improvements in the connectivity of intergrain.

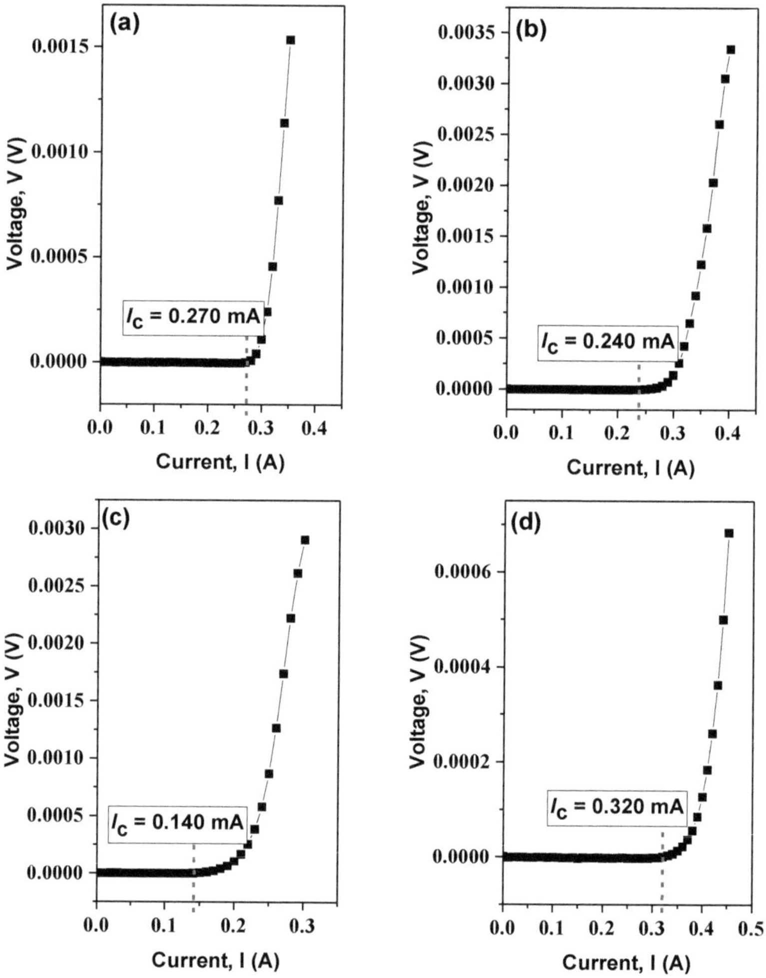

Figure 4. V-I curve in self-field at 40 K for (**a**) x = 0.0 wt.%, (**b**) x = 0.3 wt.%, (**c**) x = 0.5 wt.%, and (**d**) x = 1.0 wt.%.

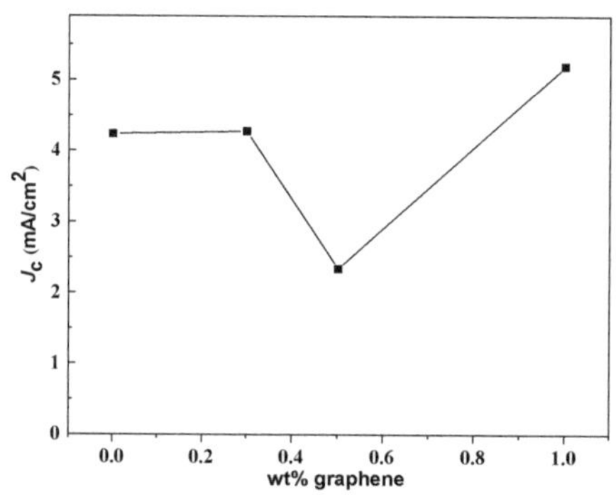

Figure 5. Variation of critical current density, J_C, against weight percentage of graphene addition content.

3.5. Microstructure Analysis

Figure 6 shows the SEM images of the surface morphology of Bi-2223 samples at 10.000X magnification. The microstructures of the samples show varying degrees of alignment of the compacted layers of thin flaky plate-like grains. The grains for samples with graphene are better aligned than those without graphene. The grains for the 1.0 wt.% are mostly aligned parallel to the *ab*-plane compared to other samples. The existence of graphene in Bi-2223 showed the different parts of the structure and distributed uniformly on the grain boundaries in great amounts. According to the analysis depicted in Figure 6, it was observed that the pure sample displayed irregular shapes that were randomly distributed. In contrast, the 1.0 wt.% sample demonstrated a denser grain structure with reduced porosity, indicating a more compacted arrangement. This indicates that the superconducting phase expanded at the expense of the surrounding phase, supporting our explanation of the higher critical temperature (T_c). Figure 7 shows a schematic illustration of how the broad flat surface of the BSCCO grains is always stacking parallel to its *ab*-plane. This observation coincides with the results from four-point probe measurements that the 1.0 wt.% also has the highest critical current density, J_c, as presented in Table 4. The inclusion of graphene nanoparticles resulted in larger grain sizes characterized by a tightly packed arrangement. This facilitates grain growth and plays a role in enhancing intergranular transport currents [26]. Possible explanations for the current transport of the grain morphology of Bi-2223 tapes can be found in the brick wall (BW) model [63,64] and railway switch (RS) model [65,66], which explain the behavior of c-axis [001] twist grain boundaries (GBs) in Bi-2223 samples. According to the BW model, the c-axis GBs create a highly connected supercurrent path when the *ab*-plane is blocked. This occurs due to the large surface area of basal-plane-faced GBs, which are formed by high aspect ratio grains. This configuration allows for a strong linkage of c-axis supercurrents. In contrast, the RS model proposes a different mechanism. It suggests that the *ab*-plane supercurrent path is primarily established through low-angle, strongly coupled, and obliquely intersecting GBs called small-angle, c-axis tilt (SCTILT) GBs. These GBs contribute to improving the texture of the samples. The presence of these well-defined supercurrent paths in both models improves the transport current occupancy of the Bi-2223 tape's cross-section. As a result, the need to rely on intrinsically lower critical current density (J_c) c-axis paths between dominant *ab*-plane paths is reduced. Ultimately, this improved connectivity and transport current occupancy enhance the J_c of the Bi-2223 tapes [67,68].

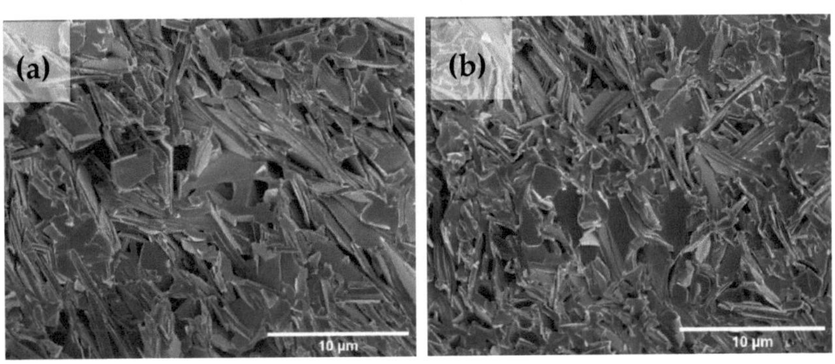

Figure 6. *Cont.*

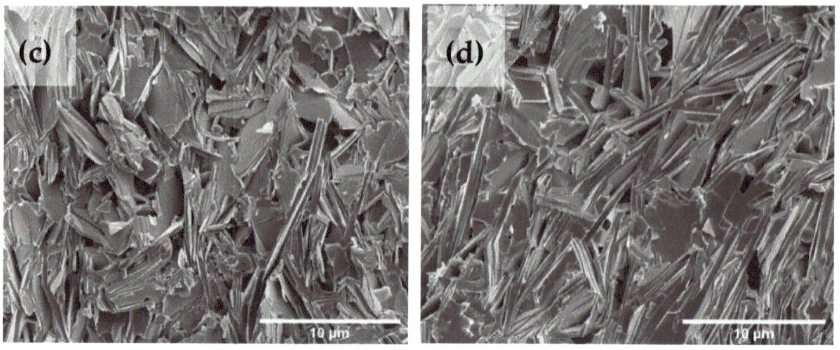

Figure 6. SEM micrographs of Bi-2223 samples with (**a**) x = 0.0 wt.%, (**b**) x = 0.3 wt.%, (**c**) x = 0.5 wt.%, and (**d**) x = 1.0 wt.% of graphene.

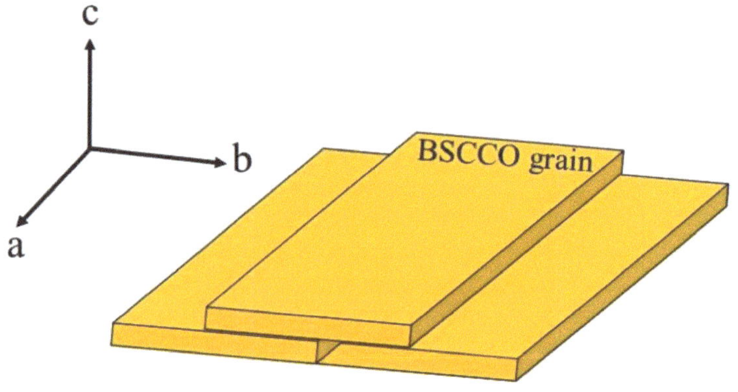

Figure 7. Schematic diagram demonstrating how the BSCCO grain is parallel to its *ab*-plane.

Table 4. Critical current density, J_c, of Bi-2223 with the addition of graphene 0.0 wt.%, 0.3 wt.%, 0.5 wt.%, and 1.0 wt.% samples.

Bi-2223 + x wt.% of Graphene	I_c (A)	J_c (mA/cm^2)
0.0	0.270	4235
0.3	0.240	4271
0.5	0.140	2340
1.0	0.320	5197

Data presented in Table 5 from EDX analysis indicate that the atomic ratio of Bi, Sr, Ca, and Cu is 2:2:2:3, with different oxygen ratios obtained for all samples. This confirms that all samples have Bi-2223 element composition, which supports the XRD results of all peaks indexed to Bi-2223 phases. The atomic percentage for C also increases, proving that graphene was added to the samples, as shown in Table 6. However, it can be seen that increasing graphene addition showed the results of unreacted graphene and localized inhomogeneity within the samples.

The information provided suggests that the elements of carbon were detected in all samples with a percentage range of 7.57% to 10.54%. The presence of graphene (C) in Bi-2223 was confirmed through elemental mapping using energy dispersive X-ray spectroscopy (EDX) as shown in Figure 8 for samples with different graphene concentrations (x = 0.3 wt.%, 0.5 wt.%, and 1.0 wt.%). This verifies the existence of carbon, which was not observed in the X-ray diffraction (XRD) patterns. Furthermore, the atomic percentage of carbon elements increased from 32.07% to 39.29% as the graphene nanoparticle content increased from x = 0.3 wt.% to x = 1.0 wt.% in the Bi-2223 samples. This finding supports a previous study that demonstrated increased graphene content with increasing graphene nanoparticle concentration in Bi-2223 samples [69].

Table 5. EDX analysis of the atomic percentage (at.%) for Bi-2223 with the addition of the graphene of 0.0 wt.%, 0.3 wt.%, 0.5 wt.%, and 1.0 wt.%.

Bi-2223 + x wt.% of Graphene (wt.%)	Elements (%)						
	Bi	Pb	Sr	Ca	Cu	O	C
0.0 (2.0:0.8:2.2:2.0:3.5:6.3)	11.98	4.86	12.93	12.11	20.79	37.33	-
0.3 (2.0:0.3:1.8:1.7:3.5:5.4)	9.23	1.39	8.53	7.68	16.34	24.77	32.07
0.5 (2.0:0.4:2.0:1.7:3.5:6.0)	8.51	1.54	8.31	7.41	14.87	25.71	33.66
1.0 (2.0:0.4:2.0:1.8:3.4:6.6)	7.5	1.52	7.31	6.71	12.84	24.83	39.29

Table 6. EDX analysis of the weight percentage (wt.%) for Bi-2223 with the addition of the graphene of 0.0 wt.%, 0.3 wt.%, 0.5 wt.%, and 1.0 wt.%.

Bi-2223 + x wt.% of Graphene (wt.%)	Elements (%)						
	Bi	Pb	Sr	Ca	Cu	O	C
0.0	35.54	14.28	16.08	6.89	18.75	8.48	-
0.3	37.90	5.64	14.68	6.04	20.39	7.78	7.57
0.5	36.43	6.52	14.91	6.08	19.35	8.42	8.28
1.0	35.02	7.04	14.30	6.01	18.22	8.87	10.54

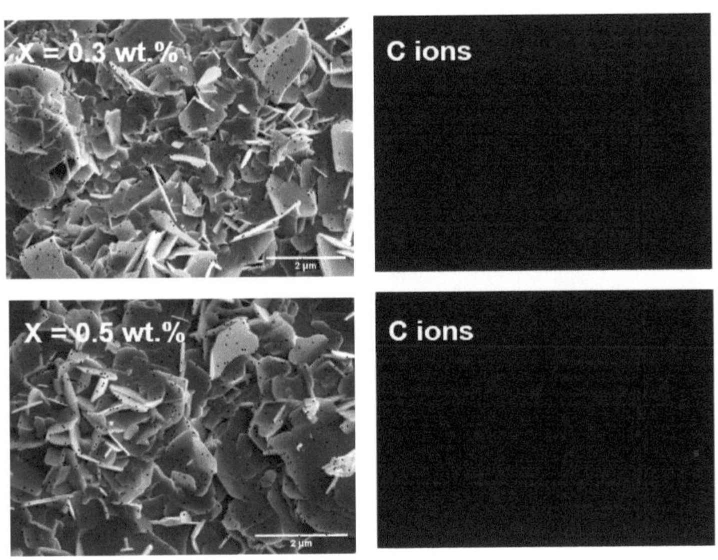

Figure 8. *Cont.*

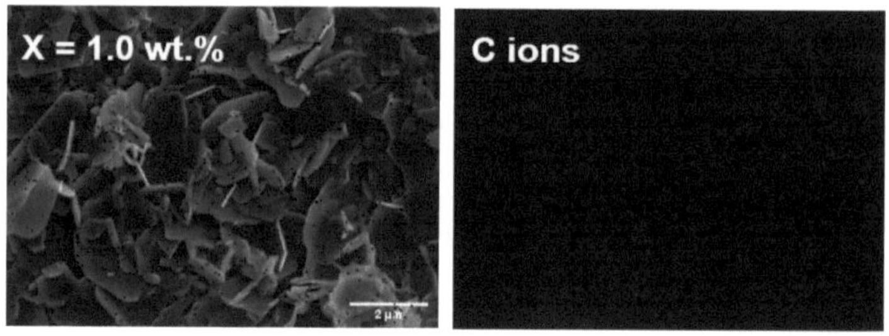

Figure 8. FESEM images and EDX mapping C ions (red dots) of the compound Bi-2223 with added graphene nanoparticles of up to x = 0.3 wt.%, 0.5 wt.%, and 1.0 wt.%, indicating the presence of the expected elements.

4. Conclusions

$Bi_{1.6}Pb_{0.4}Sr_2Ca_2Cu_3O_y$ superconducting samples with the addition of graphene nanoparticles at weight percentages of 0.0, 0.3, 0.5, and 1.0 wt.% were successfully synthesized using the co-precipitation method. A variation in T_c and J_c values was observed due to the addition of graphene content in Bi-2223. Elemental analysis via EDX showed that stoichiometry ratio of Bi-2223 superconducting compounds were obtained. The XRD patterns exhibited the primary phase indexed to Bi-2223, corresponding to a tetragonal structure, and a second phase which belonged to Bi-2212. ACS measurements established that the onset critical temperature, $T_{c\text{-onset}}$; coupling peak temperature, T_p; and intergranular critical current density, T_{cj}, decreased with the increase in graphene nanoparticle addition. The decrease in the Josephson current, I_o, with the increased addition of graphene was due to the degradation of samples' grain connection due to the amount of unreacted graphene at the grain boundary increase in the samples. The sample with 1.0 wt.% of graphene produced the highest J_c of all the samples. SEM results proved that the microstructure for samples with 1.0 wt.% was the best aligned and the grain showed a much better formation of Bi-2223 superconductors. Based on these results, adding graphene nanoparticles to Bi-2223 samples could be able to improve the pinning center and hence enhanced the critical current density, J_c.

Author Contributions: Conceptualization, S.N.A., M.M.A.K. and M.M.; methodology, S.N.A., M.M.A.K., A.N.K. and N.E.S.; formal analysis, S.N.A. and M.M.A.K.; investigation, S.N.A. and M.M.A.K.; resources, S.N.A. and M.M.A.K.; writing—original draft preparation, S.N.A.; writing—review and editing, M.M.A.K., A.N.K., Z.A.T., H.B., C.S.K., L.K.P., M.K.A.K., M.K.S. and A.H.; supervision, M.M.A.K. and A.H.S.; project administration, M.M.A.K.; funding acquisition, M.M.A.K. All authors have read and agreed to the published version of the manuscript.

Funding: This research was funded by the Ministry of Higher Education (MOHE) Malaysia under FRGS grant FRGS/1/2017/STG02/UPM/02/4.

Data Availability Statement: The data presented in this study are available on request from the corresponding author. The data are not publicly available due to ethical restrictions.

Acknowledgments: This work was partly supported by Japan Science and Technology (JST) for the Advanced Project Based Learning (aPBL), Shibaura Institute of Technology. The authors would like to thank the valuable support and discussions of all Superconductor and Thin Films Laboratory members of Universiti Putra Malaysia.

Conflicts of Interest: The authors declare no conflict of interest.

References

1. Maeda, H.; Tanaka, Y.; Fukutomi, M.; Asano, T.A. New High-T_c Oxide Superconductor without a Rare Earth Element. In *Ten Years of Superconductivity: 1980–1990. Perspectives in Condensed Matter Physics*; Ott, H.R., Ed.; A Critical Reprint Series; Springer: Berlin/Heidelberg, Germany, 1993; pp. 303–304.
2. Tarascon, J.M.; McKinnon, W.R.; Barboux, P.; Hwang, D.M.; Bagley, B.G.; Greene, L.H. Preparation, structure, and properties of the superconducting compound series $Bi_2Sr_2Ca_{n-1}Cu_nO_y$ with *n*=1,2, and 3. *Phys. Rev. B Condens. Matter.* **1988**, *38*, 8885–8892. [CrossRef] [PubMed]
3. Frank, J.O.; Charles, P.P.J. *The New Superconductors*; Stuart Wolf Naval Research Laboratory: Washington, DC, USA, 1996.
4. Hayashi, K. Commercialization of Bi-2223 Superconducting Wires and Their Applications. *Sei Tech. Rev.* **2020**, *91*, 68–74.
5. Kikuchi, M.; Ayai, N.; Ishida, T.; Tatamidani, K.; Hayashi, K.; Kobayashi, S. Development of new types of DI-BSCCO wire. *Sei Tech. Rev.* **2008**, *66*, 73–80.
6. Yumura, H.; Ashibe, Y.; Itoh, H.; Ohya, M.; Watanabe, M.; Masuda, T. Phase II of the Albany HTS cable project. *IEEE Trans. Appl. Supercond.* **2009**, *19*, 1698–1701. [CrossRef]
7. Yumura, H.; Ashibe, Y.; Ohya, M.; Itoh, H.; Watanabe, M.; Masuda, T. Update of Yokohama HTS cable project. *IEEE Trans. Appl. Supercond.* **2013**, *23*, 5402306. [CrossRef]
8. Elschner, S.; Demencik, E.; Douine, B.; Grilli, F.; Kudymow, A.; Stemmle, M. New experimental method for investigating AC losses in concentric HTS power cables. *IEEE Trans. Appl. Supercond.* **2014**, *25*, 5900105. [CrossRef]
9. Sytnikov, V.E.; Bemert, S.E.; Ivanov, Y.V.; Kopylov, S.I.; Krivetskiy, I.V.; Rimorov, D.S. HTS DC Cable line project: On-Going activities in Russia. *IEEE Trans. Appl. Supercond.* **2013**, *23*, 5401904. [CrossRef]
10. Tomita, M.; Fukumoto, Y.; Ishihara, A.; Suzuki, K.; Akasaka, T.; Caron, H. Train Running Test Transmitted by Superconducting Feeder Cable and Study as an Example of Line in Japan and France. *IEEE Trans. Appl. Supercond.* **2019**, *30*, 5400107. [CrossRef]
11. Sato, K.I.; Kobayashi, S.I.; Nakashima, T. Present status and future perspective of bismuth-based high-temperature superconducting wires realizing application systems. *Jpn. J. Appl. Phys.* **2011**, *51*, 010006. [CrossRef]
12. Yamasaki, H.; Endo, K.; Kosaka, S.; Utneda, M.; Misawa, S.; Yoshida, S. Magnetic-Field Angle Dependence of the Critical Current Density in High-Quality $Bi_2Sr_2Ca_2Cu_30_X$ Thin Films. *IEEE Trans. Appl. Supercond.* **1993**, *1*, 1536–1539. [CrossRef]
13. Patnaik, S.; Feldmann, D.M.; Polyanskii, A.A.; Yuan, Y.; Jiang, J.; Cai, X.Y. Local Measurement of Current Density by Magneto-Optical Current Reconstruction in Normally and Overpressure Processed Bi-2223 Tapes. *IEEE Trans. Appl. Supercond.* **2003**, *13*, 2930–2933. [CrossRef]
14. Arlina, A.; Halim, S.A.; Awang Kechik, M.M.; Chen, S.K. AC Losses in YBCO-Added $Bi_{1.6}Pb_{0.4}Sr_2Ca_2Cu_3O_\delta$ Superconductors. *J. Supercond. Nov. Magn.* **2015**, *28*, 1953–1958. [CrossRef]
15. Bagiah, H.; Halim, S.A.; Chen, S.K.; Lim, K.P.; Awang Kechik, M.M. Effects of rare earth nanoparticles (M = Sm_2O_3, Ho_2O_3, Nd_2O_3) addition on the microstructure and superconducting transition of $Bi_{1.6}Pb_{0.4}Sr_2Ca_2Cu_3O_{10+\delta}$ Ceramics. *Sains Malays.* **2016**, *45*, 643–651.
16. Azman, N.J.; Abdullah, H.; Abd-Shukor, R. Transport critical current density of $(Bi_{1.6}Pb_{0.4})Sr_2Ca_2Cu_3O_{10}$ ceramic superconductor with different nanosized Co_3O_4 addition. *Adv. Condens. Matter Phys.* **2014**, *2014*, 498747. [CrossRef]
17. Abd-Shukor, R.; Awang Kechik, M.M.; Halim, S.A. Transport critical current density of Bi-Sr-Ca-Cu-O/Ag superconductor tapes with addition of Fe_3O_4 as flux pinning center. *J. Phys. Conf. Ser.* **2008**, *97*, 2–8. [CrossRef]
18. Khalid, N.A.; Kong, W.; Kong, I.; Kong, C.; Awang Kechik, M.M.; Abd-Shukor, R. Significance of Cobalt Ferrite Nanoparticles on Superconducting Properties of Tl-1212 High Temperature Superconductor. *Nano Hybrids Compos.* **2021**, *31*, 1–5. [CrossRef]
19. Khalid, N.A.; Awang Kechik, M.M.; Baharuddin, N.A.; Chen, S.K.; Baqiah, H.; Lim, K.P. Carbon nanofibers addition on transport and superconducting properties of bulk $YBa_2Cu_3O_{7-\delta}$ material prepared via co-precipitation. *J. Mater. Sci. Mater. Electron.* **2020**, *31*, 16983–16990. [CrossRef]
20. Wei, K.; Abd-Shukor, R. Superconducting and transport properties of (Bi-Pb)-Sr-Ca-Cu-O with nano-Cr_2O_3 additions. *J. Electron. Mater.* **2007**, *36*, 1648–1651. [CrossRef]
21. Bae, S.; Kim, S.J.; Shin, D.; Ahn, J.H.; Hong, B.H. Towards industrial applications of graphene electrodes. *Phys. Scr.* **2012**, *T146*, 014024. [CrossRef]
22. Sato, S. Application of graphene to electronic devices. In Proceedings of the International Workshop on Active-Matrix Flatpanel Displays and Devices (AM-FPD), Kyoto, Japan, 4–7 July 2017; pp. 90–93.
23. Singh, S.; Hasan, M.R.; Sharma, P.; Narang, J. Graphene nanomaterials: The wondering material from synthesis to applications. *Sens. Int.* **2022**, *3*, 100190. [CrossRef]
24. Sahoo, B.; Routray, K.L.; Samal, D.; Behera, D. Effect of artificial pinning centers on YBCO high temperature superconductor through substitution of graphene nano-platelets. *Mater. Chem. Phys.* **2019**, *223*, 784–788. [CrossRef]
25. Maio, A.; Pibiri, I.; Morreale, M.; La Mantia, F.P.; Scaffaro, R. An Overview of Functionalized Graphene Nanomaterials for Advanced Applications. *Nanomaterials* **2021**, *11*, 1717. [CrossRef] [PubMed]
26. Dadras, S.; Falahati, S.; Dehghani, S. Effects of graphene oxide doping on the structural and superconducting properties of $YBa_2Cu_3O_{7-\delta}$. *Phys. C Supercond. Appl.* **2018**, *548*, 65–67. [CrossRef]
27. Shoushtari, M.Z.; Akbari, M.; Hajati, Y. Study of $YBa_2Cu_3O_{7-\delta}$ Superconductor/Graphene Oxide Composite. *J. Supercond. Nov. Magn.* **2018**, *31*, 2733–2739. [CrossRef]

28. Falahati, S.; Dadras, S.; Mosqueira, J. Investigation of the Magnetic and Transport Properties of $YBa_2Cu_3O_{7-\delta}$ High-Temperature Superconductor Doped with Graphene Oxide. *J. Supercond. Nov. Magn.* **2019**, *32*, 3755–3760. [CrossRef]
29. Kamarudin, A.N.; Awang Kechik, M.M.; Abdullah, S.N.; Baqiah, H.; Chen, S.K.; Karim, M.K.A.; Ramli, A.; Lim, K.P.; Halim, S.A.; Miryala, M.; et al. Effect of Graphene Nanoparticles Addition on Superconductivity of $YBa_2Cu_3O_{7-\delta}$ Synthesized via the Thermal Treatment Method. *Coatings* **2020**, *12*, 91. [CrossRef]
30. Gaffoor, M.Z.; Jarvis, A.L.L.; Young, E.A.; Dorrell, D. Comparison of the Effect of Graphene and Graphene Oxide Doping on YBCO. *J. Phys. Conf. Ser.* **2020**, *1559*, 012028. [CrossRef]
31. Wang, J.; Zhang, Q.; Liu, C.; Zhang, X. Improvement of the pinning property in $YBa_2Cu_3O_{7-X}$ films below 35 K by doping with graphene oxide. *AIP Adv.* **2019**, *9*, 015118. [CrossRef]
32. Wei, K.; Ing, K.; Hamdan, M.S.; Radiman, S.; Abd-Shukor, R. AC Susceptibility and Superconducting Properties of Graphene Added $YBa_2Cu_3O_{7-d}$. *J. Supercond. Nov. Magn.* **2018**, *31*, 2699–2703. [CrossRef]
33. Sahoo, B.; Singh, A.K.; Behera, D. Graphene oxide modified superconducting and elastic parameters of YBCO superconductor. *Mater. Chem. Phys.* **2020**, *240*, 122252. [CrossRef]
34. Sudesh, K.N.; Das, S.; Bernhard, C.; Varma, G.D. Effect of graphene oxide doping on superconducting properties of bulk MgB_2. *Supercond. Sci. Technol.* **2013**, *26*, 095008. [CrossRef]
35. Sudesh, K.N.; Das, S.; Bernhard, C.; Varma, G.D. Enhanced superconducting properties of rare-earth oxides and graphene oxide added MgB_2. *Phys. C Supercond. Appl.* **2014**, *505*, 32–38. [CrossRef]
36. De Silva, K.S.B.; Gambhir, S.; Wang, X.L.; Xu, X.; Li, W.X.; Officer, D.L. The effect of reduced graphene oxide addition on the superconductivity of MgB_2. *J. Mater. Chem. A* **2012**, *22*, 13941–13946. [CrossRef]
37. Kong, W.; Kong, I.; Awang Kechik, M.M.; Abd-Shukor, R. Effect of graphene addition on the transport critical current density of bulk $(Tl_{0.85}Cr_{0.15})Sr_2CaCu_2O_{7-\delta}$ superconductor. *Mater. Today* **2018**, *5*, 3176–3184. [CrossRef]
38. Bilgili, Ö.; Yurddaskal, M. Effects of Graphene Oxide Doping on Magnetic and Structural Properties of $Bi_{1.6}Pb_{0.4}Sr_2Ca_2Cu_3O_y$ Superconductor. *J. Electron. Mater.* **2021**, *50*, 4999–5006. [CrossRef]
39. Azhan, H.; Hawa, J.S.; Azman, K.; Hidayah, H.N.; Yusainee, S.Y.S. Superconducting properties of $Bi_{1.6}Pb_{0.4}Sr_2Ca_{2-x}Dy_xCu_3O_y$ prepared via co-precipitation method. *Adv. Mater. Res.* **2013**, *622*, 177–181.
40. Azhan, H.; Hawa, J.S.; Azura, C.M.; Azman, K.; Syamsyir, S.A. Structural and electrical properties of high and low-density Yb-doped Bi(Pb)-2223 superconductor. *J. Teknol.* **2016**, *78*, 7–12. [CrossRef]
41. Hapipi, N.M.; Chen, S.K.; Shaari, A.H.; Awang Kechik, M.M.; Tan, K.B.; Lim, K.P. Superconductivity of Y_2O_3 and $BaZrO_3$ nanoparticles co-added $YBa_2Cu_3O_{7-\delta}$ bulks prepared using co-precipitation method. *J. Mater. Sci. Mater. Elec.* **2018**, *29*, 18684–18692. [CrossRef]
42. Cui, L.J.; Zhang, P.X.; Li, J.S.; Yan, G.; Liu, X.H.; Feng, Y. Optimization of Bi-2223 precursor powder prepared by co-precipitation method. In Proceedings of the 2015 IEEE International Conference on Applied Superconductivity and Electromagnetic Devices (ASEMD), Shanghai, China, 20–23 November 2015; pp. 617–618.
43. Hamadneh, I.; Halim, S.A.; Lee, C.K. Characterization of $Bi_{1.6}Pb_{0.4}Sr_2Ca_2Cu_3O_y$ ceramic superconductor prepared via coprecipitation method at different sintering time. *J. Mater. Sci.* **2006**, *41*, 5526–5530.
44. Nur Haleeda, M.; Awang Kechik, M.M.; Abd-Shukor, R. Effect of Yb_2O_3 Nanoparticle Addition on Superconducting Properties of BSCCO (2223)/Ag Tapes by Acetate Precipitation Method. *Pertanika J. Sci. Technol.* **2016**, *2*, 1–11.
45. Hapipi, N.M.; Lim, J.K.; Chen, S.K.; Lee, O.J.; Shaari, A.H.; Awang Kechik, M.M. Comparative study on AC susceptibility of $YBa_2Cu_3O_{7-\delta}$ added with $BaZrO_3$ nanoparticles prepared via solid-state and co-precipitation method. *Crystals* **2019**, *9*, 655. [CrossRef]
46. Hapipi, N.M.; Chen, S.K.; Shaari, A.H.; Awang Kechik, M.M.; Tan, K.B.; Lim, K.P. AC Susceptibility of $BaZrO_3$ Nanoparticles Added $YBa_2Cu_3O_{7-\delta}$ Superconductor Prepared via Coprecipitation Method. *J. Supercond. Nov. Magn.* **2019**, *32*, 1191–1198. [CrossRef]
47. Knauf, N.; Harnischmacher, J.; Müller, R.; Borowski, R.; Roden, B.; Wohlleben, D. Preparation and characterisation of single-phase Bi-Pb-Sr-Ca-Cu-O high temperature superconductors. *Phys. C Supercond.* **1991**, *173*, 414–424. [CrossRef]
48. Popa, M.; Calderon-Moreno, J.M.; Crisan, D.; Zaharescu, M. Study of Fe addition on the thermal decomposition of coprecipitated oxalates for the Bi-based superconductor synthesis. *J. Therm. Anal. Calorim.* **2000**, *62*, 633–645. [CrossRef]
49. Durrani, S.K.; Qureshi, A.H.; Qayyum, S.; Arif, M. Development of superconducting phases in BSCCO and Ba-BSCCO by sol spray process. *J. Therm. Anal. Calorim.* **2009**, *95*, 87–91. [CrossRef]
50. Polasek, A.; Majewski, P.; Serra, E.T.; Rizzo, F.; Aldinger, F. Phase relations study on the melting and crystallization regions of the Bi-2223 high temperature superconductor. *Mater. Res.* **2004**, *7*, 393–408. [CrossRef]
51. Torsoni, G.B.; Cena, C.R.; Freitas, G.Q.; Carvalho, C.L. Synthesis of (Bi, Pb)-2223 superconductor powder by Pechini method. *Rev. Bras. Física Tecnológica Apl.* **2018**, *5*, 42–53. [CrossRef]
52. Van Driessche, I.; Buekenhoudt, A.; Konstantinov, K.; Bruneel, E.; Hoste, S. Evaluation of the phase composition of BPSCCO bulk samples by XRD- and susceptibility analysis. *IEEE Trans. Appl. Super.* **1996**, *4*, 185–190. [CrossRef]
53. Matsushita, T.; Suzuki, A.; Kishida, T.; Okuda, M.; Naito, H. The effect of Ag on the superconductivity of $Bi_{2-x}Pb_xSr_2Ca_2Cu_3O_y$ superconductors prepared by an optimum thermal procedure. *Supercond. Sci. Technol.* **1994**, *7*, 222–226. [CrossRef]
54. Halim, S.A.; Khawaldeh, S.A.; Mohamed, S.B.; Azhan, H. Superconducting properties of $Bi_{2-x}Pb_xSr_2Ca_2Cu_3O_y$ system derived via sol-gel and solid-state routes. *Mater. Chem. Phys.* **1999**, *61*, 251–259. [CrossRef]

55. Kamarudin, A.N.; Kechik, M.M.A.; Muralidhar, M.; Pinmangkorn, S.; Murakami, M.; Chen, S.K.; Baqiah, H.; Ramli, A.; Lim, K.P.; Halim, S.A. Microstructural, Phase Formation, and Superconducting Properties of Bulk $YBa_2Cu_3O_y$ Superconductors Grown by Infiltration Growth Process Utilizing the $YBa_2Cu_3O_y$ + $ErBa_2Cu_3O_y$ + $Ba_3Cu_5O_8$ as a Liquid Source. *Coatings* **2021**, *11*, 377. [CrossRef]
56. Langford, J.I.; Wilson, A.J.C. Scherrer after sixty years: A survey and some new results in the determination of crystallite size. *J. Appl. Crystallogr.* **1978**, *11*, 102–113. [CrossRef]
57. Mote, V.D.; Purushotham, Y.; Dole, B. Williamson-Hall analysis in estimation of lattice strain in nanometer-sized ZnO particles. *J. Theor. Appl. Phys.* **2012**, *6*, 6.
58. Kameli, P.; Salamati, H.; Eslami, M. The effect of sintering temperature on the intergranular properties of Bi2223 superconductors. *Solid State Commun.* **2006**, *137*, 30–35. [CrossRef]
59. Deac, I.; Burzo, E.; Pop, A.; Pop, V.; Tetean, R.; Kovacs, D. Intergranular properties of $(Y_{1-x-y}Zr_xCa_y)Ba_2Cu_3O_{7-\delta}$ compounds. *Int. J. Mod. Phys. B* **1999**, *13*, 1645–1654. [CrossRef]
60. Ambegaokar, V.; Baratoff, A. Tunneling Between Superconductors. *Phys. Rev. Lett.* **1963**, *10*, 486–489. [CrossRef]
61. Clem, J.R. Granular and superconducting-glass properties of the high-temperature superconductors. *Phys. C Supercond.* **1988**, *153–155*, 50–55. [CrossRef]
62. Gao, W.; Li, Z.; Sammes, N. *An Introduction to Electronic Materials for Engineers*; World Scientific Publishing Company: Hackensack, NJ, USA, 2011.
63. Bulaevskii, L.N.; Clem, J.R.; Glazman, L.I.; Malozemoff, A.P. Model for the low-temperature transport of Bi-based high-temperature superconducting tapes. *Phys. Rev. B* **1992**, *45*, 2545–2548. [CrossRef]
64. Bulaevskii, L.N.; Daemen, L.L.; Maley, M.P.; Coulter, J.Y. Limits to the critical current in high-T_c superconducting tapes. *Phys. Rev. B* **1993**, *48*, 13798–13816. [CrossRef]
65. Hensel, B.; Grivel, J.C.; Jeremie, A.; Perin, A.; Pollini, A.; Flükiger, R. A model for the critical current in $(Bi,Pb)_2Sr_2Ca_2Cu_3O_x$ silver-sheathed tapes. The role of small-angle c-axis grain boundaries and of the texture. *Phys. C Supercond.* **1993**, *205*, 329–337. [CrossRef]
66. Hensel, B.; Grasso, G.; Flükiger, R. Limits to the critical transport current in superconducting $(Bi,Pb)_2Sr_2Ca_2Cu_3O_{10}$ silver-sheathed tapes: The railway-switch model. *Phys. Rev. B* **1995**, *51*, 15456–15473. [CrossRef] [PubMed]
67. Oz, Y.; Jiang, J.; Matras, M.; Oloye, T.A.; Kametani, F.; Hellstrom, E.E. Conundrum of strongly coupled supercurrent flow in both under- and overdoped Bi-2212 round wires. *Phys. Rev. Mater.* **2021**, *5*, 074803. [CrossRef]
68. Kametani, F.; Jiang, J.; Matras, M.; Abraimov, D.; Hellstrom, E.E.; Larbalestier, D.C. Comparison of growth texture in round Bi2212 and flat Bi2223 wires and its relation to high critical current density development. *Sci. Rep.* **2015**, *5*, 8285. [CrossRef] [PubMed]
69. Parinov, I.A. *Microstructure and Properties of High-Temperature Superconductors*; Springer: Berlin/Heidelberg, Germany, 2007.

nanomaterials

MDPI

Article

Polypyrrole Solid-State Supercapacitors Drawn on Paper

Antonella Arena [1], Graziella Scandurra [1,*], Caterina Branca [2], Mariangela Ruggeri [2], Mauro Federico [2], Valentino Romano [3], Giovanna D'Angelo [2] and Carmine Ciofi [1]

1 Department of Engineering, University of Messina, 98166 Messina, Italy; arenaa@unime.it (A.A.); carmine.ciofi@unime.it (C.C.)
2 Department of Mathematical and Computational Sciences, Physical Science and Earth Science, University of Messina, 98166 Messina, Italy; caterina.branca@unime.it (C.B.); mariangela.ruggeri@studenti.unime.it (M.R.); mauro.federico@unime.it (M.F.); giovanna.dangelo@unime.it (G.D.)
3 Department of Physics, Politecnico di Milano, Piazza Leonardo da Vinci 32, 20133 Milano, Italy; valentino.romano@polimi.it
* Correspondence: gscandurra@unime.it

Abstract: Solid-state supercapacitors with areal capacitance in the order of 100 $mF \cdot cm^{-2}$ are developed on paper substrates, using eco-friendly, low-cost materials and a simple technology. The electrochemically active material used as the electrode is prepared from a stable water-based ink, obtained by doping commercial polypyrrole (PPY) powder with dodecylbenzene sulfonic acid (DBSA), and characterized by optical and electrical measurements, Raman investigation and Atomic Force Microscopy. The PPY:DBSA ink can be directly applied on paper by means of rechargeable water pens, obtaining, after drying, electrically conducting solid state tracks. The PPY:DBSA layers are then interfaced to one another through a polymer gel based on potassium hydroxide and chitosan, acting both as the ion-conducting medium and as the separator. The areal capacitance of the devices developed by following such a simple rule can be improved when the PPY:DBSA ink is applied in combination with other nanostructured carbon material.

Keywords: supercapacitor; paper substrates; polypyrrole

check for updates

Citation: Arena, A.; Scandurra, G.; Branca, C.; Ruggeri, M.; Federico, M.; Romano, V.; D'Angelo, G.; Ciofi, C. Polypyrrole Solid-State Supercapacitors Drawn on Paper. *Nanomaterials* **2023**, *13*, 3040. https://doi.org/10.3390/nano13233040

Academic Editor: Muralidharan Paramsothy

Received: 1 November 2023
Revised: 24 November 2023
Accepted: 27 November 2023
Published: 28 November 2023

1. Introduction

In recent decades, the increasing regard toward the environmental protection and preservation of our ecosystem has produced a strong collective effort, from the manufacturers to the end consumers, to replace, whenever possible, polluting material such as plastic with natural fibers and cellulose. For the same reasons, there has been unprecedented growth in the use of renewable energy sources, known to have a lower impact on the environment when compared to fossil fuels. It is, however, worth underlining how, as far as certain kinds of energy storage systems are concerned, there are still unresolved problems. In addition to safety concerns [1–4], there is in fact the issue of the disposal of spent devices, often containing toxic elements [5–7]. There is also the awareness that the reserve of strategic materials such as lithium and cobalt, often used in batteries, is not limitless and that their extraction from mines comes with high costs in terms of human safety and exploitation [8–10]. Finding metal-free, eco-friendly, reliable alternatives to commercially available energy storage devices can therefore be considered amongst the most fascinating scientific challenges for the future.

Among the energy storage systems, supercapacitors (SCs), namely devices whose electrical performances are a bridge between those of capacitors and batteries, are the ideal candidates towards which design innovation can be addressed, to reach a point where these devices can be considered possible replacements for batteries in the near future. Unlike batteries, SCs, which essentially consist of a couple of conducting electrodes interfaced by an ion-conducting medium and a separator, not only already exist in flexible [11], wearable [12], and stretchable forms [13], but they can also be entirely developed using inkjet-

and screen-printing technologies [14] suitable for low-cost mass production. Although a sharp demarcation line between the two classes does not exist, according to their operation principle, SCs can be classified as Electrochemical Double-Layer (EDL) supercapacitors or as pseudocapacitors. The capacitance of EDL supercapacitors arises from the electrically driven absorption–desorption of ions at the electrical double layers that form at the interfaces between the electrolyte and the electrodes, the latter usually consisting of nanostructured, carbon-based materials. Compared to EDL SCs, pseudocapacitors have more in common with batteries, as their electrical capacitance originates from redox transitions involving the electrodes, which are therefore required to be electrochemically active. Due to their ability to reversibly commute between an electrically conducting oxidized state and an electrically insulating reduced state, conjugated polymers such as polyaniline [15,16], polypyrrole [17], and polyethelendioxide [18] are among the most commonly used electrode materials in pseudocapacitors. Other common choices of electrodes include a variety of composites containing electrochemically active species, often metal oxides. While capacitance arising from faradaic transition at electrodes is higher compared to that originating from EDL, pseudocapacitors have unresolved cycling stability issues and their electrochemically active electrodes usually have lower conductivity compared to that of the EDL devices. One way to address these problems is to employ electrodes that support both the faradaic and the EDL mechanisms. In such a framework, amongst many other researchers throughout the world, in the past, we developed flexible supercapacitors achieving a few hundred $mF \cdot cm^{-2}$ areal capacitance, using gold-coated plastic transparency foils and a nafion membrane containing a lithium compound as substrates and as an ion transporting medium, respectively [19]. As an attempt to reformulate the design excluding lithium, gold, and plastic, here we show how eco-friendly solid-state supercapacitors with lower but acceptable capacitance can be easily developed by using simple technology. The devices are developed by applying, on paper substrates, water-dispersible conducting polymers acting as electrodes and using a chitosan gel containing potassium hydroxide as a spacer and as an ion-conducting medium.

2. Materials and Methods

Ionic surfactant dodecylbenzenesulfonic acid (DBSA) 70% in isopropanol, undoped polypyrrole loaded with 20% carbon black, multiwalled carbon nanotubes (CNTs), polyethylene oxide (PEO, average molecular weight 1,000,000), chitosan (medium molecular weight), acetic acid, and potassium hydroxide were purchased from Aldrich (Milan, Italy).

One gram of PPY, which comes as an insoluble insulating powder, was added to DBSA in isopropanol at a ratio of 1:5 in weight, and the ensemble was vigorously mixed to obtain a homogeneous paste. Then, 40 mL of deionized water was added to the mixture, which was ultrasonicated for a few hours. The developed water-based PPY:DBSA dispersion was used to fill rechargeable pens that could be used for months.

CNT dispersions were obtained by vigorously mixing carbon nanotubes and either PEO or DBSA at a ratio of 4:1 by weight until homogeneous pastes were obtained, to which a suitable amount of deionized water was added. The developed pastes, diluted with deionized water (100 mg/50 mL), were used to fill rechargeable pens in order to apply tracks on paper to be used as electrodes.

For comparison purposes, the electrical conductivity of the carbon nanotube-based pastes and inks and the PPY:DBSA inks was tested by applying samples of the material on top of the same kind of alumina substrates having platinum interdigitated electrodes on top.

Chitosan is a biopolymer soluble in dilute aqueous solutions of acetic acid that, due to its functional groups, is not only able to host but is also able to interact with inorganic salts, yielding biodegradable ion-conducting gels [20,21]. The ion conductivity of solid-state membranes derived from chitosan strongly depends on the preparation conditions, for instance on the presence of a plasticizer. Moreover, as aging affects the crosslinking degree of the gel, it has a critical impact on the mechanical stability of the gel-derived membranes.

The electrolyte used in this work was prepared by adding 40 mL of deionized water to one gram of chitosan powder, previously wet with 5 mL of acetic acid. After magnetic stirring, KOH, at a 1:0.25 ratio by weight to chitosan, was added to the mixture. After the chitosan addition, the mixture was ultrasonicated for a few hours and stored for a few days before use. The obtained gel could be either applied on the top of the electrodes, or it could be deposited on glasses, resulting in self-standing ion-conducting membranes, once dried and detached from the substrates.

The capacitors were developed in two different geometries. In the first one, the KOH–chitosan gel was drop-deposited on the gap between interdigitated PPY:DBSA and other kinds of electrodes and applied onto a single sheet of paper by rechargeable pens (Molotov empty marker size 2 mm, and Koh-I-Noor Penna China Professional 04) (Figure 1a). Molotov pens were mainly used to draw geometries by hand, while Koh-I-Noor pens were used in combination with a Graphtec pen plotter MP4300 to demonstrate the possibility of obtaining reproducible and possibly complex electrode geometries. A second geometry, sketched in Figure 1b, was achieved by sandwiching a couple of paper electrodes, obtained by cutting two rectangularly shaped pieces of paper, each having a conducting track of PPY:DBSA and other kinds of electrodes applied on the top, and interposing a KOH–chitosan layer in the middle. To improve the electrical conductivity, samples of the devices have been developed by applying the electrode material on top of layers previously drawn on paper by means of a rechargeable pen filled with commercial carbon-based conducting paste (Electric Paint by Bare Conductive®, London, UK, purchased online www.bareconductive.com), diluted with water. For comparison, samples were prepared by using PPY: DBSA, PEO/CNT, and DBSA/CNT as electrode materials.

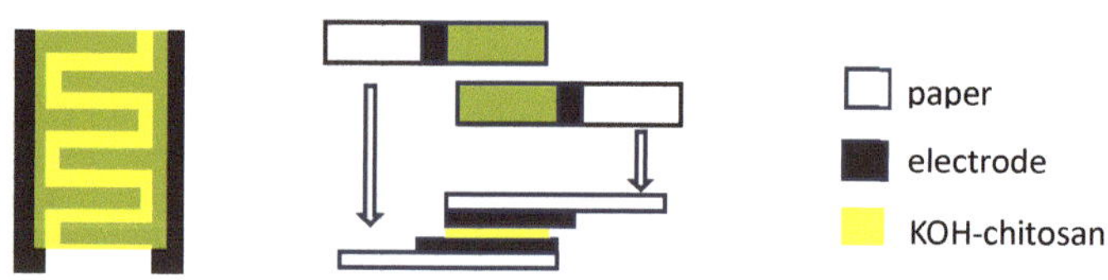

(a) (b)

Figure 1. Capacitor structures used in this study: (**a**) interdigitated electrodes are first drawn on paper and the KOH–chitosan gel is drop-deposited on top of the interdigitated section; (**b**) two strips of paper with the electrode and the KOH–chitosan gel are joined together with one strip facing the other.

The morphology of films deposited from PPY:DBSA ink and CNT dispersions was investigated by AFM measurements, which were conducted by means of a Nanosurf FlexAFM equipped with a C3000 controller (Nanosurf AG, Liestal, Switzerland). Optical measurements were performed using an HR4000 Ocean Optics microspectrophotometer. Raman spectra were recorded at room temperature using a LabRam HR800 (Horiba Italia SRL, Rome, Italy) spectrometer with a 532 nm excitation wavelength, an 1800 gr/mm grating, and a liquid nitrogen-cooled CCD camera. The measurements were performed with a 50× objective and a laser power of 1 mW to avoid any heating-induced degradation effects. The electrical measurements were performed at room temperature and at a relative humidity of 58% by means of a Keithley 2400 source meter (Tektronix, Beaverton, OR, USA).

3. Results and Discussion

Polypyrrole and its derivatives are well-known conjugated polymers that have a variety of technological applications, going from the development of electrochromic windows [22–24] to electrochemical gas sensors [25–27] and actuators [28], all based on their possibility of existing in an electrically conducting oxidized state and in an electrically insulating reduced state. Doped PPY can be achieved by either chemical or electrochemical polymerization of the pyrrole monomer, carried out in the presence of suitable dopants. The insertion of positive charges provided by the dopant cations, delocalized along the polymer backbone, oxidizes the polymer and produces an increase in electrical conductivity compared to the neutral undoped state. In addition, the presence of the dopant causes a structural rearrangement of the polymer chains and introduces additional electronic states into the bandgap. Such modifications manifest themselves with the appearance of optical absorption bands in the visible–near-infrared region and changes in the infrared and Raman spectra of the doped polymer compared to the undoped state. From the applicational point of view, doped PPY, commercialized as a powder, is virtually insoluble due to strong polymer interchain interactions and poor interactions of the material with the solvents. In a previous paper [19], we succeeded in obtaining stable dispersion of a commercially available doped polypyrrole in water, with the aid of the ionic surfactant DBSA. The excellent quality of the film deposited from such dispersions led to the conclusion that the surfactant, rather than acting as a dopant, not only affects the strong PPY interchain interaction, favoring the dispersion, but forms noncovalent bonds with the polymer backbone. Here, we find that adding suitable amounts of DBSA to commercially available insoluble, undoped polypyrrole powder results in stable inks from which conducting tracks can be deposited on a variety of substrates. Undoubtedly, the surfactant in the present case acts as a dispersing agent for the undoped PPY powder. In addition, since the PPY to be dispersed is undoped, DBSA could also play the role of the doping agent, providing the delocalized cation that oxidizes the polymer. To prove this point, the results of optical spectroscopy in the visible–infrared range and Raman measurements are shown in Figure 2a,b. According to the scientific literature, while polypyrrole in its undoped reduced state has very little optical absorption in the visible region, the material, as the doping level increases, shows absorption peaks arising in the spectral range between 450 nm and the near-infrared range [29]. Such a mechanism is responsible for the polymer's electrochromic behavior, namely a color change exhibited by the material in response to an externally applied voltage. The presence of a broad band, likely subtending more than one peak, observed in the visible region of the absorption spectrum measured on a thin film deposited on glass from the PPY:DBSA ink (Figure 2a), is, therefore, a demonstration that the surfactant DBSA actually behaves as a dopant for PPY. The change induced in the undoped PPY by the presence of DBSA is also confirmed by the results of Raman measurements. Figure 2b compares the spectrum of the untreated polypyrrole powder with that of PPY:DBSA. The two main peaks positioned at about 1580 cm^{-1} (attributed to C=C in-ring and C-C inter-ring stretching) and 1330 cm^{-1} (attributed to C-C in-ring and anti-symmetric C-N stretching) in the spectrum of the undoped PPY powder are modified in the doped sample, confirming the integration of DBSA in the PPY structure. In particular, the lowest band in doped PPY appears strongly decreased in intensity, slightly shifted in frequency, and enlarged (higher full width at half maximum, FWHM) with respect to the PPY, indicating the structural lattice distortion and higher degree of disorder of the PPY rings [30–32].

To investigate the electrochemical behavior of the PPY:DBSA ink developed as it was previously described, a simple electrochemical cell has been prepared by interfacing a couple of ITO-coated transparent glasses to each other, one having a thin PPY:DBSA layer applied on the top, and the other having a layer of KOH-doped chitosan on the top. The device, sketched in Figure 3a, exhibits a color change in response to different applied voltages, which are responsible for the change in the oxidation PPY:DBSA film, as shown in the two photographs in Figure 3b,c.

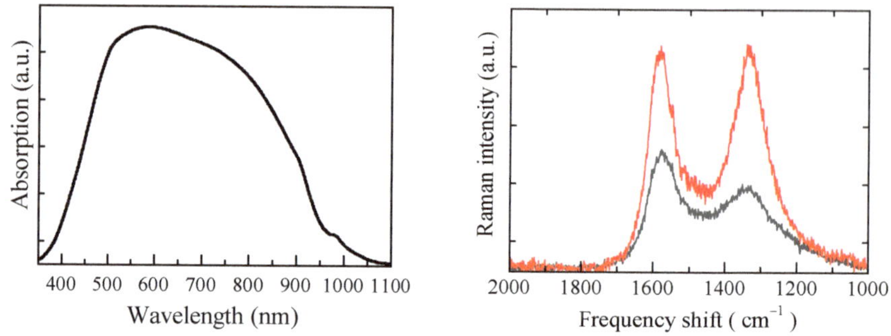

(**a**) (**b**)

Figure 2. Absorption spectrum of a PPY: DBSA film in the VIS–NIR (**a**); Raman spectrum of PPY (red line) powder compared to PPY:DBSA (black line) (**b**).

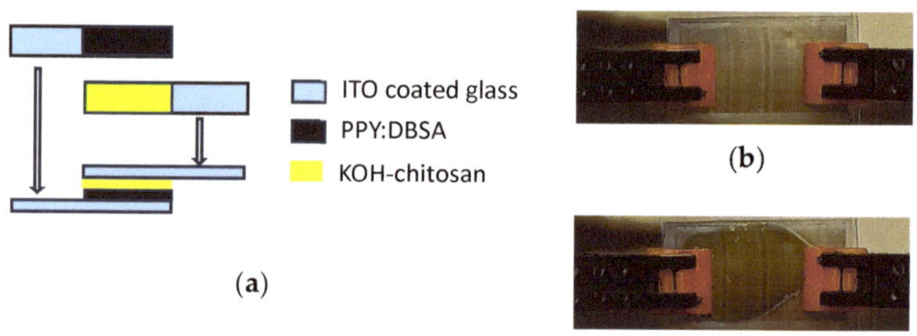

(**a**) (**b**) (**c**)

Figure 3. Schematic view of the electrochemical cells used to characterize the PPY:DBSA films (**a**). Photos of a typical electrochemical cell, with the PPY films in the reduced (**b**) and oxidized (**c**) state.

The electrical behavior of the same device has also been investigated by means of current–voltage measurements performed at different scan speeds over a voltage window ranging between −1 V and +1 V, shown in Figure 4.

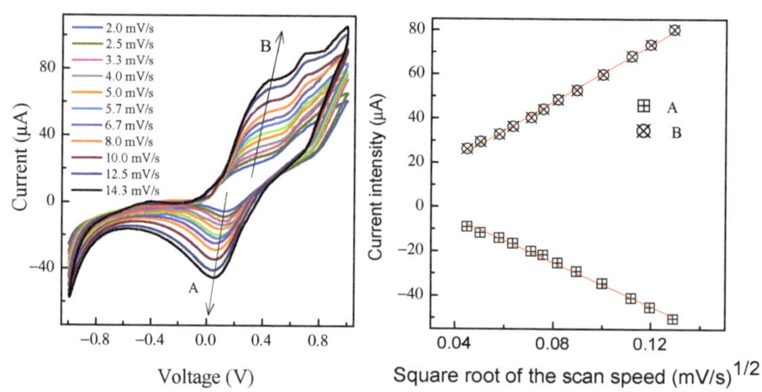

(**a**) (**b**)

Figure 4. Current–voltage loops measured at different scan speed on a typical device of the kind sketched in Figure 3a (**a**); current intensity measured in correspondence of peak A and peak B, plotted as a function of the scan speed (**b**).

As reported by Vidanapathirana et al. [33], redox processes occurring in PPY when DBSA is used as a dopant involve the ions provided by the electrolyte interfaced to the doped polymer rather than surfactant ions, which are trapped in the polymer matrix. Accordingly, the current peaks marked by the arrows A and B in the clockwise loops of Figure 4a, the intensities and positions of which change with the scan speed, originate from potassium ions provided by the KOH–chitosan layer, which are either inserted into the polymer matrix when in the reduced state or expelled from it when in the oxidized state. Figure 4b shows that the intensity of peaks marked as A and B in Figure 4a linearly depends on the square root of the scan speed, indicating that the peaks arise from diffusion-limited redox processes at the PPY:DBSA electrodes.

Figure 5 shows how the electrical resistance of a 2 cm long–2 mm wide track applied on paper decreases as the number of layers applied by means of a pen filled with the PPY:DBSA ink increases. It can be noticed that after the first three layers have been applied, the resistance becomes almost constant. The inset shows the real part of the impedance of the same conducting track as a function of frequency (the imaginary part, being two orders of magnitude lower than the real one, is not reported).

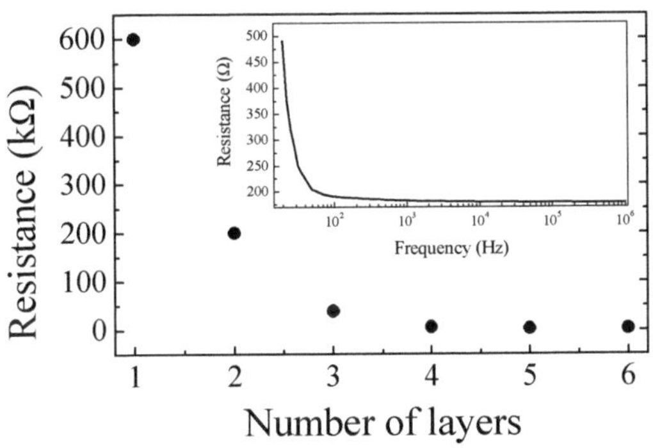

Figure 5. Resistance measured on a track applied on paper as a function of the number of layers of PPY:DBSA ink. The inset shows how the resistance after the application of 6 layers behaves as a function of frequency.

In order to test the possibility of using automated drawing systems for producing complex, geometrically consistent, and reproducible patterns, the ink was used to fill up the Koh-I-Noor Penna China Professional 04 that, by means of a proper adapter, can replace a drawing pen for the Graphtec pen plotter. An example of interdigitated test patterns drawn on paper is shown in Figure 6a. AFM is used to measure the step of crossing line patterns drawn on a microscope slide (Figure 6b shows the AFM tip positioned in correspondence with a couple of crossing lines), finding an average thickness of about 0.8 μm for lines obtained after five applications of the ink.

From the morphological point of view, the PPY:DBSA films deposited from the ink have a granular consistency, with hundred-nanometer-size oblong granules. The sample imaged in the AFM picture of Figure 7 is a single track applied onto a silicon substrate by means of a Molotov pen filled with the PPY:DBSA ink. As we found earlier in other mixtures containing the DBSA [19,34], the film appears to have a rather ordered structure, with a certain degree of alignment of the granules along a preferential direction.

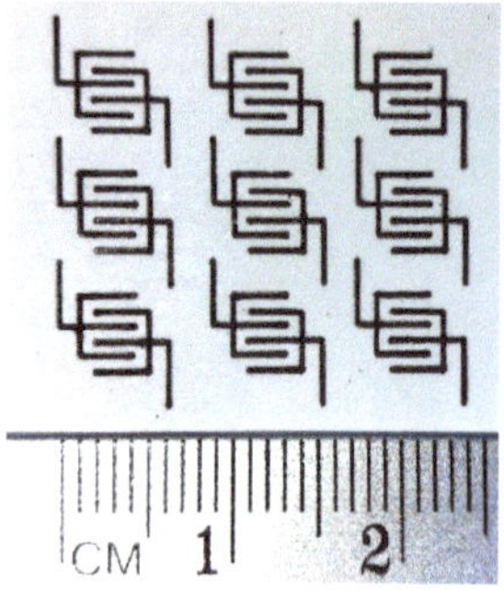

(**a**)

(**b**)

Figure 6. Interdigitated electrodes plotted on paper using the PPY:DBSA ink (**a**); magnified view of a couple of about 0.8 μm thick crossing lines, drawn using the PPY:DBSA ink below the AFM tip (**b**).

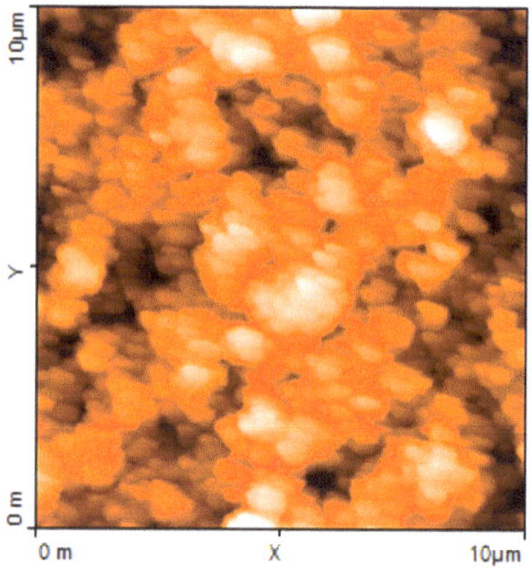

Figure 7. AFM image (area 10 μm × 10 μm) of a typical PPY:DBSA film.

As mentioned before, aimed at improving the capacitance of the electrochemically active electrodes, the faradaic contribution provided by the redox transitions can be combined with the good charge transport features and porous nature of nanostructured carbon, which favors the formation of electrical double layers at the interfaces with the electrolyte. For this reason, CNT dispersion in PEO and DBSA has been developed by us as described in a previous section, and characterized by means of Raman and morphological AFM analyses. Raman spectra of the pristine CNTs, DBSA-modified CNTs, and PEO/CNT hybrid nanocomposites are reported and compared in Figure 8. No peaks related to the DBSA or PEO matrix were revealed. Only the MWCNTs contribute to the Raman spectrum, mainly with the characteristic bands of CNTs: the D band (defect) at ~1350 cm^{-1}, the G band (graphite band) at ~1580 cm^{-1}, and the G′ band (D overtone) at ~2700 cm^{-1}.

The D band defines the in-plane breathing vibration of the aromatic ring. It is activated in the presence of disorder or deviation from the perfect graphitic lattice. The G band arises from the in-plane stretching vibrations of the sp^2-bonded carbon atoms. Thirdly, the second-order Raman band G′ arises from a double electron–phonon resonance mechanism and can be used to distinguish the number of layers of CNT. At first glance, the spectra are quite similar [35]. But, a careful analysis of the bands and their intensity ratios ID/IG, IG′/IG, and IG′/ID reveals the assessment of the relative influence of DBSO and PEO on the pristine CNTs. For the doped CNT and PEO/CNT hybrid nanocomposites, the G′ band is shifted to higher wavenumber: pristine CNTs: 2694 cm^{-1}; DBSA/CNT: 2705 cm^{-1};

PEO/CNT: 2703 cm^{-1} (Figure 9). This indicates a charge transfer interaction between the DBSA or PEO and CNTs [36]. Anyway, in all samples, the G′ peak was well fitted by a sharp Lorentzian peak, suggesting the single nature of the CNTs. The intensity ratio of D and G (ID/IG) is commonly used to acquire information on the structure of functionalized CNTs and to estimate the number of defects in the graphitic structure. Indeed, the interaction between the functional groups and the CNT wall is expected to generate a higher number of sp^3-hybridized carbon with a consequent increase in the ID/IG ratio [37]. Furthermore, the intensity ratio of the G′ and G bands (IG′/IG) and of the G′ and D bands (IG′/ID) was also determined. The IG′/IG ratio indicates the crystalline quality: the higher it is, the more extended the graphitic order is. Finally, the IG′/ID ratio is related to the degree of crystallinity [38].

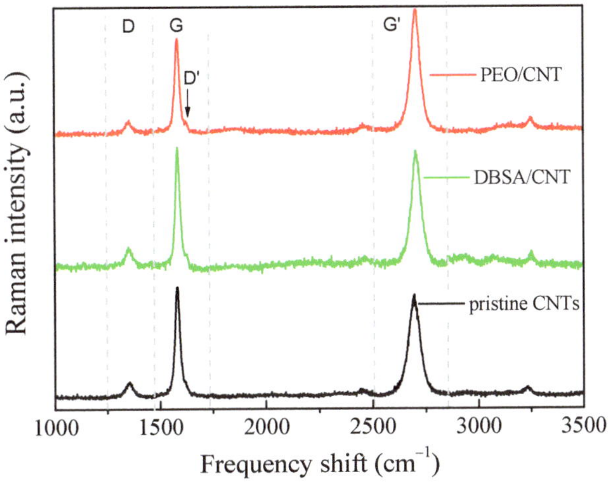

Figure 8. Raman spectra of pristine CNTs, DBSA/CNT, and PEO/CNT hybrid nanocomposites.

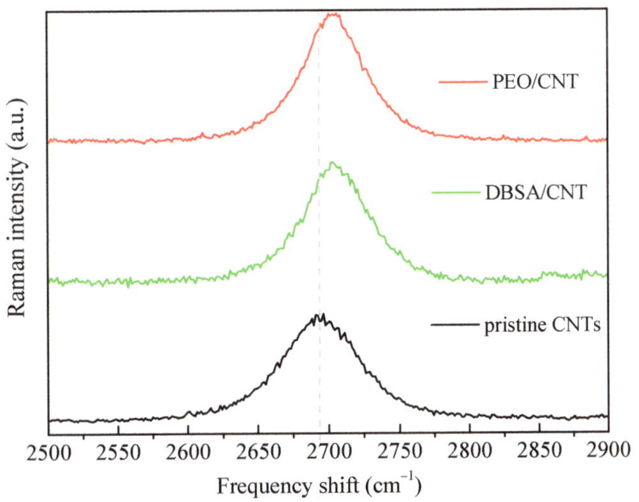

Figure 9. Raman spectra of pristine CNTs, DBSA/CNT, and PEO/CNT hybrid nanocomposites in the spectral region around the G′ band.

The intensity ratios ID/IG, IG′/IG, and IG′/ID are shown in Figure 10. The slight increase in the ID/IG ratio of the doped CNTs (ID/IG = 0.15) compared with that of the pristine CNTs (ID/IG = 0.13) confirms the successful introduction of functional groups onto the CNT surface and that the outer layers of the CNTs were somewhat chemically modified [39]. However, because the ID/IG ratio is small, we can assess that the CNTs have few defects in the modified samples.

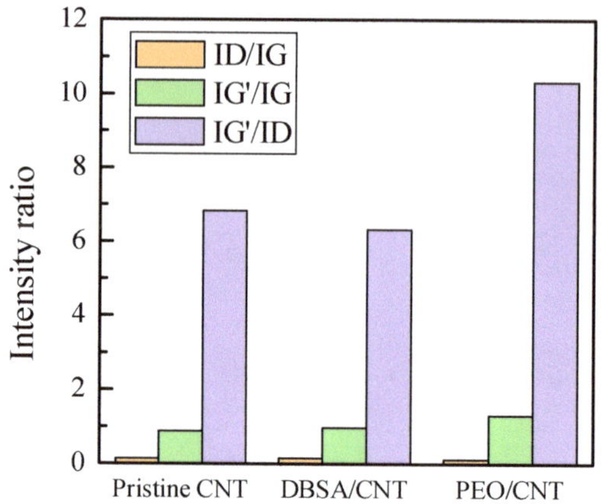

Figure 10. Relative intensity ratios of D/G, G′/G, and G′/D peaks of pristine CNTs, DBSA/CNT, and PEO/CNT hybrid nanocomposites.

The IG′/IG ratio is 0.88, 0.97, and 1.30, whereas the IG′/D ratio is 6.83, 6.33, and 10.32 for the pristine CNTs, modified CNTs, and PEO/CNT hybrid nanocomposites, respectively. The increase in the IG′/IG ratio of the modified CNTs compared with that of the pristine CNTs reveals a lower density of lattice defects in the acid-treated DBSA/CNT and hybridized PEO/CNTs. Furthermore, the hybridized PEO/CNT exhibits a high degree of crystallinity. Interestingly, the G′ peak arises from a two-phonon process; thus, it is expected that its intensity increases as the sample becomes more ordered, enabling the two-phonon coupling process [40]. The higher order in hybridized PEO/CNTs is also confirmed by a decreased full width at half-maximum (FWHM) of the G′ band: pristine CNTs: 65 cm^{-1}; DBSA/CNT: 53 cm^{-1}; PEO/CNT: 51 cm^{-1}.

The difference between PEO's and DBSA's carbon nanotube dispersions detected by Raman spectroscopy becomes more apparent when comparing the morphology of a couple of typical samples applied on silicon substrates and investigated by means of AFM measurements. Figure 11a shows the filiform morphology of the pristine carbon nanotubes powder used by us, pressed onto a silicon substrate. Wider filaments of what could be identified as polymer-wrapped nanotubes are recognizable in the AFM image of a film of carbon nanotubes dispersed in PEO deposited on silicon (Figure 11b). On the contrary, the morphology of the carbon nanotubes dispersed in DBSA (Figure 11c), which shows no filaments, resembles that of the PPY:DBSA film shown in Figure 7, confirming the peculiar role of the surfactant when used as a host on the morphologies of carbon material dispersions [19,34].

From the electrical transport point of view, it is found that samples of the pastes containing carbon nanotubes dispersed in PEO and DBSS, applied on alumina substrates with platinum interdigitated electrodes on the top, yielded lower electrical resistance compared to that of the PPY:DBSA sample. Furthermore, compared to that of PPY:DBSA, the real part of the impedance of both the PEO- and DBSA-based pastes showed little frequency dependence in the range between 20 Hz and 1 MHz.

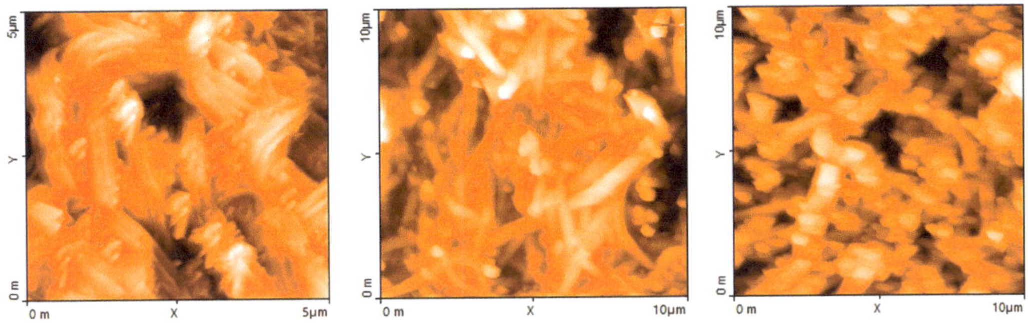

(**a**) (**b**) (**c**)

Figure 11. AFM micrograph of pressed CNT powder (**a**) and of CNTs dispersed in PEO (**b**) and in DBSA (**c**).

Figure 12 shows the results of current–voltage measurements performed at a scan speed of 10 mV/s on interdigitated electrodes applied on paper, the geometry of which is sketched in Figure 1a, with a KOH–chitosan layer on top. The inner cycle refers to a sample having electrodes deposited using the ink based on commercial conducting mentioned before. The other cycles refer to samples obtained by underlying the commercial ink-conducting tracks with ten layers of PEO/CNT, DBSA/CNT, and o PPY:DBSA, respectively. It can be noticed that better performances in terms of current intensity are achieved by using PPY:DBSA.

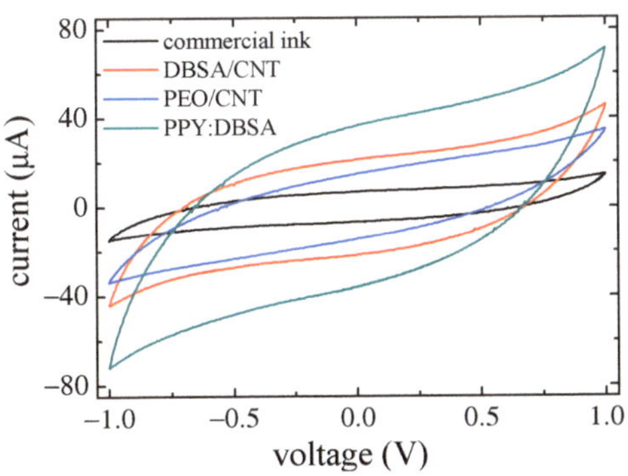

Figure 12. I–V measurements performed on samples having different kind of interdigitated electrodes with the same geometry, applied on paper substrates, with a KOH–chitosan layer on top.

Although real devices such as the ones we are investigating are far from behaving as ideal capacitors, a complex model should be developed for interpreting the results of electrical measurements to derive the value of their aptitude to serve as energy storage devices, a rough estimation of the order of magnitude of their capacitance can be derived from time current–voltage loops and from charge–discharge curves. Since the current through an ideal capacitor is given by its capacitance C multiplied by the time derivative of the voltage, an estimation C_{es1} of C can be extracted from the IV curve as the ratio between the current at zero voltage (where the current due to the resistive component in parallel is zero) and the absolute value of time rate of voltage change:

$$C_{es1} = \left. \frac{I}{dV/dt} \right|_{V=0} \tag{1}$$

where I is the current and dV/dt is the scan speed.

Another possibility of the estimation of the capacitance comes from the observation that the finite voltage change ΔV over a finite time interval Δt across a capacitor supplied with a constant current I is given by the product of the current and the time interval divided by the capacitance itself, and therefore we can obtain the estimate C_{es2} as:

$$C_{es2} = \frac{I\Delta t}{\Delta V} \tag{2}$$

from the observation of charge–discharge curves vs. time at constant current. In the case of an ideal capacitor with no dissipative components, the estimates C_{es1} and C_{es2} coincide and are independent of the test conditions. This is not the case in systems such as the ones we are investigating. However, expressions such as the ones in Equations (1) and (2) can still be used to obtain a rough estimate of the capacitance and as an index of comparison among different materials and technologies. Often, in the case in which the active surface A of the device can be clearly identified, as in the case of the devices in Figure 1b, estimates are given in terms of the areal capacitance $C_A = C/A$ and the current density I/A, rather than in terms of the absolute capacitance and the current, as this provides an information that is independent of the area of the actual device used for testing. In a similar fashion, we can reach approximate estimates for areal energy density E_A and areal power density P_A, per given voltage changes ΔV across the capacitor, as:

$$E_A = \frac{C_A(\Delta V)^2}{2}; \quad P_A = \frac{E_A}{\Delta t} \qquad \text{where } C_A = \frac{I\Delta t}{A\Delta V} \tag{3}$$

Tests for estimating capacitance and areal capacitance have been performed on devices built according to both the geometries described in Figure 1a,b.

The current–voltage loop of Figure 13a refers to a scan performed at 10 mV/s on the sample shown in the photograph in Figure 13d. The device uses as electrodes a mixture of the PPY:DBSA and DBSA/CNT inks, applied on the underlying interdigitated tracks (geometry in Figure 1a). Figure 13b reports the voltage across the sample when charged and discharged at two different levels of constant current. In the case of this type of device, an effective active area cannot be easily defined and therefore we used Equations (1) and (2) for capacitance estimation from Figures 13a and 13b, respectively. From Figure 13a, using Equation (1), we obtain C_{es1} = 6.5 mF; from Figure 12b, using the time interval required to reach 1 V starting from 0 V, we obtain values for C_{es2} between 2 and 4 mF, depending on the selected curve (that is, the value of the charging current). As expected, the estimates are different, although we obtain values of capacitance that are in the same order of magnitude, and this confirms that any of the methods discussed above, while not accurate, is useful in providing a rough indication of the charge storage capabilities of the system. Figure 13c shows the opposite of the imaginary part of the impedance, versus the real one, measured between 20 Hz and 1 MHz. From this graph, we can notice the rather high value of the resistive component with respect to the imaginary one, and this is, among other possible causes, the result of the large spacing between the electrodes. It is therefore reasonable to assume that with an optimized design, this aspect can be improved.

Current–voltage, charge–discharge, and impedance measurements were also performed on a typical device with the geometry in Figure 1b, with electrodes obtained from a mixture of the PPY:DBSA and DBSA/CNT inks. The results are reported in Figures 14–16. In particular, Figure 14a shows how the steady state loops of current density versus voltage enlarges as the time rate of voltage change increases. The values of the capacitance per unit area estimated for each loop, and reported as a function of the scan speed, are shown in Figure 14b. Figure 15a shows examples of the charge–discharge voltage measured at different current densities. The areal capacitance estimated from the data in Figure 14a is plotted as a function of the current density in Figure 15b. The obtained values, in the range from 40 to 170 mF/cm^2 for test current densities ranging from 6.4 down to 0.8 mA/cm^2, are in the same order of magnitude as those obtained from the current–voltage measurements shown in Figure 14b. The performance of the device under tests in terms of specific energy and specific power, derived from the data of Figure 15a using Equation (3), are summarized in the Ragone plot of Figure 15c. Figure 16 shows the opposite of the imaginary part of the impedance, versus the real one, measured between 20 Hz and 1 MHz. Also, in this case, the real part of the impedance is larger than the imaginary part. In this case, improvement can be possibly obtained by careful optimization of the ink composition.

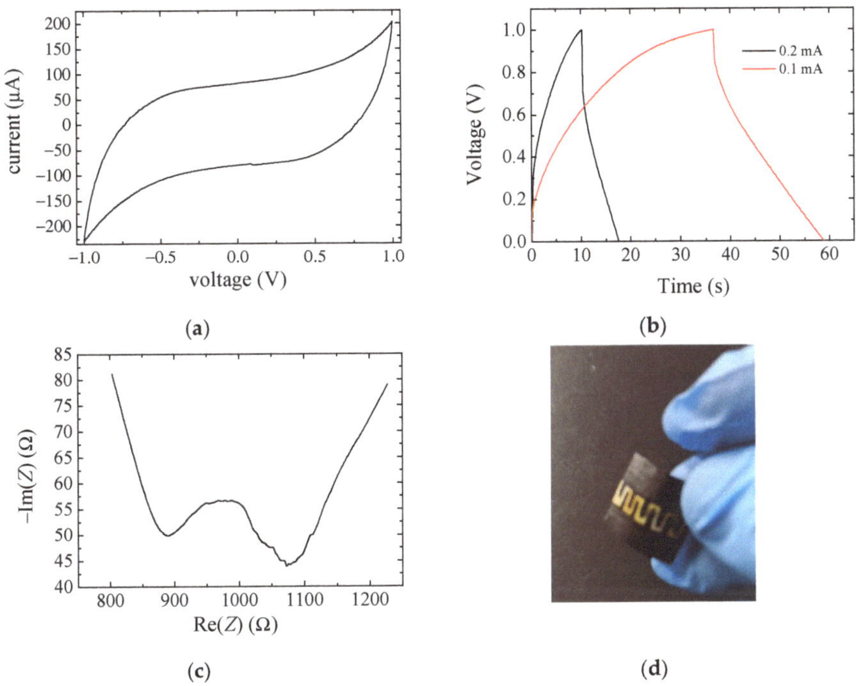

Figure 13. Current–voltage loop (**a**), voltage charge–discharge curve (**b**), frequency-dependent impedance (**c**) and photograph (**d**) of a typical device on paper, with a mixture of the PPY:DBSA and DBSA/CNT inks as electrodes and a KOH–chitosan layer on the top.

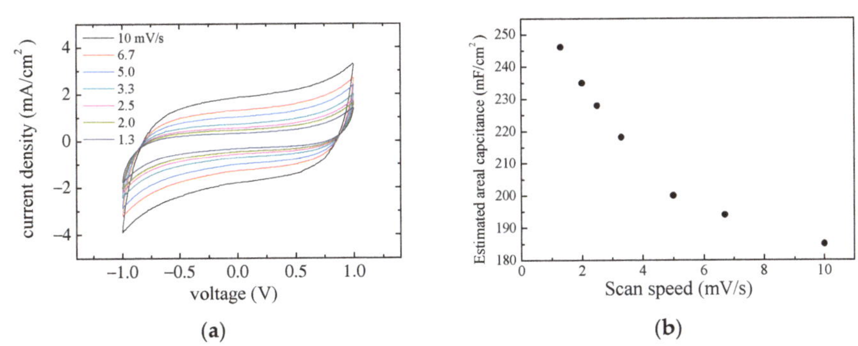

Figure 14. Current–voltage measurements performed at different scan speeds on a sample consisting of a couple of rectangularly shaped pieces of paper, each having a mixture of PPY:DBSA and DBSA/CNT applied on the top, interfaced through a KOH–chitosan layer (**a**); areal capacitance estimated from the current–voltage loops, as a function of the scan speed (**b**).

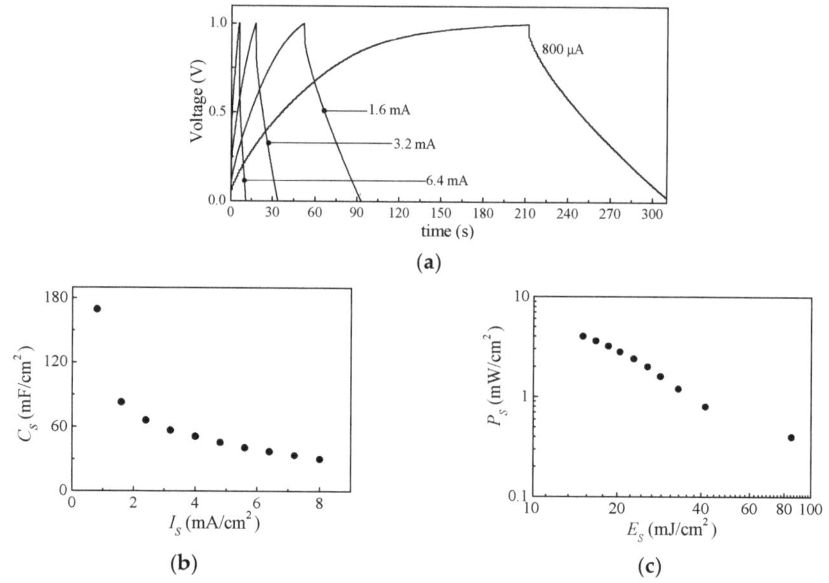

Figure 15. Voltage charge–discharge curves measured on the same device to which Figure 13 refers to (**a**); areal capacitance estimated from the charge–discharge curves, as a function of current density (**b**); power density versus energy density (**c**).

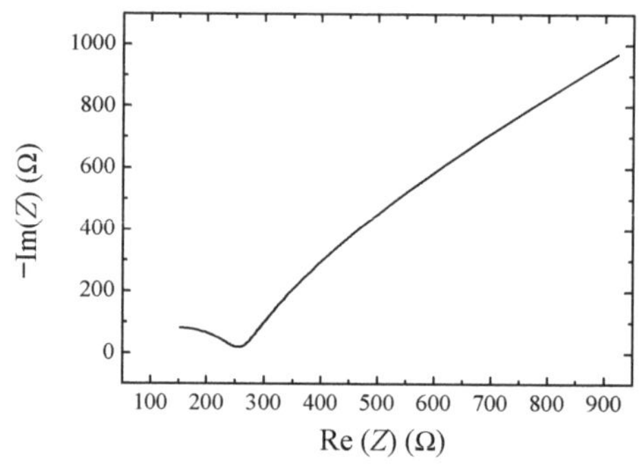

Figure 16. Cole–Cole plot of the same device to which Figures 13 and 14 refer to.

As a demonstration of the charge storage capacity of these devices, Figure 17 shows three of them connected in series (each of which having an active area of 10 mm^2), while maintaining an LED "on" for a few seconds after being initially charged to 4 V.

As far as stability is concerned, all the developed devices, when stored in air at room temperature, do not show significant changes in the current–voltage loops and in the charge–discharge behavior, even after months. As the capacitance of conjugated polymer electrodes arises from electrically driven redox transitions involving intercalation and de-intercalation of mobile ions into the polymer, such kinds of electrodes are usually the most susceptible to degradation under stress conditions. Figure 18a,b compare the results of current–voltage measurements and charge–discharge measurements, performed before and after 1000 subsequent charge–discharge cycles on the same cell, with the geometry described in Figure 1b and using only PPY:DBSA as the electrode material. It can be noticed that while the IV curve seems to be slightly affected by the stress, the charge–discharge time undergoes a 20% decrease, denoting a significant degradation of the electrode.

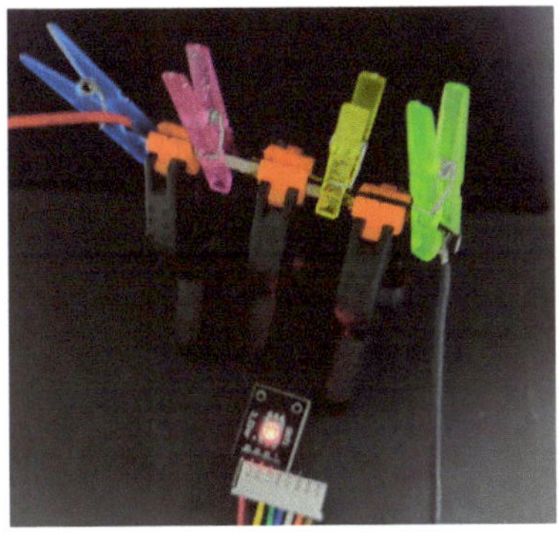

Figure 17. Snapshot of an LED being supplied by three paper devices connected in series. The series of devices was previously charged to a 4 V; the LED, with a 1 kΩ resistance in series, stays on for a few seconds after the supply is removed.

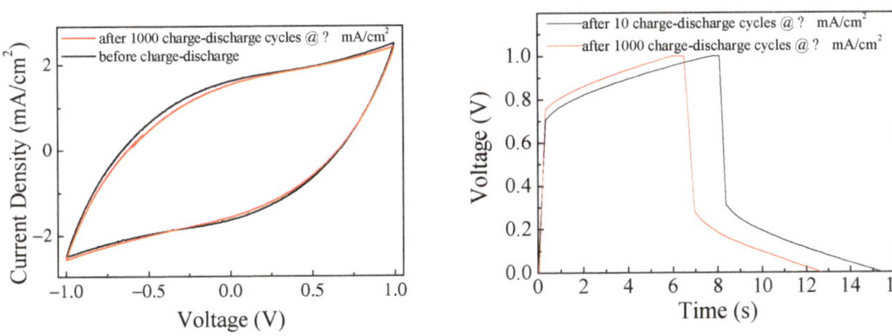

(**a**) (**b**)

Figure 18. Steady state current–voltage loop measured on a typical device with geometry as described in Figure 1b and PPY:DBSA electrodes, before and after stress (**a**); charge–discharge curve measured at constant current density on the same device, before and after stress (**b**).

4. Conclusions

The experimental results presented and discussed here demonstrate that it is possible to develop devices on paper substrates with good specific capacitance by using commercially available, eco-friendly materials such as inks and with the aid of simple and low-cost deposition techniques. The areal capacitances that have been obtained are in the range of 100 mF/cm^2 (83 mF/cm^2 at 1.6 mA/cm^2) with specific powers that can be in excess of 1 mW/cm^2. However, the performances of capacitors achieved using the approach we propose are, at present, limited by a rather large series resistance. This resistance can be, in principle, reduced by improving the conductivity of the electrodes and the electrolyte by optimizing their composition and processing, and/or by optimizing the geometry of the device by, depending on the configuration, reducing the spacing between interdigitated electrodes or increasing the overlapping surface. The performances obtained so far are encouraging, especially because we employed carbon-based electrodes without any toxic, rare, or precious chemical element or component.

Author Contributions: Conceptualization, A.A.; methodology, A.A., G.S., G.D., C.B. and C.C.; validation, formal analysis, investigation and resources: A.A, C.B., C.C., G.D., M.F., V.R., M.R. and G.S.; data curation, A.A., G.D. and C.C.; writing—original draft preparation, A.A. and C.C.; writing—review and editing, visualization: A.A, C.B., C.C., G.D., M.F., V.R., M.R. and G.S.; supervision, A.A. and G.S. All authors have read and agreed to the published version of the manuscript.

Funding: This research received no external funding.

Data Availability Statement: The data can be requested from the authors.

Conflicts of Interest: The authors declare no conflict of interest.

References

1. Yu, X.; Chen, R.; Gan, L.; Li, H.; Chen, L. Battery Safety: From Lithium-Ion to Solid-State Batteries. *Engineering* **2023**, *21*, 9–14. [CrossRef]
2. Doughty, D.H.; Roth, E.P. A General Discussion of Li Ion Battery Safety. *Electrochem. Soc. Interface* **2012**, *21*, 37.
3. Held, M.; Tuchschmid, M.; Zennegg, M.; Figi, R.; Schreiner, C.; Mellert, L.D.; Welte, U.; Kompatscher, M.; Hermann, M.; Nachef, L. Thermal runaway and fire of electric vehicle lithium-ion battery and contamination of infrastructure facility. *Renew. Sustain. Energy Rev.* **2022**, *165*, 112474–112487. [CrossRef]
4. Melchor-Martínez, E.M.; Macias-Garbett, R.; Malacara-Becerra, A.; Iqbal, H.M.N.; Sosa-Hernandez, J.E.; Parra-Saldívar, R. Environmental impact of emerging contaminants from battery waste: A mini review. *Case Stud. Chem. Environ. Eng.* **2021**, *3*, 100104–100112. [CrossRef]
5. Mrozik, W.; Rajaeifar, M.A.; Heidrichab, O.; Christensen, P. Environmental impacts, pollution sources and pathways of spent lithium-ion batteries. *Energy Environ. Sci.* **2021**, *14*, 6099–6121. [CrossRef]
6. Sun, J.; Li, J.; Zhou, T.; Yang, K.; Wei, S.; Tang, N.; Dang, N.; Li, H.; Qiu, X.; Chen, L. Toxicity, a serious concern of thermal runaway from commercial Li-ion battery. *Nano Energy* **2016**, *27*, 313–319. [CrossRef]
7. Khan, M.S.; Javed, M.; Rehman, M.D.T.; Urooj, M.; Ahmad, M.I. Heavy metal pollution and risk assessment by the battery of toxicity tests. *Sci. Rep.* **2020**, *10*, 16593. [CrossRef]
8. Sharma, S.S.; Manthiram, A. Towards more environmentally and socially responsible batteries. *Energy Environ. Sci.* **2020**, *13*, 4087–4097. [CrossRef]
9. Ponnada, S.; Kiai, M.S.; Krishnapriya, R.; Singhal, R.; Sharma, R.K. Lithium-free batteries: Needs and challenges. *Energy Fuels* **2022**, *36*, 6013–6026. [CrossRef]
10. Gourley, S.W.D.; Or, T.; Chen, Z. Breaking free from cobalt reliance in lithium-ion batteries. *Iscience* **2020**, *23*, 101505. [CrossRef]
11. Zhang, Y.; Mei, H.; Cao, Y.; Yan, X.; Yan, J.; Gao, H.; Luo, H.; Wang, S.; Jia, X.; Kachalova, L.; et al. Recent advances and challenges of electrode materials for flexible supercapacitors. *Coord. Chem. Rev.* **2021**, *438*, 213910. [CrossRef]
12. Gopi, C.V.V.M.; Vinodh, R.; Sambasivam, S.; Obaidat, I.M.; Kim, H.J. Recent progress of advanced energy storage materials for flexible and wearable supercapacitor: From design and development to applications. *J. Energy Storage* **2020**, *27*, 101035.
13. Chen, X.; Villa, N.S.; Zhuang, Y.; Chen, L.; Wang, T.; Li, Z.; Kong, T. Stretchable supercapacitors as emergent energy storage units for health monitoring bioelectronics. *Adv. Energy Mater.* **2020**, *10*, 1902769. [CrossRef]
14. Li, H.; Liang, J. Recent development of printed micro-supercapacitors: Printable materials, printing technologies, and perspectives. *Adv. Mater.* **2020**, *32*, 1805864. [CrossRef] [PubMed]
15. Vinodh, R.; Babu, R.S.; Sambasivam, S.; Gopi, C.V.V.M.; Alzahmi, S.; Kim, H.J.; Ferreira de Barros, A.L.; Obaidat, I.M. Recent advancements of polyaniline/metal organic framework (PANI/MOF) composite electrodes for supercapacitor applications: A critical review. *Nanomaterials* **2022**, *12*, 1511. [CrossRef] [PubMed]
16. Poudel, M.B.; Shin, M.; Kim, H.J. Polyaniline-silver-manganese dioxide nanorod ternary composite for asymmetric supercapacitor with remarkable electrochemical performance. *Int. J. Hydrogen Energy* **2021**, *46*, 474–485. [CrossRef]
17. Shimoga, G.; Palem, R.R.; Choi, D.S.; Shin, E.J.; Ganesh, P.S.; Saratale, G.D.; Saratale, R.G.; Lee, S.H.; Kim, S.Y. Polypyrrole-based metal nanocomposite electrode materials for high-performance supercapacitors. *Metals* **2021**, *11*, 905. [CrossRef]
18. Adekoya, G.J.; Sadiku, R.E.; Ray, S.S. Nanocomposites of PEDOT: PSS with graphene and its derivatives for flexible electronic applications: A review. *Macromol. Mater. Eng.* **2021**, *306*, 2000716. [CrossRef]
19. Arena, A.; Branca, C.; Ciofi, C.; D'Angelo, G.; Romano, V.; Scandurra, G. Polypyrrole and Graphene Nanoplatelets Inks as Electrodes for Flexible Solid-State Supercapacitor. *Nanomaterials* **2021**, *11*, 2589. [CrossRef]
20. Ramkumar, R.; Minakshi, M. Fabrication of ultrathin $CoMoO_4$ nanosheets modified with chitosan and their improved performance in energy storage device. *Dalton Trans.* **2015**, *44*, 6158–6168. [CrossRef]
21. Zallouz, S.; Le Meins, J.M.; Ghimbeu, C.M. Alkaline hydrogel electrolyte from biosourced chitosan to enhance the rate capability and energy density of carbon-based supercapacitors. *Energy Adv.* **2022**, *1*, 1051–1064. [CrossRef]
22. Camurlu, P. Pinar Camurlu Polypyrrole derivatives for electrochromic applications. *RSC Adv.* **2014**, *4*, 55832–55845. [CrossRef]
23. Boguzaite, R.; Ratautaite, V.; Mikoliunaite, L.; Pudzaitis, V.; Ramanaviciene, A.; Ramanavicius, A. Towards analytical application of electrochromic polypyrrole layers. *J. Electroanal. Chem.* **2021**, *886*, 115132. [CrossRef]
24. Loguercio, L.F.; Alves, C.C.; Thesinga, A.; Ferreira, J. Enhanced electrochromic properties of a polypyrrole–indigo carmine–gold nanoparticles nanocomposite. *Phys. Chem. Chem. Phys.* **2015**, *17*, 1234–1240. [CrossRef] [PubMed]
25. Carquignyn, S.; Sanchez, J.B.; Berger, F.; Lakard, B.; Lallemand, F. Ammonia gas sensor based on electrosynthesized polypyrrole films. *Talanta* **2009**, *78*, 199–206. [CrossRef] [PubMed]
26. Das, M.; Roy, S. Polypyrrole and associated hybrid nanocomposites as chemiresistive gas sensors: A comprehensive review. *Mater. Sci. Semicond. Process.* **2021**, *121*, 105332. [CrossRef]

27. Navale, S.T.; Mane, A.T.; Chougule, M.A.; Sakhare, R.D.; Nalage, S.R.; Patil, V.B. Highly selective and sensitive room temperature NO2 gas sensor based on polypyrrole thin films. *Synth. Met.* **2014**, *189*, 94–99. [CrossRef]
28. Hutchison, A.S.; Lewis, T.W.; Moulton, S.E.; Spinks, G.M.; Wallace, G.G. Development of polypyrrole-based electromechanical actuators. *Synth. Met.* **2000**, *113*, 121–127. [CrossRef]
29. Bredas, J.L.; Scott, J.C.; Yakushi, K.; Street, G.B. Polarons and bipolarons in polypyrrole: Evolution of the band structure and optical spectrum upon doping. *Phys. Rev. B* **1984**, *30*, 1023–1025. [CrossRef]
30. Mostany, J.; Scharifker, B.R. Impedance spectroscopy of undoped, doped and overoxidized polypyrrole films. *Synth. Met.* **1997**, *87*, 179–185.
31. Liu, Y.C.; Hwang, B.J.; Jian, W.J.; Santhanam, R. In situ cyclic voltammetry-surface-enhanced Raman spectroscopy: Studies on the doping–undoping of polypyrrole film. *Thin Solid Film.* **2000**, *374*, 85–91. [CrossRef]
32. Nguyen Thi Le, T.; Bernard, M.C.; Garcia-Renaud, B.; Deslouis, C. Raman spectroscopy analysis of polypyrrole films as protective coatings on iron. *Synth. Met.* **2004**, *140*, 287–293. [CrossRef]
33. Vidanapathirana, K.P.; Careem, M.A.; Skaarup, S.; West, K. Ion movement in polypyrrole/dodecylbenzenesulphonate films in aqueous and non-aqueous electrolytes. *Solid State Ion.* **2002**, *154–155*, 331–335. [CrossRef]
34. Arena, A.; Branca, C.; Ciofi, C.; D'Angelo, G.; Scandurra, G. Development, Characterization and Sensing Properties of Graphene Films Deposited From Platelets Mixed With Dodecyl Benzene Sulfonic Acid. *IEEE Sens. J.* **2021**, *21*, 394–402. [CrossRef]
35. Lehman, J.H.; Terrones, M.; Mansfield, E.; Hurst, K.E.; Meunier, V. Evaluating the characteristics of multiwall carbon nanotubes. *Carbon* **2011**, *49*, 2581–2602. [CrossRef]
36. Wise, K.E.; Park, C.; Siochi, E.J.; Harrison, J.S. Stable dispersion of single wall carbon nanotubes in polyimide: The role of noncovalent interactions. *Chem. Phys. Lett.* **2004**, *391*, 207–211. [CrossRef]
37. Graupner, R. Raman spectroscopy of covalently functionalized single-wall carbon nanotubes. *J. Raman Spectrosc. Int. J. Orig. Work. All Asp. Raman Spectrosc. Incl. High. Order Process. Also Brillouin Rayleigh Scatt.* **2007**, *38*, 673–683. [CrossRef]
38. Santangelo, S.; Messina, G.; Faggio, G.; Lanza, M.; Milone, C. Evaluation of crystalline perfection degree of multi-walled carbon nanotubes: Correlations between thermal kinetic analysis and micro-Raman spectroscopy. *J. Raman Spectrosc.* **2011**, *42*, 593–602. [CrossRef]
39. Zhang, L.; Hashimoto, Y.; Taishi, T.; Ni, Q.Q. Mild hydrothermal treatment to prepare highly dispersed multi-walled carbon nanotubes. *Appl. Surf. Sci.* **2011**, *257*, 1845–1849. [CrossRef]
40. Di Leo, R.A.; Landi, B.J.; Raffaelle, R.P. Purity assessment of multi-walled carbon nanotubes by Raman spectroscopy. *J. Appl. Phys.* **2007**, *101*, 064307. [CrossRef]

nanomaterials

MDPI

Article

Transition of Carbon Nanotube Sheets from Hydrophobicity to Hydrophilicity by Facile Electrochemical Wetting

Myoungeun Oh [1,2,†], Hyunji Seo [1,†], Jimin Choi [1], Jun Ho Noh [1,2], Juwan Kim [1,2], Joonhyeon Jeon [2,3,*] and Changsoon Choi [1,2,*]

[1] Department of Energy and Materials Engineering, Dongguk University, 30 Pildong-ro, 1-gil, Jung-gu, Seoul 04620, Republic of Korea; abcdeun@dongguk.edu (M.O.); amy0527330@gmail.com (H.S.); james010828@gmail.com (J.C.); shwnsgh15@dongguk.edu (J.H.N.); kjw3328@dongguk.edu (J.K.)

[2] Department of Advanced Battery Convergence Engineering, Dongguk University, 30 Pildong-ro, 1-gil, Jung-gu, Seoul 04620, Republic of Korea

[3] Division of Electronics & Electronical Engineering, Dongguk University–Seoul, 30 Pildong-ro 1-gil, Jung-gu, Seoul 04620, Republic of Korea

* Correspondence: memory@dongguk.edu (J.J.); cschoi84@dongguk.edu (C.C.)

† These authors contributed equally to this work.

Abstract: The present study delves into the transformative effects of electrochemical oxidation on the hydrophobic-to-hydrophilic transition of carbon nanotube (CNT) sheets. The paper elucidates the inherent advantages of CNT sheets, such as high electrical conductivity and mechanical strength, and contrasts them with the limitations posed by their hydrophobic nature. A comprehensive investigation is conducted to demonstrate the efficacy of electrochemical oxidation treatment in modifying the surface properties of CNT sheets, thereby making them hydrophilic. The study reveals that the treatment not only is cost-effective and time-efficient compared to traditional plasma treatment methods but also results in a significant decrease in water contact angle. Mechanistic insights into the hydrophilic transition are provided, emphasizing the role of oxygen-containing functional groups introduced during the electrochemical oxidation process.

Keywords: electrochemical wetting; carbon nanotube (CNT); supercapacitor; flexibility; wearable

Citation: Oh, M.; Seo, H.; Choi, J.; Noh, J.H.; Kim, J.; Jeon, J.; Choi, C. Transition of Carbon Nanotube Sheets from Hydrophobicity to Hydrophilicity by Facile Electrochemical Wetting. *Nanomaterials* **2023**, *13*, 2834. https://doi.org/10.3390/nano13212834

Academic Editor: Muralidharan Paramsothy

Received: 6 October 2023
Revised: 25 October 2023
Accepted: 25 October 2023
Published: 26 October 2023

1. Introduction

Forest-spun carbon nanotube (CNT) sheets exhibit multifunctionality, making them the material of choice for a wide array of flexible electronics [1]. Specifically, their low sheet resistivity (approximately 700 Ω per square in the drawing direction) and high transmittance (ranging from 95% to 98.5% for various radiation types) are highly applicable to displays, video recorders, solar cells, and solid-state lighting [1]. Moreover, the microstructure of the CNT sheet is like a network, consisting of primary CNT bundles aligned in the drawing direction and secondary branches interconnecting these main bundles. This unique network structure is formed by cooperatively rotating the CNTs in vertically oriented nanotube arrays. The networked architecture allows high retention of electrical conductivity when the sheets are mechanically bent or deformed [1], a property not found in conventional transparent conductors like indium tin oxide (ITO).

Despite such advantages, one critical limitation of CNT sheets is their intrinsic hydrophobicity originating from the graphitic carbon–carbon bonding [2–4]. Therefore, hydrophilic modification of the CNT sheets is often a critical requirement for wider applications demanding enhanced wettability and surface interactions with aqueous solutions. For instance, hydrophilic CNT sheets facilitate efficient adsorption of contaminants, thereby improving the overall performances of filtration systems. In biomedical applications, hydrophilic CNT sheets are preferred for their compatibility with biological fluids, which is essential for drug delivery or tissue engineering. Hydrophilic surfaces also promote better dispersion in polymer matrices, enhancing the mechanical and thermal properties of CNT-based composites. Most importantly, in electrochemical applications such as supercapacitors and batteries, hydrophilic CNT sheets can improve electrolyte wettability, leading to enhanced ion transport and improved charge storage capability thereof. Therefore, facile transition of CNT sheets from hydrophobicity to hydrophilicity enables new avenues for their aqueous applications in diverse technological and industrial fields.

The plasma treatment method was extensively researched in a previous study as it is one of the often-used physical methods to assign extrinsic hydrophilicity to CNTs [5–8]. Plasma treatments have been shown to alter the surface compositions of F and O atoms drastically, which in turn affect the hydrophobic and hydrophilic properties of the CNTs. Unfortunately, these treatments often involve the use of atmospheric-pressure plasma jets, high-voltage transformers, and specific gas-flow controllers, which necessitate more complex, expensive, and bulky experimental setups. Such complexity increases not only the operational challenges but also the overall cost of the treatment process, thus limiting wider application in industrial fields [5,9].

In the present work, the electrochemical wetting method is introduced as a highly effective and simple method for hydrophilic functionalization of the surfaces of CNT sheets. This method employs a straightforward three-electrode electrochemical cell setup, where the CNT sheet acts as the working electrode. A potentiostatically applied voltage to the working electrode immersed in an aqueous solution enables the facile introduction of oxygen-containing functional groups. The simplicity of this setup allows a more user-friendly as well as cost- and time-effective approach, which is particularly advantageous for both research and industrial applications. The proposed electrochemical wetting process can be completed in a relatively short time, and the duration can be easily controlled to tune the degree of surface wetting. Furthermore, electrochemical methods have been demonstrated for effective transition from hydrophobic to hydrophilic states, thereby enhancing the CNT wettability [10–14].

2. Materials and Methods

2.1. Chemicals and Materials

The CNT (carbon nanotube) sheets were drawn from well−aligned multiwalled carbon nanotube (MWNT) forests grown via chemical vapor deposition (CVD), with a height of 750 μm (NTAD 10, PDSI Corporation, Suwon, Republic of Korea). A Polyethylene terephthalate (PET) film was cut to an area of 2 (width) × 3 (length) cm^2.

In addition, 0.1 M sodium sulfate (Na_2SO_4) aqueous electrolyte was prepared by dissolving 1.42 g Na_2SO_4 in 100 mL deionized water. The solution was stirred until it was transparent.

2.2. Preparation of the Hydrophobic CNT Sheets/PET (CP) Film

Five layers of CNT sheets drawn from a 300-um-high MWNT forest (A-Tech System Co., Incheon, Republic of Korea) were uniformly and straightly into a 30-mm-long and 10-mm-wide sheet stack on the PET film. The ethanol densification process was performed, resulting in physical adhesion between the CNT sheet and the PET film. For a stable electrochemical wetting (ECW) treatment and electrochemical performance characterization, both the ends of the CNT/PET film (area: 2 cm^2) were electrically connected to Cu wires (diameter: 180 um) using silver paste and then chemically tethered using an epoxy glue (Devcon, Danvers, MA, USA). The ECW treatment was performed using a three-electrode electrochemical system consisting of Ag/AgCl (reference electrode), CNT sheets/PET film (working electrode), a Pt mesh (counter electrode), and 0.1 M Na_2SO_4 (the liquid electrolyte). Then, a potentiostatic voltage (vs. Ag/AgCl) was applied using an electrochemical analyzer (Vertex EIS, Ivium, Eindhoven, The Netherlands).

2.3. Characterization

Scanning electron microscope (SEM) images of the CNT and ECW treated CNT sheets/PET film were obtained (S−4600, Hitachi, Tokyo, Japan), along with optical images of the film using an optical camera (D750, Nikon, Tokyo, Japan). Contact and sliding angle value between CNT/PET films and water droplets were measured (Phoenix-MT(M), S.E.O., Suwon, Republic of Korea). In addition, the electrochemical performances of the CNT/PET films were evaluated using an electrochemical analyzer (Vertex EIS, Ivium, Eindhoven, The Netherlands). The chemical composition and features of the CNT sheets were determined via XPS measurements performed (Versaprobe II, ULVAC–PHI, Kazaki, Kanagawa, Japan).

3. Results and Discussion

3.1. ECW-Treated Carbon Nanotube (CNT) Sheet on a Polyethylene Terephthalate (PET) Film

Pulling a CNT sheet from a CNT forest is a fascinating and intricate procedure that has been optimized for scalability and high production rate [1]. To initiate the drawing process, the sidewalls of the CNT forest are placed in contact with metal sticks, and anisotropically aligned and self-supporting CNT sheets are subsequently hand drawn, as shown in Figure 1a. The sheets are later transferred onto a PET substrate, and five stacks of the CNT sheets are used as the working electrode for the electrochemical wetting treatment in our experiment (Figure 1b). Before treatment, the CNT sheets were cleaned and cut to the desired dimensions. Copper wires were attached to both ends of the sheets to provide electrical connections to apply the constant voltages during electrochemical treatment (Figure 1c). Before applying electrochemical wetting, a pretreatment is required to stabilize the CNT sheets, which is alcohol-based densification. As the CNT sheets are not chemically bonded to the PET substrate, they may detach or aggregate after electrochemical wetting. To prevent this, we dropped alcohol on the sheets and dried them for an hour. This pretreatment helps the sheets to attach well to the PET substrate as the sheets can be highly densified (up to 360 folds) and therefore stabilized on the substrate [1]. Electrochemical wetting of the CNT sheets was carried out in a conventional three-electrode electrochemical cell, as schematically illustrated in Figure 1d. The CNT sheets were the working electrode, while a platinum mesh and an Ag/AgCl electrode functioned as the counter and reference electrodes, respectively. An electrochemical analyzer was employed to control the applied voltage and current throughout the experiment. The working, counter, and reference electrodes were then immersed carefully in an electrolyte of 0.1 M Na_2SO_4 aqueous solution (Figure 1e). Potentiostatic voltages of 2, 3, and 4 V were applied to the working electrode, and the corresponding current densities versus wetting times were plotted (Figure 1f). The current density remained relatively stable during the early stages of the wetting process, showing plateau shapes for about 6 and 3 s at 3 and 4 V, respectively. Subsequently, a drastic decrease in current was observed, which may be attributed to the increase in sheet resistance due to excess oxygen functionalization (Figure S1a) [10].

To assess the structural integrity after electrochemical wetting, we observed the surface conditions of the CNT sheets before and after treatment (Figure 1g). From the scanning electron microscopy (SEM) images, no observable change was detected. Even at a higher magnification, highly aligned CNT bundles and their network structure were observed without significant structural collapse (Figure 1h). We expect that appropriate surface functionalization of the sheets was achieved by introducing oxygen-containing groups (e.g., hydroxyl groups), which will be discussed later (Figure 1i). The changes in the surface conditions can be simply confirmed by the large decrease in the contact angle of a water droplet placed on the surface after electrochemical wetting: the 112° contact angle of the pristine CNT decreased dramatically to 48° after wetting, as shown in Figure 1j. The change in contact angle in the image satisfies the Young's equation ($\gamma_{SV} = \gamma_{SL} + \gamma_{LV} COS(\theta)$), as mentioned by the reviewer [15].

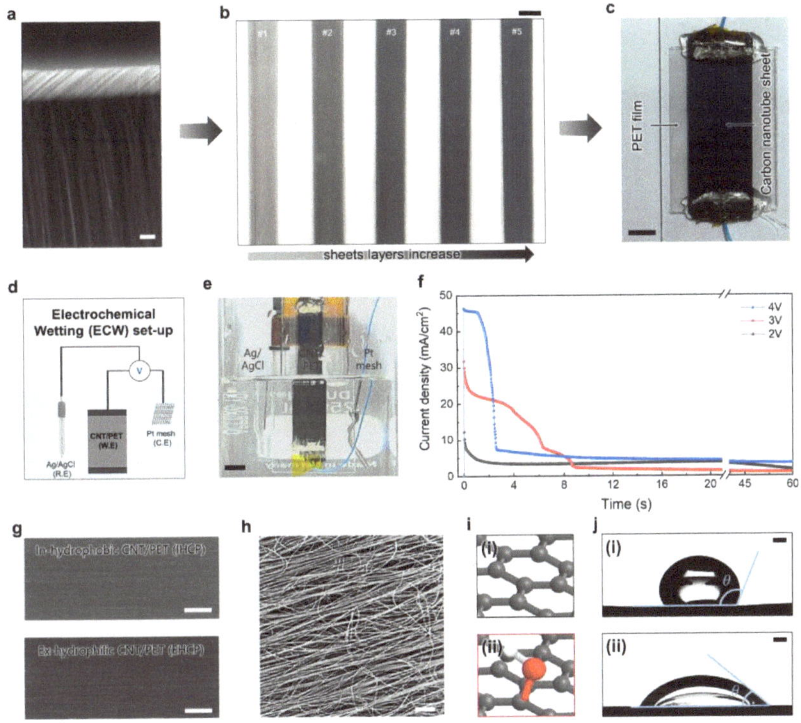

Figure 1. (**a**) Fabrication of CNT sheets/PET film. Scanning electron microscopy (SEM) image of the CNT sheet drawn from the CNT forest (scale bar = 25 μm). (**b**) Photograph showing stacking of the CNT sheets for up to five layers (scale bar = 0.5 cm). (**c**) Prepared CNT sheets loaded on PET film and electrically connected to Cu wires for electrochemical wetting (scale bar = 0.5 cm). (**d**) Schematic illustration and (**e**) photograph showing the setup for the electrochemical wetting treatment based on the three-electrode system consisting of a 20-mm-long CNT/PET film (working electrode), Ag/AgCl (reference electrode), and Pt mesh (counter electrode) in 0.1 M Na_2SO_4 electrolyte (scale bar = 1 cm). (**f**) Current density (normalized to the surface area of the CNT sheets) versus wetting time under various potentiostatically applied voltages ranging from 2.0 to 4.0 V (vs. Ag/AgCl). (**g**) SEM images of the surface before (upper panel) and after (lower panel) electrochemical wetting (scale bar = 50 um). (**h**) High-magnification SEM image showing the surface structure of the electrochemically wetted CNT sheet (scale bar = 10 μm). (**i**) Schematic images showing the graphitic atomic structures of the intrinsically hydrophobic CNT sheet (upper panel (i)) and hydroxyl group introduced by electrochemical wetting for extrinsically hydrophilic CNT sheet (lower panel (ii)). (**j**) Contact angle difference between water droplets placed on the pristine CNT sheet (upper panel (i)) and electrochemically wetted CNT sheet (lower panel (ii)); the scale bars for both images are 0.1 cm.

3.2. Changes in Contact Angle on the IHCP Surface

To inspect the electrochemical wetting effects, the contact angles between water droplets and the CNT sheets are systematically compared in Figure 2. Before electrochemical wetting, the CNT sheet exhibits a typical hydrophobic behavior characterized by a large contact angle often exceeding 110°. This intrinsic hydrophobicity of the CNT sheet is attributed to the inherent chemical structure of the CNTs, which entails symmetric bonding without any polar functional groups. Upon electrochemical wetting, the contact angle decreases progressively to about 40°, showing effective transition to extrinsic hydrophilicity.

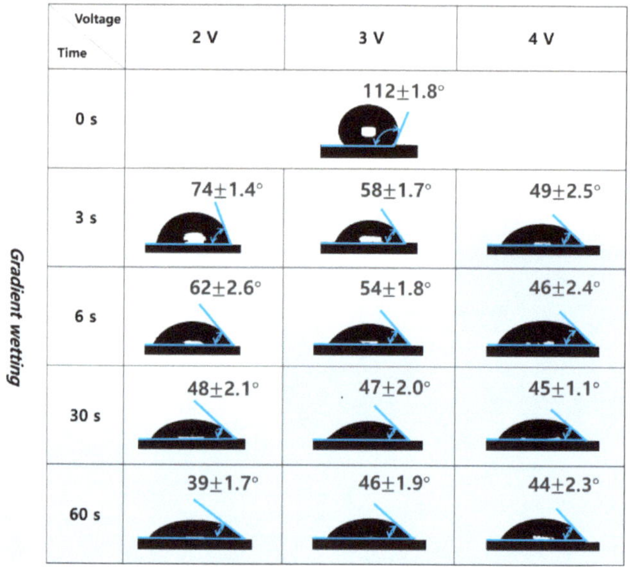

Figure 2. Photographs showing the contact angles of water droplets placed on CNT sheets that were electrochemically wetted at various voltages for various treatment times.

In the electrochemical wetting process, two key parameters, namely applied voltage and wetting time, play important roles in determining the contact angles of water droplets placed on the CNT sheets. In general, the applied voltage and wetting time are inversely related; as the applied voltage increases, the wetting time required for electrochemical CNT activation decreases. The reason behind this inverse relationship lies in the electrochemical kinetics of the process; a higher applied voltage accelerates the electrochemical reactions that introduce oxygen-containing functional groups to the CNT surfaces. Although wetting at a higher voltage induces faster transition to hydrophilicity for the CNT sheet (e.g., 49° contact angle at 4 V over 3 s), wetting at a lower voltage with a longer treatment time can produce similar effects (e.g., 48° contact angle at 2 V over 30 s). However, it should be noted that excess activation can lead to destruction of the π–π conjugate structure of the CNTs, thereby increasing the sheet resistance drastically. Therefore, the applied voltage and treatment time must be carefully optimized to achieve the desired hydrophilicity without degrading the structural integrity of the CNTs. By multiplying the applied voltage and treatment time, the optimized energy can be calculated to be approximately 0.21 J/cm^2 to fully oxidize the CNT sheets in this work.

3.3. Sliding Angle Variation for IHCP and EHCP Film

In Figure 3, based on these findings, one can also expect a substantial change in the sliding angle of the CNT sheets after electrochemical wetting. The initial hydrophobicity of the CNT sheets may likely result in a low sliding angle (42°), requiring a small tilt to initiate sliding of a water droplet. However, after electrochemical wetting, the sliding angle increases dramatically (180°) owing to the enhanced hydrophilicity of the surface, and the pinned state of the water droplet can be observed for all inspected voltages.

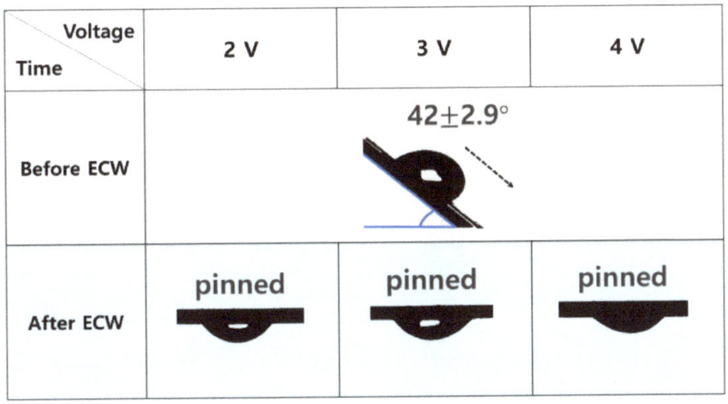

Figure 3. Photographs showing the sliding angles of water droplets placed on the CNT sheets before and after electrochemical wetting under various voltages.

As mentioned before, facile transition of the CNT sheet from hydrophobicity to hydrophilicity through electrochemical wetting is a multifaceted process that involves both chemical and structural modifications. From the chemical perspective, electrochemical wetting introduces various oxygen-containing functional groups, such as carboxyl, hydroxyl, and epoxy groups, on the surfaces of the CNTs. These polar functional groups increase the surface energy and wettability of the CNT sheet, thereby improving its hydrophilicity. From the structural perspective, electrochemical wetting involves the infiltration of electrolyte ions between adjacent CNTs within the sheets via double-layer charge injection [16,17]. Therefore, the combination of electrolyte-infiltration-induced internanotube swelling and high water interactivity of the electrochemically wetted CNT sheets contributes to a significant decrease in the contact angle [12].

3.4. *Characterization as Applied Voltage and Time Correlation among Contact, Sliding Angle and ECW Treatment*

The effects of electrochemical wetting on the changes to both contact and sliding angles are plotted in Figure 4a. The prepared pristine CNT sheets, which were densified and delaminated on the PET substrates, had average initial contact and sliding angles of 112° and 42°, which changed to 48° and 180° (pinned state), respectively, after electrochemical wetting, showing effective transition to hydrophilicity. The changes in the contact and sliding angles are plotted versus treatment time in Figure 4b. The higher applied voltage of 4 V achieves the lowest contact angle (44°) over the shortest time (3 s), whereas the lower voltage of 2 V needs more treatment time to produce a similar level of hydrophilicity. The contact angles versus calculated total energies are plotted for the applied voltages of 2, 3, and 4 V and compared (Figure 4c–e). Once a certain voltage level is applied (e.g., above 2 V) and given enough wetting time (or energy), the surface oxidation effects seem to converge regardless of the applied voltage. One of the optimum input energies for assigning extrinsic surface hydrophilicity to the CNT sheet is about 0.21 J for the given voltage range (Figure 4f). This energy value corresponds to 2 V application (30.7 mA maximum current) over 30 s, 3 V application (79.5 mA maximum current) over 6 s, and 4 V application (115.6 mA maximum current) over 3 s. The corresponding contact angles are shown in Figure S2a,b.

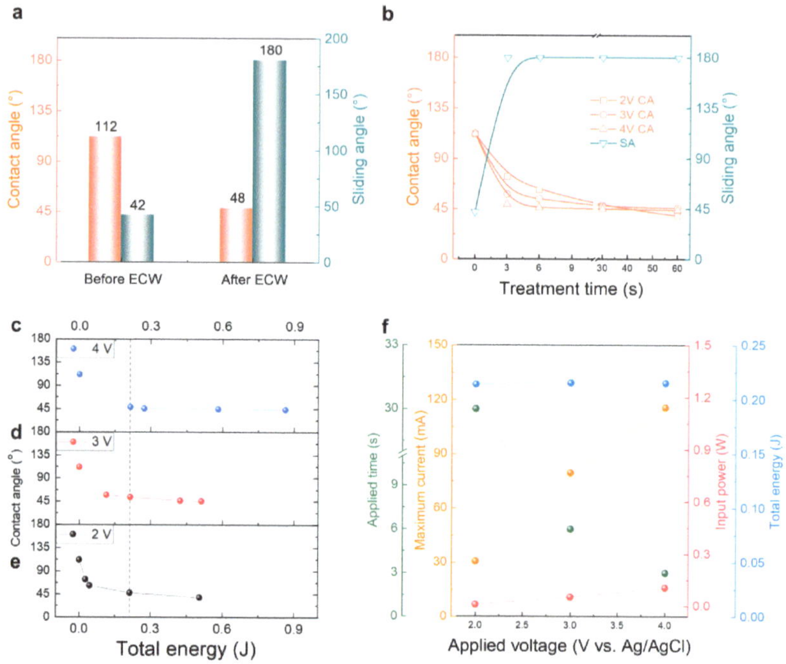

Figure 4. (**a**) Measured contact angles and sliding angle changes before and after electrochemical wetting. (**b**) Contact and sliding angles versus treatment times for various applied voltages. Measured contact angles versus total energy input for potentiostatically applied voltages of (**c**) 4 V, (**d**) 3 V, and (**e**) 2 V. (**f**) Treatment time, maximum current, input power, and total energy are compared versus the applied voltage.

3.5. Chemical and Electrochemical Characterization of the EHCP Surface after ECW Treatment Process

The electrochemical wetting method has been shown to introduce oxygen-containing functional groups on the surfaces of CNT sheets. These functional groups play pivotal roles in altering the surface properties of the materials and can be characterized using X-ray photoelectron spectroscopy (XPS). This technique provides insights into the types and quantities of functional groups introduced, thereby allowing a comprehensive understanding of the surface chemistry after electrochemical wetting. This change in surface chemistry is a primary factor contributing to the drastic reduction in the contact angle. The newly formed functional groups and altered microstructure collectively contribute to the significant decrease in contact angle, making the surface highly hydrophilic. With increasing oxidation time, the oxygen and carbon (O/C) atomic ratios rose from 0.06 to 0.25, indicating effective control over the oxygenic groups on the surfaces of carbon nanotubes (CNTs) [18]. To investigate the detailed chemical composition of CNT sheets, the XPS survey and the high-resolution C 1s profiles were investigated at different voltages and electrochemical wetting times. When the applied voltage was increased from 2.0 to 4.0 V (vs. Ag/AgCl), the wetting time decreased from 30 to 3 s, indicating that these two parameters are inversely related (each of the left side of Figure 5a–c, respectively). The gray dash line presents the original total profiles of the high resolution C 1s XPS analysis spectra. Also, other colored lines present the detailed surface functional groups content and composition were analyzed by de-convoluting the C 1s peaks (284 eV) using Gaussian peak fitting(each of the right side of Figure 5a–c, respectively). This analysis considered specific bonds: C=C (284.3 eV), C–C (284.8 eV), C–OH (286.2 eV), C–O–C (287.3 eV), and COOH (288.7 eV) [6,7,18,19]. Notably, irrespective of the applied voltage in given range, several bonds were clearly observed at

284 eV (C=C sp2), 284 eV (C–C sp3), 286 eV (C–OH), 287 eV (C–O–C), and 288 eV (COOH). The types of introduced functional groups were also observed in the high-resolution O1s profile in Figure S3 (533 eV). The formation of oxygen functional groups is clearly observed in the Raman spectroscopic analysis as well (Figure S4). Both spectra exhibited prominent peaks at ~1347 cm^{-1} (D band) and ~1584 cm^{-1} (G band). Particularly noteworthy is the substantial increase in the intensity of the D band after ECW treatment. Furthermore, the intensity ratio of the D and G bands (I_D/I_G), which indicates the extent of defect density in the graphitic carbon structure, significantly increased from 0.49 to 0.79. These results imply the presence of defects in the intrinsic carbon bonds within the CNTs as a consequence of ECW treatment.

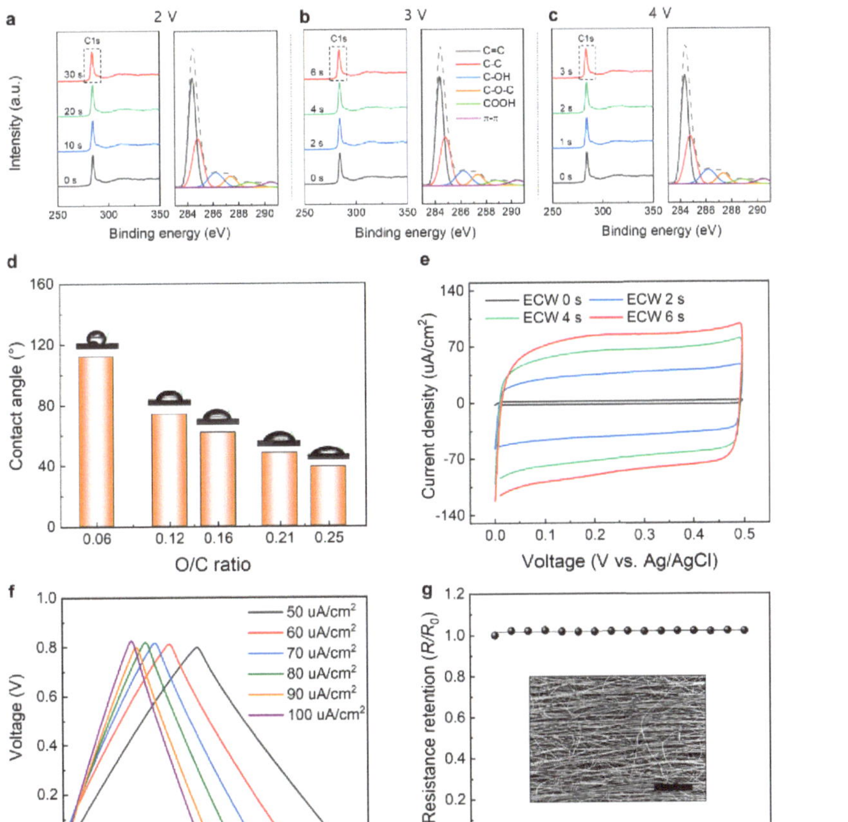

Figure 5. XPS survey spectra (left panel) and high-resolution C 1s (right panel) XPS analysis spectra of CNT sheets for various oxidation times at applied voltages of (**a**) 2 V, (**b**) 3 V, and (**c**) 4 V. (**d**) Oxygen/carbon (O/C) ratios for different applied voltages (2.0–4.0 V) and treatment times (up to 30 s). (**e**) Cyclic voltammetry curves (at 10 mV/s) with a three-electrode system and compared with ECW-treated CP films over applied time (0–6 s). (**f**) Galvanostatic charge/discharge curves of the electrochemically wetted CNT sheets over 50–100 μA/cm^2 of current density. (**g**) Resistance changes versus bending cycle (bending degree = 120°) for the electrochemically wetted CNT sheets; the inset shows the SEM image of the CNT sheet (scale bar = 10 μm).

In Figure 5d, the relationship between the water contact angles and the wetting degree of the CNT sheet is illustrated. Initially, pristine CNT sheets, characterized by an O/C ratio of 0.06, exhibited their inherent hydrophobic nature, displaying a contact angle (CA) of approximately 120°. Through the ECW treatment process, there was a proportional reduction in the CA observed between the water droplets and the sheet surface, reaching a decrease of up to 48° as the O/C ratio increased to approximately 0.25. This result signifies that the CNT sheets treated with the ECW process became significantly more hydrophilic, indicating an enhancement in wettability.

The oxygen-containing groups introduced by electrochemical wetting contribute to a pseudocapacitive behavior in an aqueous electrolyte system, significantly enhancing the capacitance of the resulting sheet-based supercapacitor [11,12,20]. Therefore, in the present case, the electrochemically wetted CNT sheet has a profound impact on capacitance improvement. Before electrochemical wetting, the energy storage mechanism of the CNT-based supercapacitor is primarily based on the electrochemical double-layer capacitance (EDLC) [20–22]. However, the specific capacitance based on the EDLC depends only on the surface ion adsorption; therefore, the values may be unsatisfactory for practical applications, especially when compared with supercapacitors containing pseudocapacitive guest materials like conducting polymers or transition metal oxides [23–31]. Upon electrochemical wetting, the CNT yarns undergo significant functionalization with oxygen-containing groups, and these groups can be electrochemically active sites for pseudocapacitive charge storage, as reported previously [10,12,13,18,20,32]. Pseudocapacitance arises from the Faradaic redox reactions that occur at the surfaces of the material, in contrast to the non-Faradaic electrostatic storage of charge in the double-layer capacitors.

Therefore, functionalization drastically enhances the performance of the sheet-based supercapacitor. For instance, the capacitance of the electrochemically wetted CNT was found to be 19 times larger than that of the neat CNT sheet, which can be confirmed from the cyclic voltammetry and galvanostatic charge/discharge curves shown in Figure 5e,f, respectively. Moreover, the oxygen-containing groups improve the wettability of the CNT sheets, which is beneficial for electrolyte infiltration. Better electrolyte infiltration ensures that a larger surface area of the CNT is accessible for charge storage, thereby enhancing the capacitance further. Therefore, electrochemical oxidation is a versatile tool for tuning the properties of CNT sheets for energy storage systems [33,34].

The unique network structure of the CNT sheets allows retention of electrical conductivity when the sheets are bent or deformed. The interconnected CNT bundles create multiple pathways for electron flow, ensuring that the electrical conductivity is maintained even when the sheets are bent. The electrochemically oxidized CNT sheets also demonstrate flexibility, where the resistance barely changes after 150 cycles of bending deformations (bending angle = 120°), as shown in Figure 5g. This is particularly important for applications requiring flexibility, such as wearable electronics, flexible batteries, or electrochemical catalysis reactions [35,36].

4. Conclusions

This study entails an in-depth exploration of the electrochemical wetting method as an effective strategy for functionalization of CNT sheets. The electrochemical wetting method is known to offer a range of advantages over other functionalization techniques, such as plasma treatment and acid wetting. While plasma treatment necessitates specialized equipment and can be both energy-intensive and time-consuming, acid wetting often involves the use of hazardous chemicals and stringent safety protocols. Electrochemical wetting, on the other hand, is characterized by its simplicity, cost-effectiveness, and efficiency; it employs a straightforward electrochemical setup and does not require specialized gases or chemicals. It is shown that this facile method induces significant transition from hydrophobic to hydrophilic behaviors, as evidenced by the dramatic decrease in the contact angles of water droplets on the CNT surfaces. The introduction of oxygen-containing functional groups and electrolyte-infiltration-induced internanotube swelling are identified as the key mechanisms behind this transition. Moreover, this study demonstrates a remarkable improvement in the supercapacitor performance of the electrochemically wetted CNT sheet.

The environmental sustainability of the electrochemical wetting method is an important consideration, especially in the current global context of growing environmental concerns. While the electrochemical wetting process is relatively straightforward and does not require hazardous chemicals, a thorough environmental impact assessment is still essential. This could involve a lifecycle analysis, evaluating the energy consumption of the electrochemical process, recyclability or disposability of the electrolyte, and any waste byproducts generated. Such assessments would offer a holistic perspective on the environmental sustainability of the electrochemical wetting method, providing valuable data for policymakers, researchers, and industry stakeholders concerned with minimizing the environmental impact.

Supplementary Materials: The following supporting information can be downloaded at: https://www.mdpi.com/article/10.3390/nano13212834/s1, Figure S1: (a) Measured current density and resistance change of the CNT working electrode versus electrochemical wetting treatment time (the mild and wild electrochemical wetting regions are indicated). (b) CV curves of the wild-ECW-treated CNT sheet/PET film electrode; Figure S2: Actual images showing contact angle at applied voltages of (a) 2 V and (b) 4 V (scale bar = 0.25 cm); Figure S3: XPS survey spectra (left panel) and high-resolution O 1s XPS analysis spectra (right panel) of CNT sheets for various oxidation times at applied voltages of 3 V; Figure S4: Raman spectra of the CNT sheets before (gray circle) and after ECW treatment (blue circle).

Author Contributions: M.O., H.S., J.J. and C.C. conceived the idea for the study. M.O., H.S. and J.C. performed the experiments. H.S., J.C. and J.H.N. analyzed the experimental results. J.K. validated the experiments. M.O., H.S. and C.C. wrote the manuscript. J.J. and C.C. edited the manuscript. M.O and H.S. contributed equally to this work. All authors have read and agreed to the published version of the manuscript.

Funding: This research received no external funding.

Data Availability Statement: Not applicable.

Acknowledgments: This research was supported by the Police–Lab 2.0 Program (www.kipot.or.kr) funded by the Ministry of Science and ICT (MSIT, Republic of Korea) and Korean National Police Agency (KNPA, Republic of Korea) [Project name: Development of fluorescent forensic nanomaterials and evidence specification technique for effective biometric evidence detection; Project number: RS-2023-00236429]. This work was also supported by the Korea Institute of Energy Technology Evaluation and Planning (KETEP) and Ministry of Trade, Industry & Energy (MOTIE) of the Republic of Korea (No. 20224000000020).

Conflicts of Interest: The authors declare that there is no conflict of interest.

References

1. Zhang, M.; Fang, S.; Zakhidov, A.A.; Lee, S.B.; Aliev, A.E.; Williams, C.D.; Atkinson, K.R.; Baughman, R.H. Strong, transparent, multifunctional, carbon nanotube sheets. *Science* **2005**, *309*, 1215–1219. [CrossRef] [PubMed]
2. Adusei, P.K.; Gbordzoe, S.; Kanakaraj, S.N.; Hsieh, Y.-Y.; Alvarez, N.T.; Fang, Y.; Johnson, K.; McConnell, C.; Shanov, V. Fabrication and study of supercapacitor electrodes based on oxygen plasma functionalized carbon nanotube fibers. *J. Energy Chem.* **2020**, *40*, 120–131. [CrossRef]
3. Naqvi, S.T.R.; Rasheed, T.; Hussain, D.; Haq, M.N.; Majeed, S.; Shafi, S.; Ahmed, N.; Nawaz, R. Modification strategies for improving the solubility/dispersion of carbon nanotubes. *J. Mol. Liq.* **2020**, *297*, 111919. [CrossRef]
4. Ham, S.W.; Hong, H.P.; Kim, J.H.; Min, S.J.; Min, N.K. Effect of oxygen plasma treatment on carbon nanotube-based sensors. *J. Nanosci. Nanotechnol.* **2014**, *14*, 8476–8481. [CrossRef]
5. Watanabe, H.; Kondo, H.; Sekine, M.; Hiramatsu, M.; Hori, M. Control of super hydrophobic and super hydrophilic surfaces of carbon nanowalls using atmospheric pressure plasma treatments. *Jpn. J. Appl. Phys.* **2012**, *51*, 01AJ07. [CrossRef]
6. Shao, Y.; Xu, F.; Marriam, I.; Liu, W.; Gao, Z.; Qiu, Y. Quasi–static and dynamic interfacial evaluations of plasma functionalized carbon nanotube fiber. *Appl. Surf. Sci.* **2019**, *465*, 795–801. [CrossRef]
7. Park, O.-K.; Kim, W.Y.; Kim, S.M.; You, N.-H.; Jeong, Y.; Lee, H.S.; Ku, B.-C. Effect of oxygen plasma treatment on the mechanical properties of carbon nanotube fibers. *Mater. Lett.* **2015**, *156*, 17–20. [CrossRef]

8. Dai, X.; Huang, X.; Yang, F.; Li, X.; Sightler, J.; Yang, Y.; Li, C. Enhanced nucleate boiling on horizontal hydrophobic-hydrophilic carbon nanotube coatings. *Appl. Phys. Lett.* **2013**, *102*, 161605. [CrossRef]
9. Lai, C.-C.; Lo, C.-T. Plasma oxidation of electrospun carbon nanofibers as supercapacitor electrodes. *RSC Adv.* **2015**, *5*, 38868. [CrossRef]
10. Son, W.; Chun, S.; Lee, J.M.; Jeon, G.; Sim, H.J.; Kim, H.W.; Cho, S.B.; Lee, D.; Park, J.; Jeon, J.; et al. Twist-stabilized, coiled carbon nanotube yarns with enhanced capacitance. *ACS Nano* **2022**, *16*, 2661–2671. [CrossRef]
11. Komatsubara, K.; Suzuki, H.; Inoue, H.; Kishibuchi, M.; Takahashi, S.; Marui, T.; Umezawa, S.; Nakagawa, T.; Nasu, K.; Hayashi, Y. Highly Oriented Carbon Nanotube Supercapacitors. *ACS Appl. Nano Mater.* **2022**, *5*, 1521–1532. [CrossRef]
12. Avasthi, P.; Kumar, A.; Balakrishnan, V. Aligned CNT Forests on Stainless Steel Mesh for Flexible Supercapacitor Electrode with High Capacitance and Power Density. *ACS Appl. Nano Mater.* **2019**, *2*, 1484–1495. [CrossRef]
13. Senokos, E.; Rana, M.; Santos, C.; Marcilla, R.; Vilatela, J.J. Controlled electrochemical functionalization of CNT fibers: Structure–chemistry relations and application in current collector-free all-solid supercapacitors. *Carbon* **2019**, *142*, 599–609. [CrossRef]
14. Liu, C.-M.; Cao, H.-B.; Li, Y.-P.; Xu, H.-B.; Zhang, Y. The effect of electrolytic oxidation on the electrochemical properties of multi-walled carbon nanotubes. *Carbon* **2006**, *44*, 2919–2924. [CrossRef]
15. Guan, D.; Zhong, J.; Xu, H.; Huang, Y.-C.; Hu, Z.; Chen, B.; Zhang, Y.; Ni, M.; Xu, X.; Zhou, W.; et al. A universal chemical-induced tensile strain tuning strategy to boost oxygen-evolving electrocatalysis on perovskite oxides. *Appl. Phys. Rev.* **2022**, *9*, 011422. [CrossRef]
16. Lima, M.D.; Li, N.; Andrade, M.J.; Fang, S.; Oh, J.; Spinks, G.M.; Kozlov, M.E.; Haines, C.S.; Suh, D.; Foroughi, J.; et al. Electrically, chemically, and photonically powered torsional and tensile actuation of hybrid carbon nanotube yarn muscles. *Science* **2012**, *338*, 928–932. [CrossRef]
17. Foroughi, J.; Spinks, G.M.; Wallace, G.G.; Oh, J.; Kozlov, M.E.; Fang, S.; Mirfakhrai, T.; Madden, J.D.W.; Shin, M.K.; Kim, S.J.; et al. Torsional carbon nanotube artificial muscles. *Science* **2011**, *334*, 494–497. [CrossRef]
18. Gao, G.; Pan, M.; Vecitis, C.D. Effect of the oxidation approach on carbon nanotube surface functional groups and electrooxidative filtration performance. *J. Mater. Chem.* **2015**, *3*, 7575. [CrossRef]
19. Yu, H.; Cheng, D.; Williams, T.S.; Severino, J.; Rosa, I.M.D.; Carlson, L.; Hicks, R.F. Rapid oxidative activation of carbon nanotube yarn and sheet by a radio frequency, atmospheric pressure, helium and oxygen plasma. *Carbon* **2013**, *57*, 11–21. [CrossRef]
20. Frackowiak, E.; Be'guin, F. Carbon materials for the electrochemical storage of energy in capacitors. *Carbon* **2001**, *39*, 937–950. [CrossRef]
21. Silva, T.D.; Damery, C.; Alkhaldi, R.; Karunanithy, R.; Gallaba, D.H.; Patil, P.D.; Wasala, M.; Sivakumar, P.; Migone, A.; Talapatra, S. Carbon nanotube based robust and flexible solid-state supercapacitor. *ACS Appl. Mater.* **2021**, *13*, 56004–56013. [CrossRef]
22. Ye, J.-S.; Liu, X.; Cui, H.F.; Zhang, W.-D.; Sheu, F.-S.; Lim, T.M. Electrochemical oxidation of multi-walled carbon nanotubes and its application to electrochemical double layer capacitors. *Electrochem. Commun.* **2005**, *7*, 249–255. [CrossRef]
23. Zhou, Y.; Wang, X.; Acauan, L.; Kalfon-Cohen, E.; Ni, X.; Stein, Y.; Gleason, K.K.; Wardle, B.L. Ultrahigh-areal-capacitance flexible supercapacitor electrodes enabled by conformal P3MT on horizontally aligned carbon-nanotube arrays. *Adv. Mater.* **2019**, *31*, 1901916. [CrossRef]
24. Ren, Y.; Liu, Y.; Wang, S.; Wang, Q.; Li, S.; Wang, W.; Dong, X. Stretchable supercapacitor based on a hierarchical PPy/CNT electrode and hybrid hydrogel electrolyte with a wide operating temperature. *Carbon Energy* **2022**, *4*, 527–538. [CrossRef]
25. Choi, C.; Kim, J.H.; Sim, H.J.; Di, J.; Baughman, R.H.; Kim, S.J. Microscopically buckled and macroscopically coiled fibers for ultra-stretchable supercapacitors. *Adv. Energy Mater.* **2017**, *7*, 1602021. [CrossRef]
26. Choi, C.; Kim, K.M.; Kim, K.J.; Lepro', X.; Spinks, G.M.; Baughman, R.H.; Kim, S.J. Improvement of system capacitance via weavable superelastic biscrolled yarn supercapacitors. *Nat. Commun.* **2016**, *7*, 13811. [CrossRef]
27. Choi, C.; Lee, J.M.; Kim, S.H.; Kim, S.J. Twistable and stretchable sandwich structured fiber for wearable sensors and supercapacitors. *Nano Lett.* **2016**, *16*, 7677–7684. [CrossRef]
28. Choi, C.; Sim, H.J.; Spinks, G.M.; Lepró, X.; Baughman, R.H.; Kim, S.J. Elastomeric and dynamic MnO_2/CNT core-shell structure coiled yarn supercapacitor. *Adv. Energy Mater.* **2016**, *6*, 1502119. [CrossRef]
29. Choi, C.; Kim, S.H.; Sim, H.J.; Lee, J.A.; Choi, A.Y.; Kim, Y.T.; Lepró, X.; Spinks, G.M.; Baughman, R.H.; Kim, S.J. Stretchable, weavable coiled carbon nanotube/MnO_2/polymer fiber solid-state supercapacitors. *Sci. Rep.* **2015**, *5*, 9387. [CrossRef] [PubMed]
30. Yu, J.; Lu, W.; Smith, J.P.; Booksh, K.S.; Meng, L.; Huang, Y.; Li, Q.; Byun, J.-H.; Oh, Y.; Yan, Y.; et al. A high performance stretchable asymmetric fiber-shaped supercapacitor with a core-sheath helical structure. *Adv. Energy Mater.* **2017**, *7*, 1600976. [CrossRef]
31. Dawoud, H.D.; Tahtamouni, T.A.; Bensalah, N. Sputtered manganese oxide thin film on carbon nanotubes sheet as a flexible and binder-ree electrode for supercapacitor. *Int. J. Energy Res.* **2019**, *43*, 1245–1254. [CrossRef]
32. Son, W.; Lee, J.M.; Kim, S.H.; Kim, H.W.; Cho, S.B.; Suh, D.; Chun, S.; Choi, C. High-power hydro-actuators fabricated from biomimetic carbon nanotube coiled yarns with fast electrothermal recovery. *Nano Lett.* **2022**, *22*, 2470–2478. [CrossRef] [PubMed]
33. Sahoo, N.G.; Cheng, H.K.F.; Cai, J.; Li, L.; Chan, S.H.; Zhao, J.; Yu, S. Improvement of Mechanical and Thermal Properties of Carbon Nanotube Composites through Nanotube Functionalization and Processing Methods. *Mater. Chem. Phys.* **2009**, *117*, 313–320. [CrossRef]
34. Pei, X.W.; Liu, W.M.; Hao, J. Functionalization of multiwalled carbon nanotube via surface reversible addition fragmentation chain transfer polymerization and as lubricant additives. *J. Polym. Sci. Polym. Chem.* **2008**, *46*, 3014–3023. [CrossRef]

35. Guan, D.; Wang, B.; Zhang, J.; Shi, R.; Jiao, K.; Li, L.; Wang, Y.; Xie, B.; Zhang, Q.; Yu, J.; et al. Hydrogen society: From present to future. *Energy Environ. Sci.* **2023**. [CrossRef]
36. Zhang, H.; Guan, D.; Hu, Z.; Huang, Y.-C.; Wu, X.; Dai, J.; Dong, C.-L.; Xu, X.; Lin, H.-J.; Che, C.-T.; et al. Exceptional lattice-oxygen participation on artificially controllable electrochemistry-induced crystalline-amorphous phase to boost oxygen-evolving performance. *Appl. Catal.* **2021**, *297*, 120484. [CrossRef]

nanomaterials

Article

Electrochemically Oxidized Carbon Nanotube Sheets for High-Performance and Flexible-Film Supercapacitors

Jun Ho Noh [1,2,†], Jimin Choi [1,†], Hyunji Seo [1], Juwan Kim [1,2] and Changsoon Choi [1,2,*]

1 Department of Energy and Materials Engineering, Dongguk University, 30 Pildong-ro, 1-gil, Jung-gu, Seoul 04620, Republic of Korea; shwnsgh15@dongguk.edu (J.H.N.); james010828@gmail.com (J.C.); amy0527330@gmail.com (H.S.); kjw3328@dungguk.edu (J.K.)

2 Department of Advanced Battery Convergence Engineering, Dongguk University, 30 Pildong-ro, 1-gil, Jung-gu, Seoul 04620, Republic of Korea

* Correspondence: cschoi84@dongguk.edu

† These authors contributed equally to this work.

Abstract: The development of flexible, high-performance supercapacitors has been a focal point in energy storage research. While carbon nanotube (CNT) sheets offer promising mechanical and electrical properties, their low electrical double-layer capacitance significantly limits their practicability. Herein, we introduce a novel approach to address this challenge via the electrochemical oxidation treatment of CNT sheets stacked on a polyethylene terephthalate substrate. This introduces oxygen-containing functional groups onto the CNT surface, thereby dramatically enhancing the pseudocapacitive effect and improving ion adsorption. Consequently, using the material in a two-electrode system increased the capacitance by 54 times compared to pristine CNT. The results of electrochemical performance characterization, including cyclic voltammograms, galvanostatic charge/discharge curves, and capacitance retention testing data, confirm the efficacy of the electrochemical oxidation approach. Furthermore, the mechanical flexibility of the electrochemically wetted CNT sheets was validated through resistance and discharge retention testing under repetitive bending (98% capacitance retention after 1000 bending cycles). The results demonstrate that electrochemically wetted CNT sheets retain their intrinsic mechanical and electrical properties while significantly enhancing the electrochemical performance (0.59 mF/cm^2 or 97.8 F/g). This work represents a significant advancement in the development of flexible, high-performance supercapacitors with potential applicability to wearable electronics, flexible displays, and next-generation energy storage solutions.

Keywords: electrochemical oxidation; carbon nanotube; supercapacitors; flexibility; wearable

Citation: Noh, J.H.; Choi, J.; Seo, H.; Kim, J.; Choi, C. Electrochemically Oxidized Carbon Nanotube Sheets for High-Performance and Flexible-Film Supercapacitors. *Nanomaterials* **2023**, *13*, 2814. https://doi.org/10.3390/nano13202814

Academic Editor: Muralidharan Paramsothy

Received: 26 September 2023
Revised: 16 October 2023
Accepted: 20 October 2023
Published: 23 October 2023

1. Introduction

The development of efficient and sustainable energy storage systems is a critical challenge in modern society [1–4]. Among the various pertinent materials, carbon nanotube (CNT) sheets have emerged as a promising candidate for supercapacitor electrodes due to their unique combination of mechanical strength, electrical conductivity, and high surface area [5–10]. These properties make them especially apt for applications requiring flexibility and durability, such as wearable electronics and portable energy storage devices [6,10–15]. Indeed, numerous studies have confirmed the efficacy of flexible supercapacitors using CNT sheet-based electrodes drawn from CNT forests [8,14]. More importantly, forest-spinnable CNT sheets are a more advantageous electrode material than classical powder CNTs. Anisotropic alignment in the drawing direction and among the highly aligned bundled structures with branches between them results in a networked microstructure with excellent electrical, mechanical, and electromechanical properties [16]. However, the inherent low-charge storage capability of CNT sheets impedes their widespread use in supercapacitors. Given that normalized capacitance is vital for supercapacitor energy storage applications, enhancing capacitance without degrading other CNT sheet properties is a pressing research concern. The low capacitance of CNT sheets can be attributed to two main factors: their innate hydrophobicity, which limits ion adsorption in aqueous electrolytes and consequently restricts the charge storage capacity [17–20], and their charge storage mechanism, which in this case is an electrical double-layer capacitance with typically lower capacitance than pseudocapacitance [21,22]. This limitation significantly impedes harnessing the full potential of CNT sheets in high-performance supercapacitors.

Various methods have been explored to address this limitation, such as doping with heteroatoms or coating with conductive polymers [23–25]. However, these approaches often compromise the intrinsic properties of CNTs, including their mechanical strength and electrical conductivity [26]. Moreover, these methods can complicate the fabrication process, thus making it less scalable and cost-effective. Another approach is to introduce functional groups onto the surface of CNTs. However, the chemical functionalization of the inactive CNT surface frequently introduces sp3 defects that disrupt π-π conjugation and, thus, decreases the inherent mechanical properties, thermal stability, and electrical conductivity of the CNTs [27–29]. Therefore, creating CNT sheets with high charge storage capability while maintaining their mechanical properties or deformability would make it possible to fabricate flexible, practicable power sources. From this perspective, the proposed electrochemical oxidation treatment (EOT) of CNT sheets could improve the capacitance while avoiding the use of brittle pseudocapacitive materials by appropriately controlling surface oxidation.

To overcome these challenges, we introduce a groundbreaking approach to significantly enhance the capacitance of CNT sheets without compromising their intrinsic properties. We employed an electrochemical activation method commonly known as EOT to fully activate the surfaces of CNT sheets stacked on a polyethylene terephthalate (PET) substrate. The EOT process involves the application of an electrical potential to the CNT sheets submerged in an electrolytic solution to facilitate the introduction of oxygen-containing functional groups, such as carboxyl, hydroxyl, and epoxy groups, onto the surface of the CNTs. These functional groups significantly improve the CNT surface's wettability, thereby enhancing ion adsorption and electrochemical performance [30–32]. Remarkably, EOT achieves this without degrading the intrinsic properties of the CNT sheets while preserving their mechanical and electrical attributes, thus ensuring their usability in flexible supercapacitors [33–35]. This is possible because the degree of oxidation can be largely adjusted by controlling the oxidation time or applied voltage. Our method enables mild oxidation to the CNT sheets, which minimizes the disruption of π-π conjugation. Thereby, we produced a thin film-shaped supercapacitor with both highly improved capacitance (54-fold) compared to untreated CNT sheets and high mechanical flexibility (180° bendability).

While our group has previously used electrochemical activation to oxidize CNT yarns for supercapacitors [23,33] and actuators [17], the impact of EOT on CNT sheets remains unexplored. By employing EOT, our research not only addresses the critical limitation of low capacitance in pristine CNT sheets but also paves the way for their applicability to thin, flexible, high-performance supercapacitors. This scalable and efficient approach to functionalizing CNT sheets represents a significant breakthrough in the field of advanced energy storage applications.

2. Materials and Methods

2.1. Chemicals and Materials

CNT sheets were drawn from well-aligned multiwalled CNT forests with a height of 750 μm grown via chemical vapor deposition (A-Tech System Co., Incheon, Republic of Korea). Silver paste (CANS, Tokyo, Japan) and Sil Poxy (Smooth-on Inc., Macungie, PA, USA) were used to construct a CNT film supercapacitor cell for use in the electrochemical analysis. Sodium sulfate (Na_2SO_4) and polyvinyl alcohol (PVA; average molecular weight = 130,000 Da) were used to prepare the gel electrolyte (Sigma-Aldrich, St. Louis, MO, USA). PET Film (Film Bank, Kyeongkido, Republic of Korea) was used as the CNT sheet supercapacitor substrate.

2.2. Preparation of EOT CNT Sheets and the Gel Electrolyte

Five layers of CNT sheets (10 mm width and length) were drawn from a CNT forest and then carefully stacked on a 5 mm thick PET substrate. Electrical interfaces (essential for the EOT process) were established by affixing copper wires to both ends of the multi-layered CNT film. Silver paste was generously applied over the copper leads (diameter 180 μm) to maintain electrical connectivity. A non-brittle Sil Poxy layer was applied over the silver paste to ensure the bending stability of the EFAS. EOT was carried out using a three-electrode system with the pristine CNT film supercapacitor, Ag/AgCl, and platinum mesh as the working, reference, and counter electrodes, respectively. A 0.1 M aqueous Na_2SO_4 solution was utilized as the ionic medium. Subsequently, the appropriate electrochemical potential range relative to the Ag/AgCl reference electrode was administered via an electrochemical analysis system. To quantify the EOT effect, cyclic voltammetry (CV) (voltage range: 0 to 0.8 V) and galvanostatic charge/discharge (GCD) analyses were conducted under the same conditions as the EOT process. Moreover, a two-electrode system featuring symmetrical CNT and EOT CNT electrodes sandwiching a quasi-solid-state PVA-Na_2SO_4 electrolyte was constructed to test the applicability of the CNT sheet superconductor for bendable wearable devices. The gel electrolyte for both the CNT and EOT CNT sheets was produced by dissolving 6 g of PVA and 6 g of Na_2SO_4 in 60 mL of deionized water. This mixture was then stirred at 200 rpm and heated to 90 °C until it became clear.

2.3. Characterization

A cell phone camera (iPhone 11, Apple, California, USA) was used to obtain images of the CNT and EOT CNT sheets during bending. Scanning electron microscopy (SEM) images of the CNT and EOT CNT sheet supercapacitors were captured using an S-4600 instrument (Hitachi, Tokyo, Japan). Their electrochemical performances were evaluated using an electrochemical analyzer (Vertex EIS, Ivium, Noord-Brabant, The Netherlands). In addition, multi-meter probes (Model 187, Fluke Corporation, Washington, USA) were employed to measure resistance. Fourier-transform infrared spectroscopy (FTIR) was performed by using an IdentifyIR instrument (Smiths Detection, England). Raman spectroscopy (NRS-3100, JASCO, Tokyo, Japan) and X-ray photoelectron spectroscopy (XPS; ESCALAB 250XI, ThermoFisher Scientific, Waltham, MA, USA) were conducted to analyze the chemical characteristics of the CNT and EOT CNT sheets.

2.4. Electrochemical Performance Calculation

The areal capacitance (μF/cm^2) was calculated from a CV curve using the following equation:

$$Capacitance = \frac{I/(dV/dt)}{unit} \tag{1}$$

where I is the average current of the CNT sheets, and dV/dt is the scan rate during the CV analysis.

3. Results and Discussion

3.1. Full Electrochemical Activation of the EOT CNT Sheet Electrode

The experimental setup for using EOT to functionalize the surface of CNT sheets (1 cm wide) is illustrated in Figure 1a. The working electrode was prepared by mechanically drawing and stacking five layers of the CNT sheets on a PET substrate (Figure 1b). This was used along with Ag/AgCl and Pt mesh as the reference and counter electrodes, respectively, in a three-electrode system. The CNT working electrode was immersed in an aqueous electrolyte (0.1 M Na_2SO_4) to electrochemically activate the CNT sheets and to characterize the electrochemical performance of the CNT electrode. The current density and resistance of the CNT sheets were measured during the EOT process (Figure 1c). It should be noted that the current density was approximately maintained during the initial phase of the EOT process but abruptly decreased after a certain treatment time. This point of current density drop coincides with a dramatic increase in the resistance of the CNT sheets, which could be due to excessive functionalization via oxygen-containing groups. Therefore, we defined just before the abrupt increase in resistance as the activation time used in all subsequent specimen preparations. Since the activation time varies depending on the applied voltage, the current densities according to time at voltages ranging from 2 to 3.5 V are shown in Figure 1d; the higher the voltage, the shorter the activation time. Moreover, the current density, activation time, and the resulting power and energy values according to the applied voltage are depicted in Figure 1e. The total energy required to activate the CNT sheets was similar at the various current densities and voltages (approximately 0.5 J), which infers much better cost and energy efficiency compared to previous oxidation methods such as sputtering. The SEM images before and after EOT, shown in Figure 1f and 1g, respectively, reveal that EOT did not degrade the structural stability of the CNT sheets, resulting in high retention of the mechanical, electrical, and electrochemical properties of the electrodes. Moreover, the SEM image in Figure S1 shows the anisotropic alignment of the CNT bundles after the EOT process.

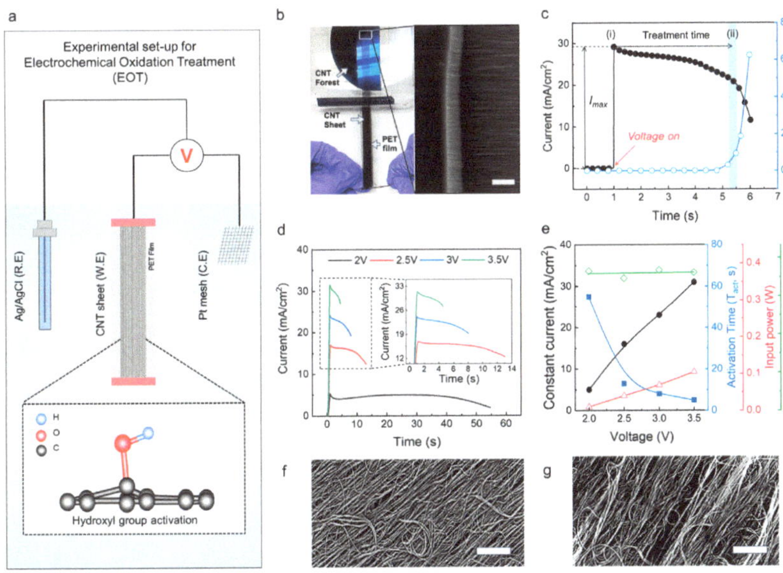

Figure 1. (**a**) A schematic illustration of the EOT setup including the CNT sheets (working electrode), Ag/AgCl (reference electrode), and Pt mesh (counter electrode) immersed in a 0.1 M Na_2SO_4 liquid electrolyte (inset: a schematic showing the expected functionalization of hydroxyl groups on the adjacent CNT surface). (**b**) A photograph showing the CNT sheet electrode preparation process. The aligned CNT sheets were mechanically drawn from the CNT forest and stacked on a PET film (scale bar = 40 mm). (**c**) Current density and sheet resistance versus oxidation time at a constant voltage of 3.5 V (versus Ag/AgCl). When the current plateaus before a sudden increase in resistance is defined as the activation time. ((i): treatment start region, (ii): treatment end region) (**d**) Current density versus oxidation time under various voltages (2, 2.5, 3, and 3.5 V) (versus Ag/AgCl). (**e**) Current density, treatment time, input power, and total energy versus the applied voltage. SEM images (**f**) before and (**g**) after EOT (scale bars = 1 μm).

3.2. *Electrochemical Performance Measurements*

One of the most compelling benefits of applying EOT to CNT sheet-based supercapacitors is its profound enhancement of the electrochemical performance compared to untreated CNT sheets. To quantify this effect, CV analysis was conducted over a voltage range from 0 to 0.8 V versus the Ag/AgCl reference electrode in a three-electrode system; CV curves for applied voltages of 2–3 V and oxidation times up to 40 s are shown in Figure 2a–c, respectively. Before EOT, the curves typically display a limited area indicative of low capacitance. However, after EOT, the area under the CV curves was dramatically enlarged, with growth roughly proportional to the treatment time. This infers a dramatic increase in charge storage capability due to EOT, which was confirmed through GCD analysis. The triangular GCD curves in Figure 2d signify an optimal capacitive charge storage mechanism while lengthening the charge/discharge time revealed capacitance augmentation via EOT. Although the main EOT parameters are applied voltage and treatment time, the influence of EOT on enhancing the capacitance seems to converge to a similar efficiency after a certain energy threshold was reached, yielding approximately a 40–50 times increase in the CV curve area (Figure 2e). Our subsequent analysis using a two-electrode system featuring symmetrical EOT CNT electrodes sandwiching a PVA-Na_2SO_4 quasi-solid-state electrolyte was conducted to validate the reliability of the capacitance improvement by the EOT effect. To this end, we performed multiple EOT experiments and plotted the resulting capacitance improvements and specific capacitances of 12 supercapacitors (Figure 2f). We also assessed the capacitance retention performance and high-rate capability of the pristine and EOT CNT sheet supercapacitors at various scan rates (from 10 to 1000 mV/s), of which the resulting areal capacitances are exhibited in Figure 2g; those of pristine and EOT CNT sheets measured at a scan rate of 10 mV/s were 18.2 and 587 μF/cm^2, respectively. Moreover, 76% of the capacitance was retained at a very high scan rate of 1000 mV/s rate, as indicated by the well-defined rectangular CV curve at high scan rates without a notable loss in the CV curve area in Figure 2h. Finally, long-term stability was assessed through 1000 repeated charge/discharge cycles (Figure 2i), during which the EOT CNT sheet supercapacitor retained 89% of its performance. We also compared the areal capacitance of the EOT CNT sheets with those of previously reported carbon-based materials used as supercapacitor electrodes in Figure S2 [36–45].

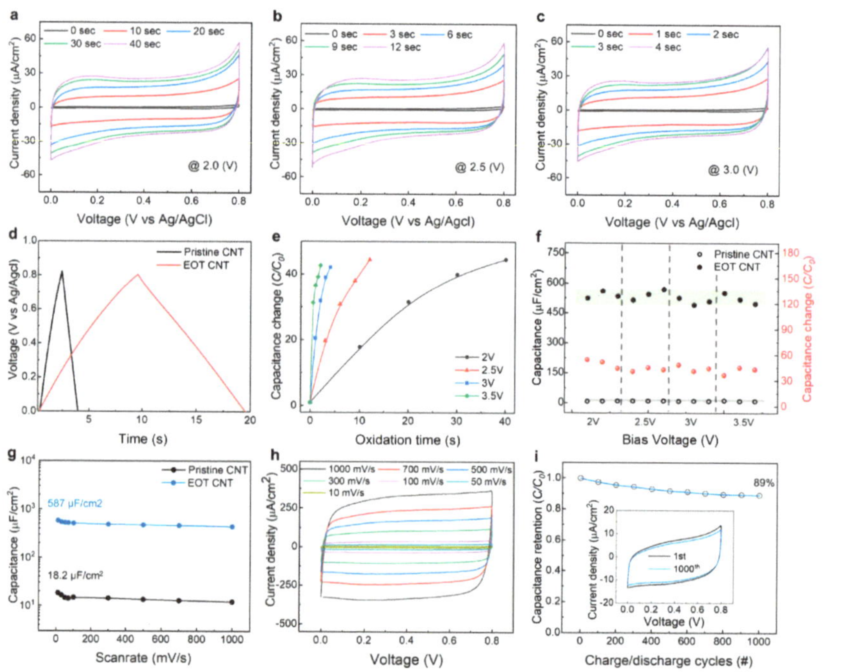

Figure 2. CV curves of EOT CNT sheets prepared at various activation times under applied voltages of (**a**) 2.0, (**b**) 2.5, or (**c**) 3.0 V at 30 mV/s. (**d**) GCD curves at a current density = 30 μA/cm^2 of CNT

sheets before and after EOT. (**e**) Capacitance change ratio versus oxidation time under applied voltages of 2, 2.5, 3, or 3.5 V. (**f**) Areal capacitance and capacitance change ratio versus applied voltage for 12 different supercapacitors in a two-electrode system consisting of two symmetrical electrodes coated with PVA/Na_2SO_4 gel electrolyte. (**g**) Areal capacitances of pristine and EOT CNT supercapacitors at scan rates from 10 to 1000 mV/s. (**h**) CV curves of the EOT CNT supercapacitor measured at scan rates from 10 to 1000 mV/s. (**i**) Capacitance retention versus charge/discharge cycles of the EOT CNT supercapacitor showing 89% retention after 1000 cycles (inset: the CV curves between the first and 1000th cycles at 30 mV/s).

3.3. The Capacitance Improvement Mechanism

The FTIR spectra in Figure 3a offer definitive evidence of certain functional groups. The pronounced absorption peak in the spectrum of EOT CNTs near 3400 cm^{-1} arises from O-H stretching vibrations attributed to hydroxyl groups (specifically O=C-H and C-OH) stemming from the EOT of the CNT layers. The resonance near 1620 cm^{-1} is due to the vibrational stretching of C=C bonds. The peak at around 2330 cm^{-1} caused by the vibrational absorption of the C-O-C bonds was higher for the EOT CNTs, further verifying the oxidation of CNTs by EOT. The absorption peak around 1735 cm^{-1} linked to the carbonyl (C=O) group hints at the stretching vibrations of carboxyl units (-COOH). The resonance at around 1100 cm^{-1} related to the C-O stretching of the carboxylic acid group was higher in the spectrum of the EOT CNTs. These results indicate the successful formation of functional groups on the surface of the EOT CNTs. Raman spectroscopy provided additional evidence of functional group formation (Figure 3b). EOT increased the intensity of the D-band located at 1350 cm^{-1} to the G-band (located at 1590 cm^{-1} (I_D/I_G) ratio from 0.51 to 0.76, suggesting a higher defect density in the graphitic carbon structure and corroborating the FTIR findings. From our previous work [33], the hydroxyl group (C-OH) content seems to be pivotal for activating CNT.

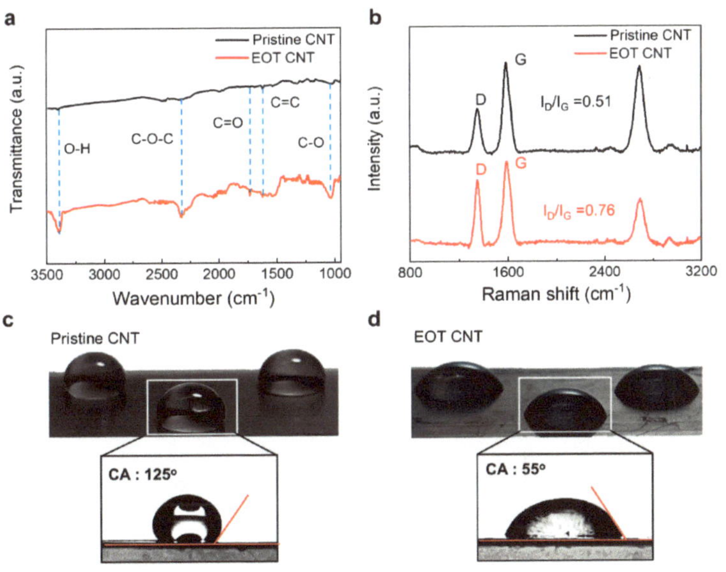

Figure 3. (**a**) FTIR and (**b**) Raman spectra of pristine and EOT CNT sheets(D: defect, G: graphite). (**c**) Optical images showing water droplets and their contact angles (CAs) on (**c**) pristine CNT sheets and (**d**) EOT CNT sheets.

XPS analysis was conducted to measure the degree of oxidation caused by EOT, with XPS spectra before and after EOT being presented in Figure S3a,b, respectively. We found that the main functional groups present were hydroxyl and epoxy. The ratio of oxygen to carbon atoms (O/C) before and after EOT increased from 8% to 28%, respectively. This 20% increase is quite mild because EOT was terminated before the drastic increase in resistance in the CNT electrode (Figure S3c), which could be due to excessive oxidation. If we apply harsh conditions by over-wetting the mild-oxidation region, the resistance drastically increases due to severe oxidation. Therefore, the internal resistance drop during the charge storage process predominates, resulting in the highly degraded CV curve shown in Figure S3d.

Functional groups significantly improve the wettability of the CNT surface, thereby increasing the electrochemically active area for ion adsorption. To validate the improved wettability of the EOT CNT sheets over untreated ones, we analyzed the contact angles before and after EOT (Figure 3c and 3d, respectively). The contact angle of water droplets on CNT sheets should be significantly different before and after EOT. Untreated CNT sheets generally exhibit a higher contact angle of approximately 125°, thereby indicating hydrophobicity due to a lack of polar functional groups on the surface (Figure 3c). On the other hand, the EOT CNT sheets had a considerably lower contact angle of approximately 55°, thereby indicating hydrophilicity and enhanced wettability (Figure 3d). Hence, the formation of oxygen-containing functional groups during EOT increased the surface polarity and consequently boosted wettability. A more hydrophilic surface enhances ion adsorption, leading to improved electrochemical performance. Therefore, the reduction in contact angle induced by EOT not only confirms the successful introduction of functional groups but also indicates an improvement in the electrochemical properties of the CNT sheets. This change in contact angle can thus act as a simple yet effective marker for the success of the EOT process in enhancing the performance of CNT-based supercapacitors.

3.4. Flexibility of the EOT CNT Sheet Supercapacitor

One of the standout features of our EOT CNT sheet-based supercapacitor is its exceptional mechanical flexibility, a property that is becoming increasingly important for wearable electronics, flexible displays, and portable energy storage devices [17]. To rigorously assess the flexibility of our supercapacitors, we fabricated elongated EOT CNT sheets (1 × 5 cm) and conducted a series of tests measuring resistance and discharge retention performances against bending curvature and the number of bending cycles. Our results demonstrate remarkable stability in both resistance and discharge retention under various bending conditions. Specifically, changes in the resistance of the supercapacitor were negligible across a range of bending curvatures (up to 0.63/cm), indicating that the electrical pathways within the CNT sheets remained largely intact, even when being bent to 180° (Figure 4a,b). Furthermore, the charge/discharge retention performance of the supercapacitor remained exceptionally stable under mechanical bending (the GCD curves measured statically while bending at 90° and 180° are shown in Figure 4c). Moreover, the dynamic charge/discharge characteristics analyzed via GCD and CV are shown in Figure 4d and 4e, respectively. From these results, it is evident that the electrochemical performance of the EOT CNTs was well preserved while being bent. Typically, mechanical stress can result in reduced electrochemical performance due to disruptions in the active material or electrical contacts. However, our EOT CNT-based supercapacitor maintained a stable discharge profile even after thousands of bending cycles, thereby underscoring its robustness and reliability (Figures 4f and S4). These findings not only affirm the mechanical flexibility of our CNT sheet-based supercapacitor but also highlight its potential for integration into a wide array of flexible and wearable devices.

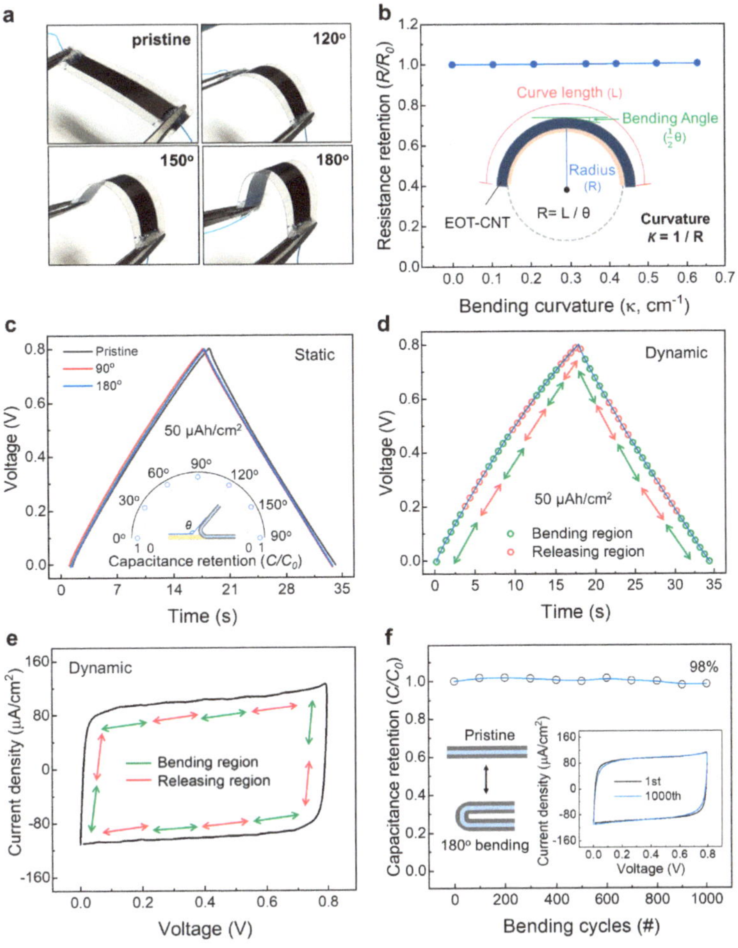

Figure 4. (**a**) Optical images showing an EOT CNT sheet electrode bent at various angles. (**b**) Resistance change ratio versus bending curvature of EOT CNT sheets at 0–180°. (**c**) A GCD curve at a current density of 50 μA/cm^2 of the bent EOT CNT sheets. The test was performed with a two-electrode system while being bent at 0°, 90°, or 180°. (**d**) GCD curves at a current density of 50 μA/cm^2 and (**e**) CV curves at 30 mV/s measured during 180° bending and releasing dynamic cycles of the EOT CNT sheets. (**f**) Capacitance retention versus 180° bending cycles of the EOT CNT sheets showing 98% capacitance retention after 1000 cycles (inset: a comparison of the CV curves between the first and 1000th bending cycles at 30 mV/s).

4. Conclusions

We presented a groundbreaking approach to enhance the electrochemical performance of CNT sheet-based supercapacitors using EOT. Our work addresses the critical limitation of low capacitance found in pristine CNT sheets by introducing oxygen-containing functional groups onto the CNT surface via EOT. This approach improved ion adsorption by making the CNT sheets hydrophilic (a contact angle decrease from 125° to 55°), which consequently improved the capacitance by 54-fold. Moreover, the EOT CNT sheets retained their intrinsic mechanical and electrical properties, including exceptional flexibility and 98% capacitance retention after 1000 bending cycles. This represents a significant advancement in the field of energy storage, particularly for applications requiring flexible and durable supercapacitors.

Our findings have far-reaching implications, not only in energy storage but also in wearable electronics, flexible displays, and other next-generation technologies.

Supplementary Materials: The following supporting information can be downloaded at: https://www.mdpi.com/article/10.3390/nano13202814/s1, Figure S1: Magnified SEM image of EOT CNT sheets; Figure S2: Performance comparison of the EOT CNT supercapacitor; Figure S3: XPS analysis of EOT process and performance of Wild EOT process. Figure S4: SEM images of the CNT working electrode after the 1000 bending cycle.

Author Contributions: J.H.N., J.C. and C.C. conceived the idea for the study. J.C. and J.H.N. performed and analyzed the experiments. H.S. designed the methodology and performed the experiments. J.K. conducted the analysis. J.H.N. and J.C. contributed to writing the manuscript. C.C. edited the manuscript. J.H.N. and J.C. equally contributed to this work. All authors have read and agreed to the published version of the manuscript.

Funding: This research received no external funding.

Data Availability Statement: Not applicable.

Acknowledgments: This research was supported by the Police-Lab 2.0 Program (www.kipot.or.kr) funded by the Korean Ministry of Science and ICT (MSIT) and Korean National Police Agency (KNPA) [Project name: Development of fluorescent forensic nanomaterials and evidence specification technique for effective biometric evidence detection; Project number: RS-2023-00236429]. This work was also supported by the Korean Institute of Energy Technology Evaluation and Planning (KETEP) and the Ministry of Trade, Industry & Energy (MOTIE) of the Republic of Korea (No. 20224000000020).

Conflicts of Interest: The authors declare no conflict of interest.

References

1. Deng, W.; Xu, Y.; Zhang, X.; Li, C.; Liu, Y.; Xiang, K.; Chen, H. (NH4)$_2$Co$_2$V$_{10}$O$_{28}$·16H$_2$O/(NH$_4$)$_2$V$_{10}$O$_{25}$·8H$_2$O heterostructure as cathode for high-performance aqueous Zn-ion batteries. *J. Alloys Compd.* **2022**, *903*, 163824. [CrossRef]
2. Deng, W.; Liu, W.; Zhu, H.; Chen, L.; Liao, H.; Chen, H. Click-chemistry and ionic cross-linking induced double cross-linking ionogel electrolyte for flexible lithium-ion batteries. *J. Energy Storage* **2023**, *72*, 108509. [CrossRef]
3. Wen, W.; Luo, J.; Xiang, K.; Zhou, W.; Zhang, C.; Chen, H. High-performance monoclinic WO_3 nanospheres with the novel NH_4^+ diffusion behaviors for aqueous ammonium-ion batteries. *Chem. Eng. J.* **2023**, *458*, 141381. [CrossRef]
4. Deng, W.D.; Li, Y.H.; Xu, D.F.; Zhou, W.; Xiang, K.X.; Chen, H. Three-dimensional hierarchically porous nitrogen-doped carbon from water hyacinth as selenium host for high-performance lithium–selenium batteries. *Rare Met.* **2022**, *41*, 3432–3445. [CrossRef]
5. Dawoud, H.D.; Tahtamouni, T.A.; Bensalah, N. Sputtered manganese oxide thin film on carbon nanotubes sheet as a flexible and binder-free electrode for supercapacitors. *Int. J. Energy Res.* **2019**, *43*, 1245–1254. [CrossRef]
6. Malik, R.; Zhang, L.; McConnel, C.; Schott, M.; Hsieh, Y.Y.; Noga, R.; Alvarez, N.T.; Shanov, V. Three-dimensional, free-standing polyaniline/carbon nanotube composite-based electrode for high-performance supercapacitors. *Carbon* **2017**, *116*, 579–590. [CrossRef]
7. Bathula, C.; Rabani, I.; Kadam, A.; Opoku, H.; Patil, S.A.; Shreshta, N.K.; Hwang, J.H.; Seo, Y.S.; Kim, H.S. Sonochemically exfoliated polymer-carbon nanotube interface for high performance supercapacitors. *J. Colloid Interface Sci.* **2022**, *606*, 1792–1799. [CrossRef]
8. Kim, J.H.; Ko, Y.I.; Kim, Y.A.; Kim, K.S.; Yang, C.M. Sulfur-doped carbon nanotubes as a conducting agent in supercapacitor electrodes. *J. Alloys Compd.* **2021**, *855*, 157282. [CrossRef]
9. Guo, Z.; Han, X.; Zhang, C.; He, S.; Liu, K.; Hu, J.; Yang, W.; Jian, S.; Jiang, S.; Duan, G. Activation of biomass-derived porous carbon for supercapacitors: A review. *Chin. Chem. Lett.* **2023**, *CCLET*, 109007. [CrossRef]
10. Ben, J.; Song, Z.; Liu, X.; Lü, W.; Li, X. Fabrication and Electrochemical Performance of PVA/CNT/PANI Flexible Films as Electrodes for Supercapacitors. *Nanoscale Res. Lett.* **2020**, *15*, 151. [CrossRef]
11. Li, L.; Nam, J.S.; Kim, M.S.; Wang, Y.; Jiang, S.; Hou, H.; Kim, I.D. Sulfur–Carbon Electrode with PEO-LiFSI-PVDF Composite Coating for High-Rate and Long-Life Lithium–Sulfur Batteries. *Adv. Energy Mater.* **2023**, *13*, 2302139. [CrossRef]
12. Zhou, W.; Zeng, G.; Jin, H.; Jiang, S.; Huang, M.; Zhang, C.; Chen, H. Bio-Template Synthesis of V_2O_3@Carbonized Dictyophora Composites for Advanced Aqueous Zinc-Ion Batteries. *Molecules* **2023**, *28*, 2147. [CrossRef]
13. Son, W.; Chun, S.; Lee, J.M.; Jeon, G.; Sim, H.J.; Kim, H.W.; Cho, S.B.; Lee, D.; Park, J.; Jeon, J.; et al. Twist-stabilized, coiled carbon nanotube yarns with enhanced capacitance. *ACS Nano* **2022**, *16*, 2661–2671. [CrossRef]
14. Sun, J.; Liu, Y.; Huang, J.; Li, J.; Chen, M.; Hu, X.; Liu, Y.; Wang, R.; Shen, Y.; Li, J.; et al. Size-refinement enhanced flexibility and electrochemical performance of MXene electrodes for flexible waterproof supercapacitors. *J. Energy Chem.* **2021**, *63*, 594–603. [CrossRef]

15. Vandeginste, V. A Review of Fabrication Technologies for Carbon Electrode-Based Micro-Supercapacitors. *Appl. Sci.* **2022**, *12*, 862. [CrossRef]
16. Zhang, M.; Fang, S.; Zakhidov, A.A.; Lee, S.B.; Aliev, A.E.; Williams, C.D.; Atkinson, K.R.; Baughman, R.H. Strong, Transparent, Multifunctional, Carbon Nanotube Sheets. *Science* **2005**, *309*, 1215–1219. [CrossRef]
17. Son, W.; Lee, J.M.; Kim, S.H.; Kim, H.W.; Cho, S.B.; Suh, D.; Chun, S.; Choi, C. High-Power Hydro-Actuators Fabricated from Biomimetic Carbon Nanotube Coiled Yarns with Fast Electrothermal Recovery. *Nano Lett.* **2022**, *22*, 2470–2478. [CrossRef]
18. Sianipar, M.; Kim, S.H.; Khoiruddin, K.; Iskandar, F.; Wenten, I.G. Functionalized carbon nanotube (CNT) membrane: Progress and challenges. *RSC Adv.* **2017**, *7*, 51175–51198. [CrossRef]
19. Wu, J.; An, A.K.; Guo, J.; Lee, E.J.; Farid, M.U.; Jeong, S. CNTs reinforced super-hydrophobic-oleophilic electrospun polystyrene oil sorbent for enhanced sorption capacity and reusability. *Chem. Eng. J.* **2017**, *314*, 526–536. [CrossRef]
20. Wang, C.; Zhou, S.; Wu, C.; Yang, Z.; Zhang, W. Janus carbon nanotube sponges for highly efficient solar-driven vapor generation. *Chem. Eng. J.* **2023**, *454*, 140501. [CrossRef]
21. Jiang, Y.; Liu, J. Definitions of Pseudocapacitive Materials: A Brief Review. *Energy Environ. Mater.* **2019**, *2*, 30–37. [CrossRef]
22. Wu, N.; Bai, X.; Pan, D.; Dong, B.; Wei, R.; Naik, N.; Patil, R.R.; Guo, Z. Recent Advances of Asymmetric Supercapacitors. *Adv. Mater. Interfaces* **2021**, *8*, 2001710. [CrossRef]
23. Iyer, M.S.; Rajangam, I. Graphene variants and heteroatom doped nanocomposites as electrode material for symmetric supercapacitors. *Colloids Interface Sci. Commun.* **2022**, *46*, 100577. [CrossRef]
24. Sajjad, L.; Ali, G.; Mansoor, M.A.; Khan, M.F. Facile synthesis and electrochemical study of $NiMnO_3$/CNT/PANI ternary hybrid as a high-performance supercapacitor electrode material. *J. Energy Storage* **2023**, *72*, 108351. [CrossRef]
25. Pour, G.B.; Ashourifar, H.; Aval, L.F.; Solaymani, S. CNTs-Supercapacitors: A Review of Electrode Nanocomposites Based on CNTs, Graphene, Metals, and Polymers. *Symmetry* **2023**, *15*, 1179. [CrossRef]
26. Chun, S.; Son, W.; Lee, G.; Kim, S.H.; Park, J.W.; Kim, S.J.; Pang, C.; Choi, C. Single-Layer Graphene-Based Transparent and Flexible Multifunctional Electronics for Self-Charging Power and TouchSensing Systems. *ACS Appl. Mater. Interfaces* **2019**, *11*, 9301–9308. [CrossRef]
27. Pei, X.; Liu, W.; Hao, J. Functionalization of Multiwalled Carbon Nanotube via Surface Reversible Addition Fragmentation Chain Transfer Polymerization and as Lubricant Additives. *J. Phys. Chem. C* **2007**, *111*, 2947–2952. [CrossRef]
28. Sahoo, N.G.; Cheng, H.K.F.; Cai, J.; LI, L.; Chan, S.H.; Zhao, J.; Yu, S. Improvement of mechanical and thermal properties of carbon nanotube composites through nanotube functionalization and processing methods. *Mater. Chem. Phys.* **2009**, *117*, 313–320. [CrossRef]
29. Beese, A.M.; Sarkar, S.; Nair, A.; Naraghi, M.; An, Z.; Moravsky, A.; Loutfy, R.O.; Buehler, M.J.; Nguyen, S.T.; Espinosa, H.D. Bio-Inspired Carbon NanotubePolymer Composite Yarns with Hydrogen Bond-Mediated Lateral Interactions. *ACS Nano* **2013**, *7*, 3434–3446. [CrossRef]
30. Banks, C.E.; Davies, T.J.; Wildgoode, G.G.; Compton, R.G. Electrocatalysis at graphite and carbon nanotube modified electrodes: Edge-plane sites and tube ends are the reactive sites. *Chem. Commun.* **2005**, 829–841. [CrossRef]
31. Lota, G.; Tyczkowski, J.; Kapica, R.; Lota, K.; Frackowiak, E. Carbon materials modified by plasma treatment as electrodes for supercapacitors. *J. Power Sources* **2010**, *195*, 7535–7539. [CrossRef]
32. Xiao, X.; Li, T.; Peng, Z.; Jin, H.; Zhong, Q.; Hu, Q.; Yao, B.; Luo, Q.; Zhang, C.; Gong, L.; et al. Freestanding functionalized carbon nanotube-based electrode for solid-state asymmetric supercapacitors. *Nano Energy* **2014**, *6*, 1–9. [CrossRef]
33. Son, W.; Lee, J.M.; Chun, S.; Yu, S.; Noh, J.H.; Kim, H.W.; Cho, S.B.; Kim, S.J.; Choi, C. Enhanced Hydro-Actuation and Capacitance of Electrochemically Inner-Bundle-Activated Carbon Nanotube Yarns. *ACS Appl. Mater. Interfaces* **2023**, *15*, 13484–13494. [CrossRef] [PubMed]
34. Ranjithkumar, R.; Arasi, S.E.; Sudhahar, S.; Nallamuthu, N.; Devendran, P.; Lakshmanan, P.; Kumar, M.K. Enhanced electrochemical studies of ZnO/CNT nanocomposite for supercapacitor devices. *Phys. B Condens. Matter* **2019**, *568*, 51–59. [CrossRef]
35. Ma, J.; Yuan, J.; Ming, W.; He, W.; Zhang, G.; Zhang, H.; Cao, Y.; Jiang, Z. Non-traditional processing of carbon nanotubes: A revew. *Alex. Eng. J.* **2022**, *61*, 597–617. [CrossRef]
36. Izadi-Najafabadi, A.; Yamda, T.; Futaba, D.N.; Yudasaka, M.; Takagi, H.; Hatori, H.; Iijima, S.; Hata, K. High-Power Supercapacitor Electrodes from Single-Walled Carbon Nanohorn/Nanotube Composite. *ACS Nano* **2011**, *5*, 811–819. [CrossRef]
37. Zheng, C.; Quan, W.; Cui, C.; Zhang, Q.; Jin, Y.; Zhao, M.; Tan, P.; Wei, F. Hierarchical carbon nanotube membrane with high packing density and tunable porous structure for high voltage supercapacitors. *Carbon* **2012**, *50*, 5167–5175. [CrossRef]
38. Di, J.; Hu, D.; Chen, H.; Yong, Z.; Chen, M.; Feng, Z.; Zhu, Y.; Li, Q. Ultrastrong, Foldable, and Highly Conductive Carbon Nanotube Film. *ACS Nano* **2012**, *6*, 5457–5464. [CrossRef]
39. Li, X.; Gu, T.; Wei, B. Dynamic and Galvanic Stability of Stretchable Supercapacitors. *Nano Lett.* **2012**, *12*, 6366–6371. [CrossRef]
40. Niu, Z.; Dong, H.; Zhu, B.; Li, J.; Hng, H.H.; Zhou, W.; Chen, X.; Xie, S. Highly Stretchable, Integrated Supercapacitors Based on Single-Walled Carbon Nanotube Films with Continuous Reticulate Architecture. *Adv. Mater.* **2013**, *25*, 1058–1064. [CrossRef]
41. Yuksel, R.; Sarioba, Z.; Cirpan, A.; Hiralal, P.; Unalal, H.E. Transparent and Flexible Supercapacitors with Single Walled Carbon Nanotube Thin Film Electrodes. *ACS Appl. Mater. Interfaces* **2014**, *6*, 15434–15439. [CrossRef] [PubMed]
42. Lee, J.; Kim, W.; Kim, W. Stretchable Carbon Nanotube/Ion-Gel Supercapacitors with High Durability Realized through Interfacial Microroughness. *ACS Appl. Mater. Interfaces* **2014**, *6*, 13578–13586. [CrossRef] [PubMed]

43. Zheng, C.; Zhou, F.; Cao, H.L.; Wang, G.H.; Liu, Z.P. Edge-enriched porous graphene nanoribbons for high energy density supercapacitors. *J. Mater. Chem. A* **2014**, *2*, 7484. [CrossRef]
44. Li, H.; Tao, Y.; Zheng, X.; Li, Z.; Liu, D.; Xu, Z.; Lu, C.; Luo, J.; Kang, F.; Yang, Q.H. Compressed porous graphene particles for use as supercapacitor electrodes with excellent volumetric performance. *Nanoscale* **2015**, *7*, 18459–18463. [CrossRef]
45. Tang, X.; Lui, Y.H.; Chen, B.; Hu, S. Functionalized carbon nanotube based hybrid electrochemical capacitors using neutral bromide redox-active electrolyte for enhancing energy density. *J. Power Sources* **2017**, *352*, 118–126. [CrossRef]

nanomaterials

Article

Effect of Atmospheric Temperature on Epoxy Coating Reinforced with Carbon Nanotubes for De-Icing on Road Systems

Seung-Jun Lee [1], Yu-Jin Jung [2], Chunhee Cho [3,*] and Sung-Hwan Jang [2,4,*]

1 Department of Civil and Environmental Engineering, Hanyang University, Seoul 04763, Republic of Korea; sj5523@hanyang.ac.kr
2 Department of Smart City Engineering, Hanyang University ERICA, Ansan 15588, Republic of Korea; yujin0421@hanyang.ac.kr
3 Department of Civil and Environmental Engineering, University of Hawaii at Manoa, Honolulu, HI 98622, USA
4 Department of Civil and Environmental Engineering, Hanyang University ERICA, Ansan 15588, Republic of Korea
* Correspondence: chunhee@hawaii.edu (C.C.); sj2527@hanyang.ac.kr (S.-H.J.)

Abstract: Traffic accidents caused by road icing are a serious global problem, and conventional de-icing methods like spraying chemicals have several limitations, including excessive manpower management, road damage, and environmental pollution. In this study, the carbon nanotubes reinforced de-icing coating for the road system with a self-heating function was developed as part of the development of a new system to prevent accidents caused by road icing. The electrical characteristics of the fabricated coating were analyzed, and the carbon nanotube coating heating performance experiment was conducted to measure the temperature increments by applying a voltage to the coating at a sub-zero temperature using an environmental chamber. In addition, the coating was installed on the road pavement and the applicability was investigated through a heating test in winter. As a result of the experiment, the coating made with the higher carbon nanotube concentration presented higher heating owing to its higher electrical conductivity. In addition, the coating showed sufficient heating performance, although the maximum temperature by Joule heating decreased for the entire coating at sub-zero temperatures. Finally, field tests demonstrated the potential of electrically conductive coatings for de-icing applications.

Keywords: carbon nanotube; CNT/EP coating; de-icing; Joule heating

Citation: Lee, S.-J.; Jung, Y.-J.; Cho, C.; Jang, S.-H. Effect of Atmospheric Temperature on Epoxy Coating Reinforced with Carbon Nanotubes for De-Icing on Road Systems. *Nanomaterials* **2023**, *13*, 2248. https://doi.org/10.3390/nano13152248

Academic Editor: Muralidharan Paramsothy

Received: 10 July 2023
Revised: 31 July 2023
Accepted: 31 July 2023
Published: 3 August 2023

1. Introduction

Road systems face their greatest challenges during the winter season in many countries because snowfalls and icing create serious problems in the field of road systems. These challenges involve the mobility of vehicles and the assurance of driver safety. In addition, many road departments are not prepared to immediately remove snow and ice. As a result, snow and ice accumulation on road systems lead to significant financial setbacks. For example, persistent snowfall in northwestern Germany resulted in more than 2000 traffic accidents and a direct economic loss of 100 million Euros in November 2005, and traffic in the northeastern United States was paralyzed by a snowstorm for four days, with a direct economic loss of US $10 billion in January 1996 [1,2]. Considering the financial consequences and threats to public safety, extensive global research has been conducted on various snow and ice removal methods, including their practical implementations. Typical approaches for eliminating snow and ice involve combining the use of deicing chemicals with mechanical extraction techniques. However, these techniques come with drawbacks such as storage and acquisition expenses, extensive labor requirements, environmental harm, and potential damage to the road infrastructure [3–6]. For example, the salty runoff from road surface deicing operations is responsible for soil and water contamination and adverse health effects for human, plant, and aquatic life [7–12]. In addition, deicing substances account for the largest portion of greenhouse gas emissions related to traditional deicing techniques [13,14]. Globally, the consumption of deicing materials has continuously increased over the past few years which creates serious environmental problems [15].

To address these issues, many research efforts have been devoted to finding cleaner techniques for safely deicing road systems during winter. Some of the emerging techniques involve the application of embedding electrically heated sheet/grille elements inside the road [16] and the application of heated road systems [17–22]. In recent years, conductive composite materials based on carbon nanotubes has attracted attention to improve road systems during winter [23,24]. Dispersing these types of nanoparticles into polymeric structures can enhance thermal stability, resistance to photooxidation, and mechanical attributes, whilst also equipping the resulting nanocomposites with the capacity to exhibit functional properties [25–27]. In recent years, carbon nanotubes (CNTs) have been used in various fields as fillers and CNT-reinforced composites have been confirmed as an alternative de-icing approach. Joule-heating effects occurs when electrical current is passed through conductive composites. CNTs demonstrate a significant self-heating effect when an electric current is applied [28–30]. Joule-heating capabilities have been reported for over a decade. Jang and Park [31] suggested the use of composite materials reinforced with carbon nanotubes, designed for dual purposes like temperature detection and ice removal. Their study showed the feasibility of CNT-reinforced polymer composites for coating systems capable of detecting freezing temperature and self-heating. Yum et al. [32] proposed multi-functional road coating material substances composed of carbon nanotubes (CNTs) and a polyurethane (PU) matrix, commonly used materials in road marking. Prolongo et al. [33] and Redondo et al. [34] proposed the doping of epoxy resins with graphene nanoplatelets as coatings (GNP, 8–10 wt.%) or CNTs (0.1–0.5 wt.%) for the heating of epoxy resin by the Joule effect. In their study, CNTs exhibited greater efficiency in deicing and anti-icing applications owing to their superior electrical conductivity, achieving higher temperatures at reduced electrical voltages. Meanwhile, GNPs generated lower yet more evenly distributed heat. However, studies on carbon nanotube-based heating composites are mainly conducted at room temperature and since they are laboratory-scale studies, it is necessary to study the effect of the atmospheric environment on heating performance and consider large-scale applications.

In our research, we developed a novel coating comprising carbon nanotubes and an epoxy matrix capable of replacing conventional methods of de-icing road systems. We investigated the heating performance of the coating according to atmospheric temperature and its potential applicability through field experiments. We began by examining the electrical properties of the CNT/EP coatings in relation to varying CNT concentrations. Following that, we assessed the heating efficiency of the CNT/EP coating at room temperature, utilizing the Joule effect. To explore the impact of atmospheric temperatures, we measured the heating temperature of the CNT/EP coating under below-freezing conditions. Furthermore, we evaluated the coating's applicability by observing its performance during winter.

2. Experimental

2.1. Materials

In this study, multi-walled carbon nanotubes were adopted from Nanolab, Inc. (Waltham, MA, USA). The CNTs have a purity of higher than 85 wt.% (industrial grade), a diameter of 15 nm, and a length of 5–20 μm. Epoxy was sourced from Easy Composites Ltd. (EpoxAcast 690, Staffordshire, UK), exhibiting a density ranging between 1.12–1.18 g/cm^3 and a viscosity of 200–450 mPa·s. The dispersant used acetone with a purity of 99.7% from Samchun Pure Chemical Co., Ltd. (Pyeongtaek-si, Gyeonggi-do, Republic of Korea).

2.2. Fabrication Procedure

The CNT/EP coating was fabricated following the procedure shown in Figure 1 [35]. A mixture was created by adding 50 g of acetone and 20 g of epoxy resin to a 200 mL beaker, then manually stirring with a stick. Then, varying concentrations of CNTs (0–5 wt.%) were introduced into the beaker and mixed in the same manner (i). An ultrasonicator (Q700CA, Qsonica LLC, Newtown, CT, USA) was then employed to effectively distribute the CNTs within the solution. For this experiment, the ultrasonicator was operated in pulse mode at 90% amplitude for a duration of 30 min (ii). To prevent the acetone from evaporating because of heat, the beaker was surrounded by ice. After dispersion, the sample was placed on a hot plate (60 °C) for 24 h to completely evaporate acetone (iii). Then, 6 g of curing agent was added to the sample and mixed evenly in a 3-roll mill (TR 50M, Trilos, San Ramon, CA, USA) (iv). Following the molding process (v), the sample was positioned within a vacuum chamber for 30 min to eliminate any air bubbles present within the sample (vi).

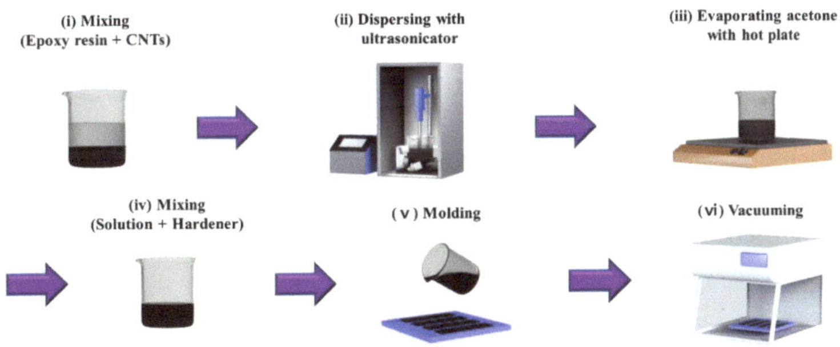

Figure 1. Fabrication procedure for CNT/EP coating.

2.3. Characterization

The heating performance due to the Joule effect is based on the electrical conductivity and the applied voltage of the application. Therefore, in our study, we have set the concentration of CNTs and the applied voltage as the main parameters. The resistance of CNT/EP coatings was gauged using a Keithley 2700 (Tektronix, Beaverton, OR, USA) for standard resistance, and a Keithley 2450 (Tektronix, Beaverton, OR, USA) for high resistance levels exceeding 10^9 Ω. Electrical resistance was assessed based on the current–voltage curves acquired by applying voltages ranging from −10 V to +10 V. The test coating samples were prepared in various CNT concentrations (0.63–5.00 wt.%) with dimensions of 50 mm × 30 mm × 2 mm. High-purity silver paint was applied to both ends of the coating samples to reduce the contact resistance between the coating and the probe point. The electrical conductivity (σ) of the specimens was calculated by $\sigma = L/RA$, where L is the length of the coating (m), R is the resistance of the coating (Ω), and A is the area of the coating (m^2) [36,37]. The microstructure of the CNT/EP coatings was examined by observing the cross-section of the sample under a scanning electron microscope (MIRA3 FE-SEMs from TESCAN, based in Brno, Czech) operating at 15 kV. The cross-section of the coating was coated with platinum using sputter coating (QUORUMQ150T S, Laughton, UK) for 10 min in preparation for a measurement with a magnification of greater than 10,000 times.

As shown in Figure 2a, a voltage-adjustable DC power supply (2260B–800–1, Tektronix, Beaverton, OR, USA) provides electrical energy which is then transformed into thermal energy via the CNT/EP coating. On both ends of the CNT/EP coatings (4-row parallel type: 100 mm × 20 mm × 5 mm), two pieces of copper tape were affixed, intended to function as electrodes, and the input voltage (5–30 V) was applied to the composite to induce heat, creating uniform heat distribution.

(**a**) (**b**)

Figure 2. Equipment and details of the heating test; (**a**) effect of atmospheric temperature test; (**b**) application test on the road in winter.

A thermal infrared camera (FLIR A655sc, Wilsonville, OR, USA) was used to record the heat distribution across the samples during Joule heating, resulting in thermal images. To investigate the effect of atmospheric temperature on the heating performance, the infrared thermal imaging camera and CNT/EP coating were set in the environmental chamber, and the temperature increments of the CNT/EP coating by applied voltage was measured with respect to the environmental temperature (−20 °C to +20 °C) as shown in Figure 2a. For the field test, CNT/EP coating (5.0 wt.%) was printed on the road. The sizes of the coating were 80 mm × 3000 mm × 5 mm or 40 mm × 3000 mm × 5 mm depending on the actual road width. To examine the heating efficiency of the coating under below-freezing conditions, a study was conducted to observe the relationship between the coating's temperature rise and the voltage applied. This experiment took place during winter with an ambient temperature of approximately −8 °C. Figure 2b presented the experimental setup for the heating performance of the CNT/EP coating application in the field test.

3. Results and Discussion

3.1. Electrical Characteristics of CNT/EP Coating

The addition of CNT in the epoxy metrics can significantly improve the electrical conductivity of the coating because of the superior electrical conductivity and high aspect ratio of the CNT [38]. The electrical conductivity of the CNT/EP coatings was measured for various CNT concentrations as shown in Figure 3a. In general, materials with electrical conductivities below about 10^{-8} S/m are insulator, and those with electrical conductivities above 10^{-8} S/m are conductive. At low concentration, lower than 0.25 wt.% of CNTs, the coatings showed non-conductivity. The electrical conductivity of the coating rapidly increased when the CNT concentration rose to between 0.25 to 1.0 wt.% based on increases in the CNT networks. This sharp increase in the electrical conductivity of CNT/EP coating is because of the formation of a percolation threshold [39–42]. In this study, the percolation threshold, which is the minimum CNT concentration in the matrix after which there is no significant change in electrical conductivity, occurred at approximately 0.63 wt.% CNTs. The effectiveness of electron transfer between CNTs is highly dependent on the CNTs spacing distance. After 1.0 wt.%, the electrical conductivity increased only gradually, showing saturation. Figure 3b shows the microstructure of the 5 wt.% coating. These distributions of the CNTs within the epoxy matrix contribute high and stable electrical conductivity of the CNT/EP coatings.

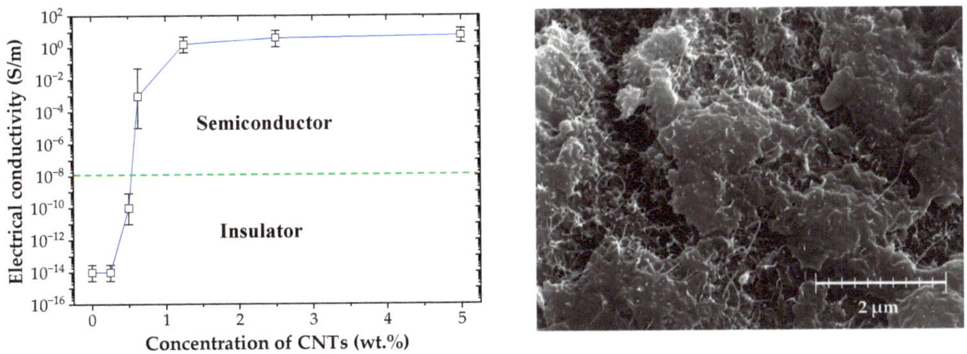

(a) (b)

Figure 3. Electrical characteristics of CNT/EP coating; (**a**) electrical conductivity; (**b**) SEM image.

3.2. Heating Characteristics of CNT/EP Coating at Room Temperature

According to Joule's law, the CNT/EP coating converts electric energy into thermal energy after loading voltage, which mainly shows rapidly increasing temperature stage of the coating [43,44]. Figure 4a–d shows the time–temperature curve of the CNT/EP coating at room temperature (20 °C). As observed, a fast temperature increase of the coatings is seen in the first five minutes. Following this, the temperature gradient in time is reduced, indicating stability of the heat transfer process. Furthermore, the higher the CNT concentration, the higher the temperature response. The results show that the coatings showed high heating performance as the CNT concentration and voltage increased. In particular, increasing the CNT concentration to 5.00 wt.% CNT/EP coating results in a heating temperature range of 20.1–184.4 °C under 5–30 V. Figure 4e showed the temperature increment and distribution with CNT concentrations at room temperature. This figure depicts all CNT concentration heating tests in which a fixed voltage of 30 V was applied. It shows an even temperature distribution of the coatings according to the applied voltage. These results suggest that CNTs are well-dispersed in the epoxy matrix [45]. Figure 4g shows the maximum temperature of the CNT/EP coating at room temperature as a function of applied voltage. The maximum temperature of the coating increases nonlinearly as the voltage and the CNT concentration increase. These results are supported by the Joule effect ($Q = I^2Rt = V^2t/R$, where Q is Joule heat, I is current, R is resistance, V is applied voltage and t is operating time). Another important aspect of the Joule heating system is the heating ratio [45,46]. Figure 4f shows the heating ratio of the coating during the first 30 s. The heating ratio increases rapidly with the increase of CNT concentration and the applied voltage. For example, the heating ratio is 0.17, 0.51, and 1.61 °C/s for coatings with 1.25, 2.50, and 5.00 wt.% coatings when applied 30 V, respectively. This high initial heating ratio of the CNT/EP coating is suitable for rapid de-icing.

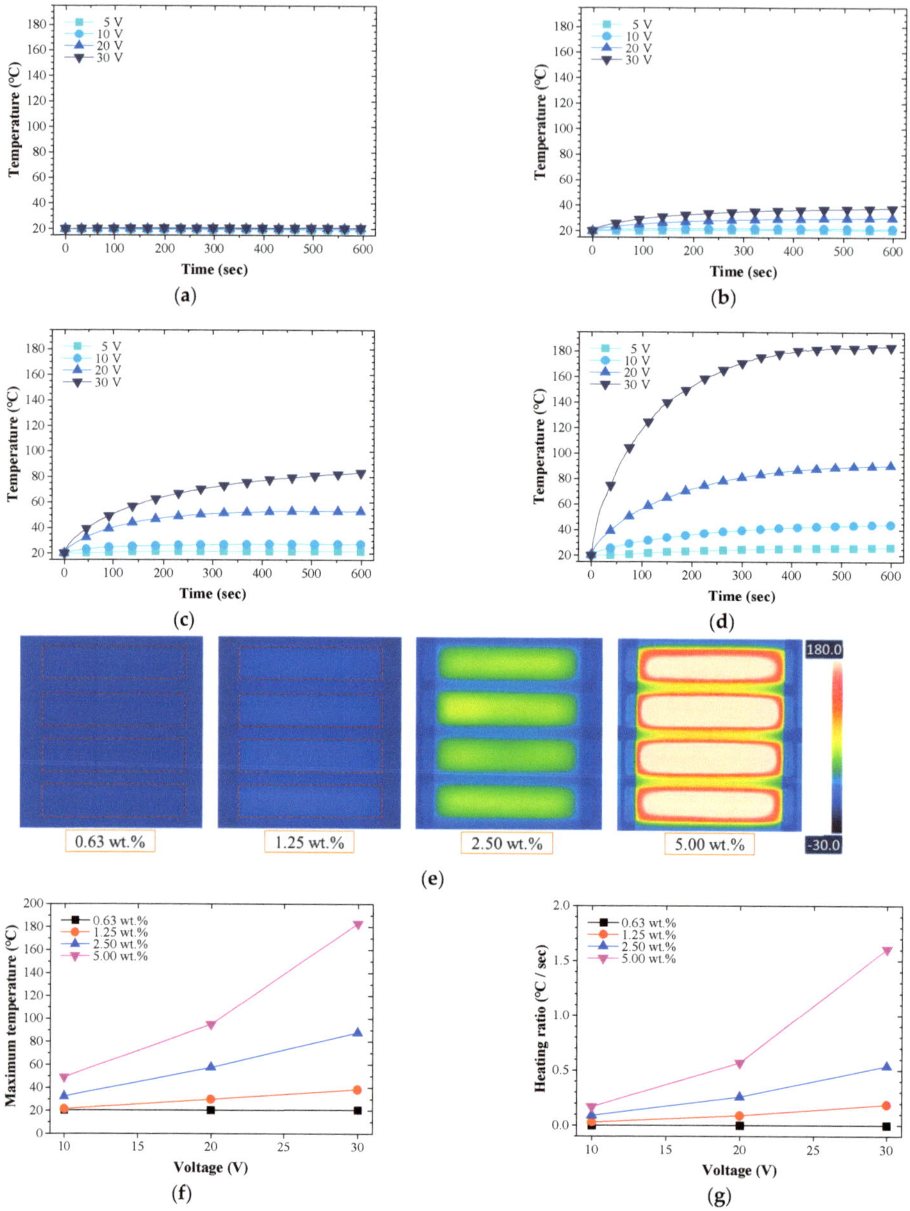

Figure 4. Heating characteristics of the CNT/EP coating at room temperature; (**a**) 0.63 wt.%; (**b**) 1.25 wt.%; (**c**) 2.50 wt.%; (**d**) 5.00 wt.%; (**e**) heating distribution; (**f**) maximum temperature; (**g**) heating ratio during the first 30 s.

To evaluate the applicability of CNT/EP coatings based on Joule heating, a deeper exploration of thermal efficiency is required, as it signifies the de-icing capability in relation to the energy consumed. Figure 5 shows the energy characteristics of CNT/EP coating according to the applied voltage and CNT concentration. The electric power is calculated by $P = V^2/R$. The relationship between the applied voltage and the power is comparable to that between the maximum temperature of the coating and the voltages shown in Figure 4f. This indicates that most of the electric power applied to the coatings was transformed into heat during the electric heating experiments [34,47]. To evaluate the energy efficiency, the power density (W·cm^{-3}) can be calculated as electric power per unit volumetric [48,49]. Figure 5a shows the temperature increments of CNT/EP coating by applied voltages as a function of power density. The temperature increase produced by the CNT/EP coating is linearly related to the power density, indicating high compliance with Joule law [50]. The slope of achieved temperature versus power density provides thermal efficiency, an important factor to evaluate the heating performance of a heater [51–53]. A larger slope value indicates that less energy is required to increase the temperature, thereby being indicative of higher thermal efficiency. Figure 5b shows the results of the thermal efficiency of the CNT/EP coating. The 1.25 wt.% coating had the highest thermal efficiency, and the thermal efficiency gradually decreased with CNT concentration. This result can be explained by the higher maximum temperature intensifying the heat exchange with the coating and its surroundings. Importantly, the thermal efficiency of the coating is quite high compared with similar systems reported in other literature [54,55]. This result shows that the CNT/EP coating has high thermal efficiency by attaining maximum temperatures with relatively low electric power.

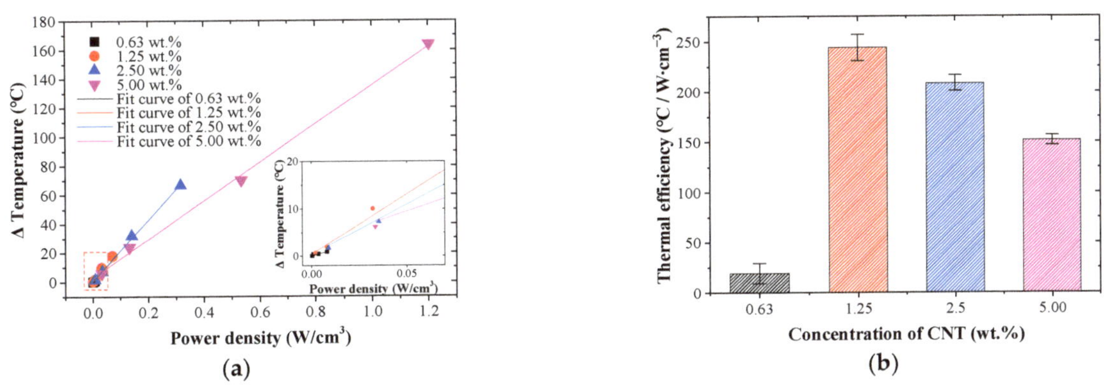

Figure 5. Energy characteristics for heating of CNT/EP coating; (**a**) temperature increase by power density; (**b**) thermal efficiency.

3.3. *Effect of Atmospheric Temperatures on Heating Performance of CNT/EP Coating*

Many studies related to Joule heating performance of CNT composites have been conducted [56–59]. However, most studies have investigated the heating performance of CNT composites at room temperature or at a fixed temperature level. Only a few studies on the effects of the atmospheric temperature have studied heating performance. In this study, the heating performance of the CNT/EP coatings was investigated at various environmental temperatures to evaluate the effect of atmospheric temperature. Figure 6a shows the temperature change on the coating (1.25–5.00 wt.%) increases with respect to the atmospheric temperature when a voltage of 30 V is applied. On the contrary, the temperature change on the coating decreased significantly as the atmospheric temperature decreased. In the case of 5.00 wt.%, the maximum temperature increment of the coating was reduced by about 29% at −20 °C compared to room temperature (20 °C). Low atmospheric temperature affected not only the maximum heat temperature but also the heating ratio. Figure 6b shows the temperature of the coatings over time when a voltage of 30 V was applied at room temperature and −20 °C. The heating ratio also significantly decreased by 31–43% according to the lower atmospheric temperature compared to the heating ratio at room temperature. Figure 6c shows the heating test results at −20 °C and revealed an even temperature distribution like the room temperature test results, but the overall temperature increment was reduced. As a result, some results did not reach the zero temperature. For example, the 1.25 wt.% coating had the highest thermal efficiency from previous results but did not reach the temperature required for de-icing of −20 °C. This result indicates coatings too low in concentration are not suitable as CNT/EP coatings in low temperature environments. The decrease in temperature is due to various influences, one of which is the creation of the frost as the atmospheric temperature decreases. When the surface temperature of the coating is below the water freezing temperature, the transferred water vapor may condense and then freeze on the cold surface [60,61]. This frost affects the lowering of Joule heating [62]. Another factor could be the resistance properties of the coating according to temperature. Figure 6d shows the result of coating resistance with decreasing atmospheric temperature. The CNT/EP coating showed a negative temperature coefficient with increasing resistance of all coatings with decreasing environmental temperature [31,63]. This increase in resistance causes a decrease in the electrical power, which can decrease in thermal efficiency. Figure 6e shows the thermal efficiency of the coating according to the atmospheric temperature. As the atmospheric temperature decreased, the thermal efficiency of all coatings decreased from about 30–37%. Although the heating performance deteriorated owing to the environmental temperature, the CNT/EP coating still showed high heating performance. These results mean that the CNT/EP coating can be applied even in extreme environments. Furthermore, to apply de-icing systems using Joule heating as well as the proposed CNT/EP coating, we must investigate the effect of atmospheric temperature on the heating performance.

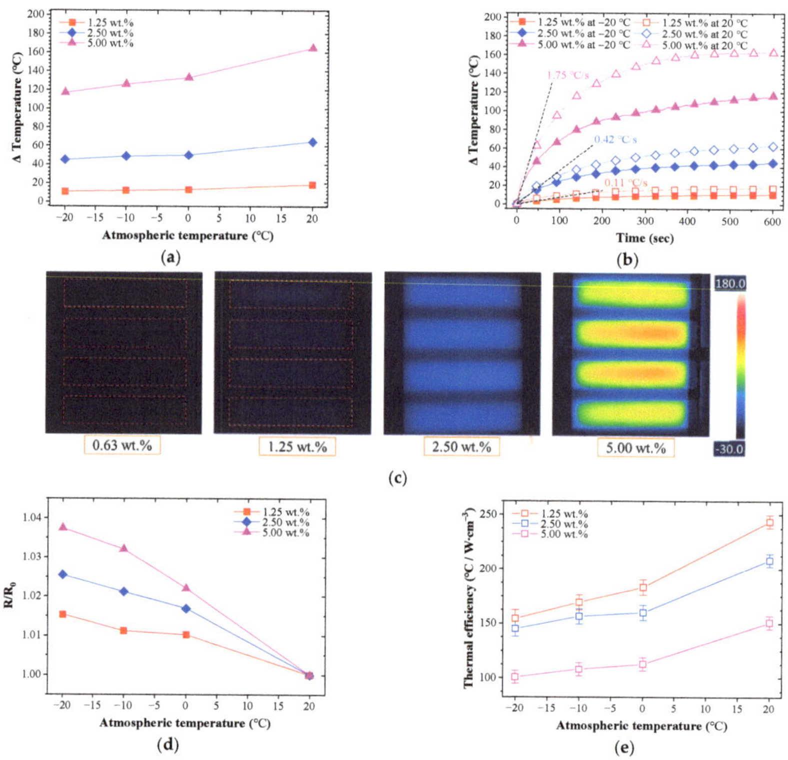

Figure 6. Effect of the atmospheric temperature on CNT/EP coating; (**a**) maximum temperature increments by 30 V; (**b**) comparison of heating performance for −20 °C and room temperature; (**c**) infrared image at −20 °C; (**d**) normalized resistance; (**e**) thermal efficiency.

3.4. Application of Road Heating System

To demonstrate the feasibility of the proposed method, we placed CNT/EP coating on the road pavements as shown in Figure 7a. The 5.00 wt.% coating was used as the highest heating value. After installation, heating performance of CNT/EP coating was measured by 220 V at −8 °C air temperature. This is to compare the heating performance by applying power like the experiment in which 30 V was applied to the 5.0 wt.% concentration of the laboratory-scale sample (about 50 W). Figure 7b showed the temperature increments of CNT/EP coating by applied voltage for 20 min. Figure 7c,d shows the thermal images of the CNT/EP coating at times t_0 and t_1, respectively. The resistance of EC01 and EC02 was 960 Ω and 1960 Ω at 20 °C, respectively. The resistance of EC02 is about twice that of EC01. The experiment was carried out during an evening in winter 2021. As a result of the heating test, the surface temperature of the coating increased above 10 °C from about −5 °C. After the voltage was applied, it heated up rapidly for 200 s, and the EC01 sample showed a higher temperature than EC02. It should show a similar heating pattern for the same applied voltage, but the EC01 sample showed a higher and more even temperature. This is because the dispersion of the CNTs in EC01 sample was more evenly distributed than that of the EC02 sample which can be confirmed by comparing the heating distribution of the coating as shown in Figure 7d. As the coating size increases, the thermal efficiency is lower than that of laboratory-scale samples, but it is possible to heat above zero depending on the applied voltage from sub-zero temperature. In addition, it is possible to increase power efficiency by optimizing the coating standard, and, evidenced through these studies, can be used as a road heating coating. Figure 8 shows a concept of the CNT/EP coating system for road de-icing. It comprises a heating coating that removes icing and a control system that applies voltage when icing occurs. Through this system, it is possible to achieve active, efficient, and eco-friendly ice removal unlike conventional methods such as spraying chemicals.

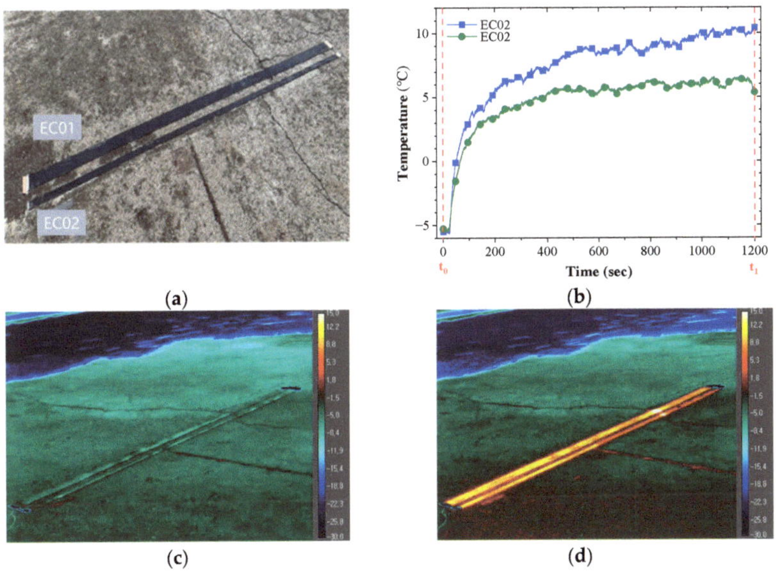

Figure 7. Application for CNT/EP coating on the road; (**a**) painting the coating on the road; (**b**) temperature increments; (**c**) surface temperature at t_0; (**d**) surface temperature at t_1.

Figure 8. A concept of the CNT/EP coating system for road de-icing.

4. Conclusions

The aim of this study was to examine the relationship between the concentration of CNTs and supply voltage with respect to the heating and electrical performance of CNT/EP coating, and to assess the impact of atmospheric temperature on this relationship. The goal was to provide an alternative solution to the issues caused by chemical de-icing agents in road systems. By dispersing CNTs in an epoxy matrix, we have successfully fabricated highly electrically conductive coatings. Measurement of the variation of electrical conductivity with CNT concentration showed that the percolation threshold occurred when CNT concentration was between 0.5 wt.% and 1.0 wt.%. To evaluate the applicability of CNT/EP coatings based on Joule heating, thermal efficiency was investigated at room temperature, as it signifies the de-icing capability in relation to the energy consumed. Results show that the CNT/EP coating has high thermal efficiency by attaining maximum temperatures with relatively low electric power. As a result of the heating test of the CNT/EP coating according to atmospheric temperature, it was found that a decrease in atmospheric temperature reduced the heating performance of the coating. Nevertheless, it was confirmed that the heating performance of the coating remained above freezing even under sub-zero temperatures. Finally, application tests confirmed that the de-icing coating developed and verified in this study present good heating performance and applicability. As a result of the application test at sub-zero temperature, it was proved that the coating can reach the de-icing temperature and it is possible to respond according to the installation environment. Through these results, it was confirmed that the developed coating could be used not only to prevent ice on the road system as an eco-friendly coating but also may be applied to structures in various fields.

Author Contributions: Conceptualization, S.-H.J.; methodology, S.-H.J.; formal analysis, S.-H.J.; investigation, S.-J.L. and S.-H.J.; writing—original draft preparation, S.-J.L.; writing—review and editing, C.C. and S.-H.J.; visualization, S.-J.L. and Y.-J.J.; supervision, C.C. and S.-H.J.; project administration, S.-H.J.; funding acquisition, S.-H.J. All authors have read and agreed to the published version of the manuscript.

Funding: This work was supported by the research fund of Hanyang University (HY–202200000002564) and the National Research Foundation of Korea (NRF) grant funded by the Korea government (MSIT) (NRF–2020R1C1C1005273).

Institutional Review Board Statement: Not applicable.

Informed Consent Statement: Not applicable.

Data Availability Statement: Not applicable.

Conflicts of Interest: The authors declare no conflict of interest.

References

1. Vitaliano, D.F. An economic assessment of the social costs of highway salting and the efficiency of substituting a new deicing material. *J. Policy Anal. Manag.* **1992**, *11*, 397–418. [CrossRef]
2. Yu, W.; Yi, X.; Guo, M.; Chen, L. State of the art and practice of pavement anti-icing and de-icing techniques. *Sci. Cold Arid Reg.* **2014**, *6*, 14–21.
3. Arabzadeh, A.; Ceylan, H.; Kim, S.; Gopalakrishnan, K.; Sassani, A. Super-hydrophobic coatings on asphalt concrete surfaces. *Transport. Res. Rec. J. Transport. Res. Board.* **2016**, *2551*, 10–17. [CrossRef]
4. Arabzadeh, A.; Ceylan, H.; Kim, S.; Gopalakrishnan, K.; Sassani, A. Fabrication of polytetrafluoroethylene-coated asphalt concrete biomimetic surfaces: A nanomaterials based pavement winter maintenance approach. In Proceedings of the 2016 International Conference on Transportation and Development, Houston, TX, USA, 26–29 June 2016; pp. 54–64.
5. Lai, Y.; Liu, Y.; Ma, D. Automatically melting snow on airport cement concrete pavement with carbon fiber grille. *Cold Reg. Sci. Technol.* **2014**, *103*, 57–62. [CrossRef]
6. Ramakrishna, D.M.; Viraraghavan, T. Environmental impact of chemical deicers—A review. *Water Air Soil Pollut.* **2005**, *166*, 49–63. [CrossRef]
7. Christopher, A.; Strong, J.E.; Mosher, P.A. Effect of deicing salts on metal and organic matter mobilization in roadside soils. *Environ. Sci. Technol.* **1992**, *26*, 703–709.
8. Bäckström, M.; Karlsson, S.; Bäckman, L.; Folkeson, L.; Lind, B. Mobilisation of heavy metals by deicing salts in a roadside environment. *Water Res.* **2004**, *38*, 720–732. [CrossRef]
9. Sanzo, D.; Hecnar, S.J. Effects of road de-icing salt (NaCl) on larval wood frogs (*Rana sylvatica*). *Environ. Pollut.* **2006**, *140*, 247–256. [CrossRef]
10. Czerniawska-Kusza, I.; Kusza, G.; Duzynski, M. Effect of deicing salts on urban soils and health status of roadside trees in the Opole Region. *Environ. Toxicol.* **2004**, *19*, 296–301. [CrossRef]
11. Cunningham, M.A.; Snyder, E.; Yonkin, D.; Ross, M.; Elsen, T. Accumulation of deicing salts in soils in an urban environment. *Urban Ecosyst.* **2008**, *11*, 17–31. [CrossRef]
12. Karraker, N.E.; Gibbs, J.P.; Vonesh, J.R. Impacts of road deicing salt on the demography of vernal pool-breeding Amphibians. *Ecol. Appl.* **2016**, *18*, 724–734. [CrossRef] [PubMed]
13. Shen, W.; Ceylan, H.; Gopalakrishnan, K.; Kim, S.; Taylor, P.C.; Rehmann, C.R. Life cycle assessment of heated apron pavement system operations. *Transport. Res. Road Environ.* **2016**, *48*, 316–331. [CrossRef]
14. Wang, H.; Thakkar, C.; Chen, X.; Murrel, S. Life-cycle assessment of airport pavement design alternatives for energy and environmental impacts. *J. Clean. Prod.* **2016**, *133*, 163–171. [CrossRef]
15. Xu, H.; Tan, Y. Modeling and operation strategy of pavement snow melting systems utilizing low-temperature heating fluids. *Energy* **2015**, *80*, 666–676. [CrossRef]
16. Chang, C.; Ho, M.; Song, G.; Mo, Y.-L.; Li, H. A feasibility study of self-heating concrete utilizing carbon nanofiber heating elements. *Smart Mater. Struct.* **2009**, *18*, 127001. [CrossRef]
17. Farcas, C.; Galao, O.; Navarro, R.; Zornoza, E.; Baeza, F.J.; Del Moral, B.; Pla, R.; Garcés, P. Heating and de-icing function in conductive concrete and cement paste with the hybrid addition of carbon nanotubes and graphite products. *Smart Mater. Struct.* **2021**, *30*, 045010. [CrossRef]
18. Gomis, J.; Galao, O.; Gomis, V.; Zornoza, E.; Garces, P. Self-heating and deicing conductive cement. Experimental study and modeling. *Constr. Build. Mater.* **2015**, *75*, 442–449. [CrossRef]
19. Pan, P.; Wu, S.; Xiao, Y.; Liu, G. A review on hydronic asphalt pavement for energy harvesting and snow melting. *Renew. Sustain. Energy Rev.* **2015**, *48*, 624–634. [CrossRef]
20. Tuan, C.Y. *Implementation of Conductive Concrete for Deicing (Roca Bridge)*; University of Nebraska: Lincoln, NE, USA, 2008.
21. Wang, H.; Liu, L.; Chen, Z. Experimental investigation of hydronic snow melting process on the inclined pavement. *Cold Reg. Sci. Technol.* **2010**, *63*, 44–49. [CrossRef]
22. Xu, H.; Wang, D.; Tan, Y.; Zhou, J.; Oeser, M. Investigation of design alternatives for hydronic snow melting pavement systems in China. *J. Clean. Prod.* **2018**, *170*, 1413–1422. [CrossRef]
23. Han, B.G.; Yu, X.; Kwon, E. A self-sensing carbon nanotube/cement composite for traffic monitoring. *Nanotechnology* **2009**, *20*, 44. [CrossRef] [PubMed]
24. Maurin, L.; Boussoir, J.; Rougeault, S.; Bugaud, M.; Ferdinand, P.; Landrot, A.G.; Grunevald, Y.H.; Chauvin, T. FBG-based smart composite bogies for railway applications. In Proceedings of the 2002 15th Optical Fiber Sensors Conference Technical Digest. OFS 2002 (Cat. No. 02EX533), Portland, OR, USA, 10 May 2002; pp. 91–94.
25. Pujar, N.V.; Nanjundaradhya, N.V.; Sharma, R.S. Effect of graphene oxide nano filler on dynamic behaviour of GFRP composites. In Proceedings of the AIP Conference Proceedings, Lausanne, Switzerland, 19–21 March 2018; Volume 1943.
26. Fouda, H.; Guo, L. The mechanical properties of CF/Epoxy resin composite with adding different types of CNTs. *Int. J. Eng. Res. Technol.* **2017**, *6*, 311–315.
27. Mostovoy, A.; Yakovlev, A.; Tseluikin, V.; Lopukhova, M. Epoxy nanocomposites reinforced with functionalized carbon nanotubes. *Polymers* **2020**, *12*, 1816. [CrossRef] [PubMed]
28. Isaji, S.; Bin, Y.; Matsuo, M. Electrical conductivity and self-temperature-control heating properties of carbon nanotubes filled polyethylene films. *Polymer* **2009**, *50*, 1046–1053. [CrossRef]

29. Kumar, S.; Pimparkar, N.; Murthy, J.Y.; Alam, M.A. Self-consistent electrothermal analysis of nanotube network transistors. *J. Appl. Phys.* **2011**, *109*, 014315. [CrossRef]
30. Chien, A.T.; Cho, S.; Joshi, Y.; Kumar, S. Electrical conductivity and Joule heating of polyacrylonitrile/carbon nanotube composite fibers. *Polymer* **2014**, *55*, 6896–6905. [CrossRef]
31. Jang, S.H.; Park, Y.L. Carbon nanotube-reinforced smart composites for sensing freezing temperature and deicing by CNT/EP. *Nanomater. Nanotechnol.* **2018**, *8*, 1847980418776473. [CrossRef]
32. Yum, S.G.; Yin, H.; Jang, S.H. Toward multi-functional road surface design with the nanocomposite coating of carbon nanotube modified polyurethane: Lab-scale experiments. *Nanomaterials* **2020**, *10*, 1905. [CrossRef]
33. Prolongo, S.G.; Moriche, R.; Del Rosario, G.; Jiménez-Suárez, A.; Prolongo, M.G.; Ureña, A. Joule effect self-heating of epoxy composites reinforced with graphitic nanofillers. *J. Polym. Res.* **2016**, *23*, 189. [CrossRef]
34. Redondo, O.; Prolongo, S.G.; Campo, M.; Sbarufatti, C.; Giglio, M. Anti-icing and de-icing coatings based Joule's heating of graphene nanoplatelets. *Compos. Sci. Technol.* **2018**, *164*, 65–73. [CrossRef]
35. Lee, S.J.; Jung, Y.J.; Park, J.W.; Jang, S.H. Temperature Detectable Surface Coating with Carbon Nanotube/Epoxy Composites. *Nanomaterials* **2022**, *12*, 2369. [CrossRef] [PubMed]
36. Fang, Y.; Li, L.Y.; Jang, S.H. Calculation of electrical conductivity of self-sensing carbon nanotube composites. *Compos. B Eng.* **2020**, *199*, 108314. [CrossRef]
37. Jang, S.H.; Li, L.Y. Self-sensing carbon nanotube composites exposed to glass transition temperature. *Materials* **2020**, *13*, 259. [CrossRef] [PubMed]
38. Jang, S.H.; Kim, D.; Park, Y.L. Accelerated curing and enhanced material properties of conductive polymer nanocomposites by joule heating. *Materials* **2018**, *11*, 1775. [CrossRef]
39. Kim, J.K.; Mai, Y.W. (Eds.) *Engineered Interfaces in Fiber Reinforced Composites;* Elsevier: Amsterdam, The Netherlands, 1998.
40. Vertuccio, L.; Guadagno, L.; Spinelli, G.; Lamberti, P.; Zarrelli, M.; Russo, S.; Iannuzzo, G. Smart coatings of epoxy based CNTs designed to meet practical expectations in aeronautics. *Compos. Part B Eng.* **2018**, *147*, 42–46. [CrossRef]
41. Xin, X.; Liang, M.; Yao, Z.; Su, L.; Zhang, J.; Li, P.; Sun, C.; Jiang, H. Self-sensing behavior and mechanical properties of carbon nanotubes/epoxy resin composite for asphalt pavement strain monitoring. *Constr. Build. Mater.* **2020**, *257*, 119404. [CrossRef]
42. Cataldo, A.; Biagetti, G.; Mencarelli, D.; Micciulla, F.; Crippa, P.; Turchetti, C.; Pierantoni, L.; Bellucci, S. Modeling and electrochemical characterization of electrodes based on epoxy composite with functionalized nanocarbon fillers at high concentration. *Nanomaterials* **2020**, *10*, 850. [CrossRef]
43. Cortés, A.; Romate, X.F.S.; Jiménez-Suárez, A.; Campo, M.; Prolongo, M.G.; Ureña, A.; Prolongo, S.G. 3D printed anti-icing and de-icing system based on CNT/GNP doped epoxy composites with self-curing and structural health monitoring capabilities. *Smart Mater. Struct.* **2020**, *30*, 025016. [CrossRef]
44. Liang, S.; Wang, H.; Tao, X. Multiwalled carbon nanotube/cationic cellulose nanofibril electrothermal films: Mechanical, electrical, electrothermal, and cycling performances. *Wood Sci. Technol.* **2021**, *55*, 1711–1723. [CrossRef]
45. Farcas, C.; Galao, O.; Vertuccio, L.; Guadagno, L.; Romero-Sánchez, M.D.; Rodríguez-Pastor, I.; Garcés, P. Ice-prevention and de-icing capacity of epoxy resin filled with hybrid carbon-nanostructured forms: CNT/EP by Joule effect. *Nanomaterials* **2021**, *11*, 2427. [CrossRef]
46. Lee, T.W.; Jeong, Y.G. Regenerated cellulose/multiwalled carbon nanotube composite films with efficient electric heating performance. *Carbohydr. Polym.* **2015**, *133*, 456–463. [CrossRef]
47. An, J.E.; Jeong, Y.G. Structure and electric heating performance of graphene/epoxy composite films. *Eur. Polym. J.* **2013**, *49*, 1322–1330. [CrossRef]
48. Lee, H.B.; Veerasubramani, G.K.; Lee, K.S.; Lee, H.; Han, T.H. Joule heating-induced faradaic electrode-decorated graphene fibers for flexible fiber-shaped hybrid supercapacitor with high volumetric energy density. *Carbon* **2022**, *198*, 252–263. [CrossRef]
49. Wang, B.; Wang, G.F.; Jiang, S.S.; Zhang, K.F. Effect of pulse current on thermal performance and deep drawing of SiCp/2024Al composite sheet. *Int. J. Adv. Manuf. Technol.* **2013**, *67*, 623–627. [CrossRef]
50. Zhou, B.; Han, X.; Li, L.; Feng, Y.; Fang, T.; Zheng, G.; Wang, B.; Dai, K.; Liu, C.; Shen, C. Ultrathin, flexible transparent Joule heater with fast response time based on single-walled carbon nanotubes/poly(vinyl alcohol) film. *Compos. Sci. Technol.* **2019**, *183*, 107796. [CrossRef]
51. Zhang, X.; Wang, X.; Lei, Z.; Wang, L.; Tian, M.; Zhu, S.; Xiao, H.; Tang, X.; Qu, L. Flexible MXene-decorated fabric with interwoven conductive networks for integrated Joule heating, electromagnetic interference shielding, and strain sensing performances. *ACS Appl. Mater. Interfaces* **2020**, *12*, 14459–14467. [CrossRef]
52. Zhou, B.; Li, Z.; Li, Y.; Liu, X.; Ma, J.; Feng, Y.; Zhang, D.; He, C.; Liu, C.; Shen, C. Flexible hydrophobic 2D Ti3C2Tx-based transparent conductive film with multifunctional self-cleaning, electromagnetic interference shielding and joule heating capacities. *Compos. Sci. Technol.* **2021**, *201*, 108531. [CrossRef]
53. Sun, Y.; Ding, R.; Hong, S.Y.; Lee, J.; Seo, Y.K.; Nam, J.D.; Suhr, J. MXene-xanthan nanocomposite films with layered microstructure for electromagnetic interference shielding and Joule heating. *Chem. Eng. J.* **2021**, *410*, 128348. [CrossRef]
54. Jeong, Y.G.; An, J.E. Effects of mixed carbon filler composition on electric heating behavior of thermally-cured epoxy-based composite films. *Compos. Part A Appl. Sci. Manuf.* **2014**, *56*, 1–7. [CrossRef]

55. Wang, F.X.; Liang, W.Y.; Wang, Z.Q.; Yang, B.; He, L.; Zhang, K. Preparation and property investigation of multi-walled carbon nanotube (MWCNT)/epoxy composite films as high-performance electric heating (resistive heating) element. *Express Polym. Lett.* **2018**, *12*, 285–295. [CrossRef]
56. Tang, P.; Zhang, R.; Shi, R.; Bin, Y. Synergetic effects of carbon nanotubes and carbon fibers on electrical and self-heating properties of high-density polyethylene composites. *J. Mater.* **2015**, *50*, 1565–1574. [CrossRef]
57. Prolongo, S.G.; Moriche, R.; Jiménez-Suárez, A.; Delgado, A.; Ureña, A. Printable self-heating coatings based on the use of carbon nanoreinforcements. *Polym. Compos.* **2020**, *41*, 271–278. [CrossRef]
58. Chu, H.; Zhang, Z.; Liu, Y.; Leng, J. Self-heating fiber reinforced polymer composite using meso/macropore carbon nanotube paper and its application in deicing. *Carbon* **2014**, *66*, 154–163. [CrossRef]
59. Xu, C.; Xu, S.; Eticha, R.D. Experimental investigation of thermal performance for pulsating flow in a microchannel heat sink filled with PCM (paraffin/CNT composite). *Energy Convers. Manag.* **2021**, *236*, 114071. [CrossRef]
60. Brian, P.T.; Reid, R.C.; Shah, Y.T. Frost deposition on cold surfaces. *Ind. Eng. Chem. Fund.* **1970**, *9*, 375–380. [CrossRef]
61. Na, B.; Webb, R.L. A fundamental understanding of factors affecting frost nucleation. *Int. J. Heat Mass Transf.* **2003**, *46*, 3797–3808. [CrossRef]
62. Liebscher, M.; Tzounis, L.; Junger, D.; Dinh, T.T.; Mechtcherine, V. Electrical Joule heating of cementitious nanocomposites filled with multi-walled carbon nanotubes: Role of filler concentration, water content, and cement age. *Smart Mater. Struct.* **2020**, *29*, 125019. [CrossRef]
63. Li, Q.; Xue, Q.Z.; Gao, X.L.; Zheng, Q.B. Temperature dependence of the electrical properties of the carbon nanotube/polymer composites. *Express Polym. Lett.* **2009**, *3*, 769–777. [CrossRef]

nanomaterials

MDPI

Article

A Comparison of Shell Theories for Vibration Analysis of Single-Walled Carbon Nanotubes Based on an Anisotropic Elastic Shell Model

Matteo Strozzi [1], Isaac E. Elishakoff [2,*], Michele Bochicchio [1], Marco Cocconcelli [1], Riccardo Rubini [1] and Enrico Radi [1]

[1] Department of Sciences and Methods for Engineering, University of Modena and Reggio Emilia, 42122 Reggio Emilia, Italy; matteo.strozzi@unimore.it (M.S.); marco.cocconcelli@unimore.it (M.C.); riccardo.rubini@unimore.it (R.R.); enrico.radi@unimore.it (E.R.)
[2] Department of Ocean and Mechanical Engineering, Florida Atlantic University, Boca Raton, FL 33431, USA
* Correspondence: elishako@fau.edu

Abstract: In the present paper, a comparison is conducted between three classical shell theories as applied to the linear vibrations of single-walled carbon nanotubes (SWCNTs); specifically, the evaluation of the natural frequencies is conducted via Donnell, Sanders, and Flügge shell theories. The actual discrete SWCNT is modelled by means of a continuous homogeneous cylindrical shell considering equivalent thickness and surface density. In order to take into account the intrinsic chirality of carbon nanotubes (CNTs), a molecular based anisotropic elastic shell model is considered. Simply supported boundary conditions are imposed and a complex method is applied to solve the equations of motion and to obtain the natural frequencies. Comparisons with the results of molecular dynamics simulations available in literature are performed to check the accuracy of the three different shell theories, where the Flügge shell theory is found to be the most accurate. Then, a parametric analysis evaluating the effect of diameter, aspect ratio, and number of waves along the longitudinal and circumferential directions on the natural frequencies of SWCNTs is performed in the framework of the three different shell theories. Assuming the results of the Flügge shell theory as reference, it is obtained that the Donnell shell theory is not accurate for relatively low longitudinal and circumferential wavenumbers, for relatively low diameters, and for relatively high aspect ratios. On the other hand, it is found that the Sanders shell theory is very accurate for all the considered geometries and wavenumbers, and therefore, it can be correctly adopted instead of the more complex Flügge shell theory for the vibration modelling of SWCNTs.

Keywords: carbon nanotubes; vibration analysis; anisotropic elastic model; shell theories; natural frequencies

Citation: Strozzi, M.; Elishakoff, I.E.; Bochicchio, M.; Cocconcelli, M.; Rubini, R.; Radi, E. A Comparison of Shell Theories for Vibration Analysis of Single-Walled Carbon Nanotubes Based on an Anisotropic Elastic Shell Model. *Nanomaterials* **2023**, *13*, 1390. https://doi.org/10.3390/nano13081390

Academic Editor: Muralidharan Paramsothy

Received: 21 March 2023
Revised: 11 April 2023
Accepted: 11 April 2023
Published: 17 April 2023

1. Introduction

Since their discovery in 1991 in Japan in the laboratories of the NEC Corporation by Professor Sumio Iijima [1], the study of the vibrations of carbon nanotubes has represented a very demanding challenge for many researchers all over the world.

This interest in carbon nanotubes is due to their extraordinary mechanical properties, in particular the very high elastic modulus and tensile strength, together with their very small diameter, which allows them to reach natural frequencies of the THz order, and therefore leads them to be applied in several high sensitivity electro-mechanical systems, such as resonators, sensors, and oscillators [2–6].

In order to study the vibratory behaviour of carbon nanotubes, three different methods have been proposed: experimental analyses, molecular dynamics simulations, and continuous models.

The experimental analyses, conducted on the basis of resonant Raman spectroscopy, allow to obtain the natural frequencies only of peculiar modes of carbon nanotubes, the so-called "radial breathing modes", which are characterized by zero waves both longitudinal and circumferential, as undeformed axisymmetric modes [7–9]. In addition to this, the main limitation of the experimental analyses is due to their very high technical difficulty and the need to work with scanning or transmission electron microscopes with very high resolution (and therefore very high cost).

Molecular dynamics simulations take into account the discrete nature of carbon nanotubes by modelling bonds and interaction forces between the different carbon atoms based on the fundamental concepts of molecular mechanics [10]. These analyses allow to obtain the natural frequencies of both radial breathing and beam-like modes, where the latter are very important for carbon nanotubes, which have high aspect ratio (i.e., the ratio between length and radius) and, therefore, have a vibratory behaviour very similar to beam structures [11,12]. The main limitation of molecular dynamics simulations results from the high computational effort needed to carry out numerical analyses, especially in the presence of a large number of carbon atoms, a condition that usually occurs in carbon nanotubes, due to the fact that they are long and of reduced thickness (i.e., the ratio between thickness and radius), which leads them to be considered "thin-walled structures".

As an alternative to the experimental analyses and numerical simulations, researchers have proposed several continuous models, mainly of beam-type or shell-type, to study carbon nanotube vibrations via continuum mechanics [13].

In particular, the continuous beam-type models are able to properly simulate the torsional vibrations of nanotubes [14], whereas the continuous shell-type models are able to also properly simulate the flexural vibrations of nanotubes, which give the highest natural frequencies; they therefore represent more complete models [15].

In general, the main problem in the continuous modelling of carbon nanotubes is the choice of the equivalent parameters that allow to study the discrete nanotubes as continuous structures.

For the shell models, which as previously mentioned are the most accurate ones, Yakobson [16], starting from results of molecular dynamics simulations, proposed equivalent values of tensile and flexural stiffness, and therefore thickness and surface density, thus allowing the study of carbon nanotubes as continuous isotropic cylindrical shells.

Considering an isotropic elastic shell model and using these equivalent parameters, several papers have been published concerning the study of the natural frequencies of single- or multi-walled carbon nanotubes in the framework of thin shell theories; these principally include Donnell-Mushtari, Sanders-Koiter, and Lur'ye-Flügge-Byrne [17–20].

Readers interested in deepening the peculiarities of these thin shell theories can find the related strain–displacement relationships and equations of motion in the fundamental monographs by Leissa [21], Yamaki [22], Amabili [23], Soedel [24], and Ventsel [25].

An interesting comparison of thin shell theories for vibrations of circular cylindrical shells was carried out by Amabili [26]. It was derived from this work that, among the Donnell, Flügge, and Sanders shell theories, the Donnell shell theory, which has the lowest analytical complexity (i.e., the lowest number of terms in the equations of motion), presents the lowest accuracy in modelling shell vibrations. On the other hand, it was found that the Flügge shell theory, which has the highest analytical complexity, presents the highest accuracy. Finally, the Sanders shell theory, which has an intermediate analytical complexity, presents a relatively high accuracy. Moreover, an examination of cylindrical shell theories, specifically Donnell and Sanders shell theories, for the buckling of carbon nanotubes was performed by Wang et al. [27].

The same continuous models adopted to predict CNT vibrations at the nanoscopic scale can be used at the macroscopic scale to investigate the vibrations of different configurations of circular cylindrical shells, e.g., FGM shells [28–30], multi-layer composite shells [31,32], sandwich composite shells [33], and laminated orthotropic shells [34], which are considered in several mechanical components. It is important to stress that, in the passage from the nanoscopic to macroscopic scale, the continuous models lose the anisotropic behaviour due to the inherent chirality of carbon nanotubes.

However, given the intrinsically anisotropic nature of carbon nanotubes, in order to correctly study their vibratory behaviour and also to take into account the dependence of their elastic properties on chirality, it is necessary to adopt an anisotropic model.

To this aim, Chang [35,36], starting from considerations of molecular mechanics, developed a novel and very accurate anisotropic elastic shell model capable of correctly predicting the dependence of the material elastic properties on CNT chirality and dimensions, and therefore, also able to calculate the natural frequencies very similar to those obtained via molecular dynamics simulations.

Adopting this anisotropic elastic model, the linear vibrations of single-walled and multi-walled carbon nanotubes for different geometries and wave numbers were investigated separately via the Donnell [37], Sanders [38], and Flügge [39,40] thin shell theories.

To the knowledge of the authors of this paper, a study on the linear vibrations of SWCNTs based on an anisotropic elastic shell model that compares the values obtained for the natural frequencies by applying the Donnell, Sanders, and Flügge thin shell theories for different geometries and wavenumbers has not yet been published in the literature. In fact, the authors of this paper believe that it could be very useful to investigate the field of applicability and limitation of the three previously indicated thin shell theories in order to identify which of them is able to provide sufficiently accurate results with a relatively low computational effort.

To this aim, in the present paper, the natural frequencies of SWCNTs are obtained in the framework of the Donnell, Sanders, and Flügge shell theories, where the actual discrete SWCNT is modelled via a continuous homogeneous cylindrical shell considering equivalent thickness and surface density. An anisotropic elastic shell model is adopted to take into account the intrinsic chirality effects of CNTs. Simply supported boundary conditions are imposed and the complex method is used to solve the dynamic equations of motion and to obtain the natural frequencies. Vibration modes with different numbers of longitudinal and circumferential waves are studied. SWCNTs with different geometries are analysed.

Taking the more accurate but more complex Flügge theory as a reference, the main objective of this work was to establish whether the simpler Donnell or Sanders shell theories allow to obtain sufficiently accurate natural frequencies and, therefore, can be adopted instead of the Flügge theory to correctly model the linear vibrations of SWCNTS on the basis of an anisotropic elastic shell model.

2. Thin Shell Theories for SWCNTs

In the present paper, the actual discrete SWCNT of Figure 1a is modelled by means of an equivalent continuous elastic thin cylindrical shell, see Figure 1b,c with radius R, length L, and thickness h. A cylindrical coordinate system (O, x, θ, z) is adopted, where the origin O of the reference system is located at the centre of one end of the cylindrical shell. Three displacements are present: longitudinal $u(x, \theta, t)$, circumferential $v(x, \theta, t)$, and radial $w(x, \theta, t)$, where the radial displacement w is assumed as positive outward, (x, θ) are the longitudinal and angular coordinates of an arbitrary point on the middle surface of the shell, z is the radial coordinate along the thickness h, and t is the time.

In this paper, the natural frequencies obtained by considering three different thin shell theories are compared regarding the linear vibrations of SWCNTs with different geometries and wavenumbers. These theories are based on Love's first approximation assumptions [21]: (i) the thickness h of the shell is small with respect to the radius of curvature R of the middle surface; (ii) the strains are small; (iii) the transverse normal stress is small; (iv) and the normal to the undeformed middle surface remains straight and normal to the middle surface after the deformation, and undergoes no thickness stretching (Kirchhoff-Love kinematic hypothesis). The considered thin shell theories are: (a) Donnell-

Mushtari [23], (b) Sanders-Koiter [24], and (c) Flügge-Lur'ye-Byrne [25]; for all of them, both rotary inertia and shear deformations are neglected.

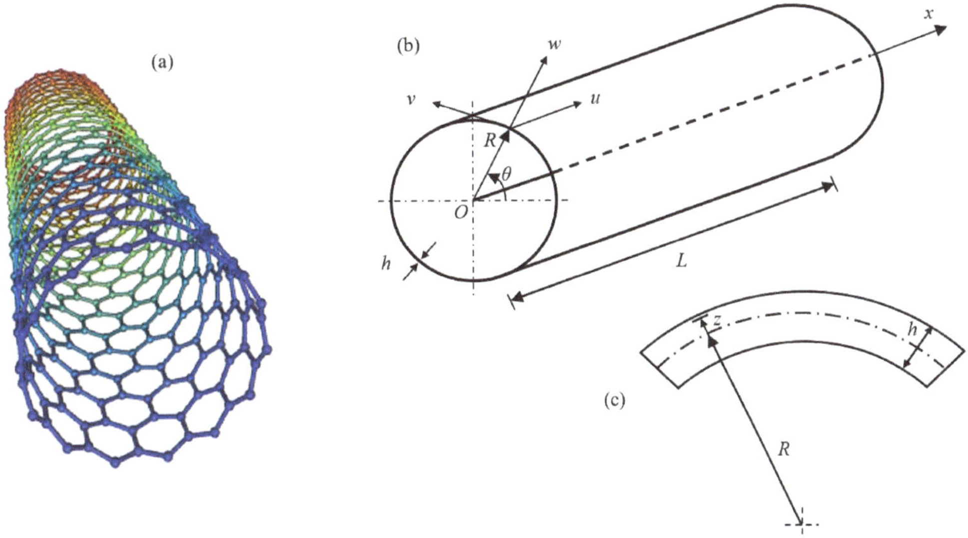

Figure 1. Continuous modelling of SWCNTs: (**a**) actual discrete SWCNT; (**b**) geometry of the equivalent continuous circular cylindrical shell; and (**c**) cross-section of the surface of the equivalent continuous circular cylindrical shell [20].

2.1. Strain–Displacement Relationships

According to the Donnell, Sanders, and Flügge shell theories, the middle surface strains $(\varepsilon_{x,0}, \varepsilon_{\theta,0}, \gamma_{x\theta,0})$ of the shell are related to the displacements (u, v, w) in the form [21]:

$$\varepsilon_{x,0} = \frac{\partial u}{\partial x} \qquad \varepsilon_{\theta,0} = \frac{1}{R}\frac{\partial v}{\partial \theta} + \frac{w}{R} \qquad \gamma_{x\theta,0} = \frac{\partial v}{\partial x} + \frac{1}{R}\frac{\partial u}{\partial \theta} \tag{1}$$

From Equation (1), it can be observed that the middle surface strains are expressed in the same form for the three different shell theories considered.

According to Donnell, Sanders, and Flügge, the middle surface changes in the curvature and torsion $(k_x, k_\theta, k_{x\theta})$ of the shell are related to the displacements (u, v, w) in the form [21]:

$$\begin{gathered} k_x = -\frac{\partial^2 w}{\partial x^2} \qquad k_\theta = \frac{1}{R^2}\left(-\frac{\partial^2 w}{\partial \theta^2} + \psi\frac{\partial v}{\partial \theta} - \varphi w\right) \\ k_{x\theta} = -\frac{2}{R}\frac{\partial^2 w}{\partial x \partial \theta} - \frac{1}{R^2}\left(\frac{1}{2}\psi + \varphi\right)\frac{\partial u}{\partial \theta} + \frac{1}{R}\left(\frac{3}{2}\psi + \varphi\right)\frac{\partial v}{\partial x} \end{gathered} \tag{2}$$

From Equation (2), it can be noted that the middle surface change in the curvature k_θ and torsion $k_{x\theta}$ are written in a different form for the three different shell theories, where parameters $(\psi = 0, \varphi = 0)$ denote the Donnell shell theory, parameters $(\psi = 1, \varphi = 0)$ denote the Sanders shell theory, and parameters $(\psi = 0, \varphi = 1)$ denote the Flügge shell theory. Since the Sanders and Flügge shell theories have more terms in the Expansion (2) than the Donnell shell theory, the first two theories can be expected to be more accurate than the third in the modelling of SWCNT linear vibrations.

2.2. Strain Components at an Arbitrary Point of the Shell Surface

According to the Donnell, Sanders and Flügge shell theories, the strain components $(\varepsilon_x, \varepsilon_\theta, \gamma_{x\theta})$ at an arbitrary point of the surface of the shell are related to the middle surface strains $(\varepsilon_{x,0}, \varepsilon_{\theta,0}, \gamma_{x\theta,0})$ and to the changes in curvature and torsion of the middle surface $(k_x, k_\theta, k_{x\theta})$ by the relationships [23]:

$$\varepsilon_x = \varepsilon_{x,0} + zk_x \qquad \varepsilon_\theta = \varepsilon_{\theta,0} + zk_\theta \qquad \gamma_{x\theta} = \gamma_{x\theta,0} + zk_{x\theta} \tag{3}$$

where z is the distance of the considered arbitrary point of the shell from the middle surface.

Substituting Equations (1) and (2) into Equation (3), the following is obtained:

$$\begin{gathered}\varepsilon_x = \frac{\partial u}{\partial x} - z\frac{\partial^2 w}{\partial x^2} \qquad \varepsilon_\theta = \frac{1}{R}\frac{\partial v}{\partial \theta} + \frac{w}{R} + \frac{z}{R^2}\left(-\frac{\partial^2 w}{\partial \theta^2} + \psi\frac{\partial v}{\partial \theta} - \varphi w\right) \\ \gamma_{x\theta} = \frac{\partial v}{\partial x} + \frac{1}{R}\frac{\partial u}{\partial \theta} - \frac{2z}{R}\frac{\partial^2 w}{\partial x \partial \theta} - \frac{z}{R^2}\left(\frac{1}{2}\psi + \varphi\right)\frac{\partial u}{\partial \theta} + \frac{z}{R}\left(\frac{3}{2}\psi + \varphi\right)\frac{\partial v}{\partial x}\end{gathered} \tag{4}$$

where Equation (4) relate the strain components at an arbitrary point of the shell surface $(\varepsilon_x, \varepsilon_\theta, \gamma_{x\theta})$ to the displacements (u, v, w).

3. Anisotropic Elastic Shell Model

Considering the molecular based anisotropic elastic shell model developed by Chang [35,36], which includes the chirality effects characteristic of SWCNTs, the stress–strain relationships can be written as:

$$\begin{aligned}\sigma_x &= \tfrac{1}{h}(Y_{11}\varepsilon_x + Y_{12}\varepsilon_\theta + Y_{13}\gamma_{x\theta}) \\ \sigma_\theta &= \tfrac{1}{h}(Y_{21}\varepsilon_x + Y_{22}\varepsilon_\theta + Y_{23}\gamma_{x\theta}) \\ \tau_{x\theta} &= \tfrac{1}{h}(Y_{31}\varepsilon_x + Y_{32}\varepsilon_\theta + Y_{33}\gamma_{x\theta})\end{aligned} \tag{5}$$

where $(\sigma_x, \sigma_\theta, \tau_{x\theta})$ are the stress components at an arbitrary point of the shell surface, and Y_{ij} are the anisotropic surface elastic constants of an arbitrary SWCNT, which are defined as [35,36]:

$$Y_{ij} = \frac{2}{3\sqrt{3}}\left(K_\rho G_{li}G_{lj} + 2\frac{K_\theta}{a^2}H_{li}H_{lj}\right),\ i,j,l = 1,2,3(\text{sum over } l) \tag{6}$$

in which a is the carbon–carbon bond length; (K_ρ, K_θ) are force constants associated with stretching and angular distortion of the carbon-carbon bond, respectively, where these constants can be obtained from quantum (ab initio) mechanics, empirical molecular potential, or fitted to experimental data; and $\left(G_{li}, G_{lj}, H_{li}, H_{lj}\right)$ are elements of matrices $\boldsymbol{G}$ and $\boldsymbol{H}$, respectively, which are given in detail in Ref. [36].

$$\begin{aligned}\sigma_x = &\tfrac{1}{h}\left\{Y_{11}\left(\tfrac{\partial u}{\partial x} - z\tfrac{\partial^2 w}{\partial x^2}\right) + Y_{12}\left[\tfrac{1}{R}\tfrac{\partial v}{\partial \theta} + \tfrac{w}{R} + \tfrac{z}{R^2}\left(-\tfrac{\partial^2 w}{\partial \theta^2} + \psi\tfrac{\partial v}{\partial \theta} - \varphi w\right)\right]\right. \\ &\left.+Y_{13}\left[\tfrac{\partial v}{\partial x} + \tfrac{1}{R}\tfrac{\partial u}{\partial \theta} - \tfrac{2z}{R}\tfrac{\partial^2 w}{\partial x \partial \theta} - \tfrac{z}{R^2}\left(\tfrac{1}{2}\psi + \varphi\right)\tfrac{\partial u}{\partial \theta} + \tfrac{z}{R}\left(\tfrac{3}{2}\psi + \varphi\right)\tfrac{\partial v}{\partial x}\right]\right\}\end{aligned} \tag{7}$$

$$\begin{aligned}\sigma_\theta = &\tfrac{1}{h}\left\{Y_{21}\left(\tfrac{\partial u}{\partial x} - z\tfrac{\partial^2 w}{\partial x^2}\right) + Y_{22}\left[\tfrac{1}{R}\tfrac{\partial v}{\partial \theta} + \tfrac{w}{R} + \tfrac{z}{R^2}\left(-\tfrac{\partial^2 w}{\partial \theta^2} + \psi\tfrac{\partial v}{\partial \theta} - \varphi w\right)\right]\right. \\ &\left.+Y_{23}\left[\tfrac{\partial v}{\partial x} + \tfrac{1}{R}\tfrac{\partial u}{\partial \theta} - \tfrac{2z}{R}\tfrac{\partial^2 w}{\partial x \partial \theta} - \tfrac{z}{R^2}\left(\tfrac{1}{2}\psi + \varphi\right)\tfrac{\partial u}{\partial \theta} + \tfrac{z}{R}\left(\tfrac{3}{2}\psi + \varphi\right)\tfrac{\partial v}{\partial x}\right]\right\}\end{aligned} \tag{8}$$

$$\begin{aligned}\tau_{x\theta} = &\tfrac{1}{h}\left\{Y_{31}\left(\tfrac{\partial u}{\partial x} - z\tfrac{\partial^2 w}{\partial x^2}\right) + Y_{32}\left[\tfrac{1}{R}\tfrac{\partial v}{\partial \theta} + \tfrac{w}{R} + \tfrac{z}{R^2}\left(-\tfrac{\partial^2 w}{\partial \theta^2} + \psi\tfrac{\partial v}{\partial \theta} - \varphi w\right)\right]\right. \\ &\left.+Y_{33}\left[\tfrac{\partial v}{\partial x} + \tfrac{1}{R}\tfrac{\partial u}{\partial \theta} - \tfrac{2z}{R}\tfrac{\partial^2 w}{\partial x \partial \theta} - \tfrac{z}{R^2}\left(\tfrac{1}{2}\psi + \varphi\right)\tfrac{\partial u}{\partial \theta} + \tfrac{z}{R}\left(\tfrac{3}{2}\psi + \varphi\right)\tfrac{\partial v}{\partial x}\right]\right\}\end{aligned} \tag{9}$$

4. Equations of Motion

The general equations of motion for an arbitrary SWCNT in terms of force $(N_x, N_\theta, N_{x\theta})$ and moment $(M_x, M_\theta, M_{x\theta})$ resultants are written in the form [22]:

$$\begin{gathered}\frac{\partial N_x}{\partial x}+\frac{1}{R}\frac{\partial N_{x\theta}}{\partial \theta}-\frac{1}{2R^2}\frac{\partial M_{x\theta}}{\partial \theta}-\rho h\frac{\partial^2 u}{\partial t^2}=0\\ \frac{1}{R}\frac{\partial N_\theta}{\partial \theta}+\frac{\partial N_{x\theta}}{\partial x}+\frac{3}{2R}\frac{\partial M_{x\theta}}{\partial x}+\frac{1}{R^2}\frac{\partial M_\theta}{\partial \theta}-\rho h\frac{\partial^2 v}{\partial t^2}=0\\ \frac{\partial^2 M_x}{\partial x^2}+\frac{2}{R}\frac{\partial^2 M_{x\theta}}{\partial x\partial \theta}+\frac{1}{R^2}\frac{\partial^2 M_\theta}{\partial \theta^2}-\frac{N_\theta}{R}-\rho h\frac{\partial^2 w}{\partial t^2}=0\end{gathered} \tag{10}$$

where ρh is the mass density per unit lateral area (i.e., the surface density) of SWCNT.

In the anisotropic elastic shell model, the force and moment resultants are defined based on the stress components in Equation (5), in the form [39]:

$$\begin{aligned}N_x=\int_{-h/2}^{h/2}\sigma_x\left(1+\varphi\frac{z}{R}\right)dz=Y_{11}\frac{\partial u}{\partial x}+\frac{Y_{12}}{R}\left(\frac{\partial v}{\partial \theta}+w\right)+Y_{13}\left(\frac{\partial v}{\partial x}+\frac{1}{R}\frac{\partial u}{\partial \theta}\right)\\ +\varphi\left[-\frac{X_{11}}{R}\frac{\partial^2 w}{\partial x^2}-\frac{X_{12}}{R}\left(\frac{1}{R^2}\frac{\partial^2 w}{\partial \theta^2}+\frac{w}{R^2}\right)+\frac{X_{13}}{R}\left(-\frac{2}{R}\frac{\partial^2 w}{\partial x\partial \theta}-\frac{1}{R^2}\frac{\partial u}{\partial \theta}+\frac{1}{R}\frac{\partial v}{\partial x}\right)\right]\end{aligned} \tag{11}$$

$$N_\theta=\int_{-h/2}^{h/2}\sigma_\theta dz=Y_{21}\frac{\partial u}{\partial x}+\frac{Y_{22}}{R}\left(\frac{\partial v}{\partial \theta}+w\right)+Y_{23}\left(\frac{\partial v}{\partial x}+\frac{1}{R}\frac{\partial u}{\partial \theta}\right) \tag{12}$$

$$\begin{aligned}N_{x\theta}=\int_{-h/2}^{h/2}\tau_{x\theta}\left(1+\varphi\frac{z}{R}\right)dz=Y_{31}\frac{\partial u}{\partial x}+\frac{Y_{32}}{R}\left(\frac{\partial v}{\partial \theta}+w\right)+Y_{33}\left(\frac{\partial v}{\partial x}+\frac{1}{R}\frac{\partial u}{\partial \theta}\right)\\ +\varphi\left[-\frac{X_{31}}{R}\frac{\partial^2 w}{\partial x^2}-\frac{X_{32}}{R}\left(\frac{1}{R^2}\frac{\partial^2 w}{\partial \theta^2}+\frac{w}{R^2}\right)+\frac{X_{33}}{R}\left(-\frac{2}{R}\frac{\partial^2 w}{\partial x\partial \theta}-\frac{1}{R^2}\frac{\partial u}{\partial \theta}+\frac{1}{R}\frac{\partial v}{\partial x}\right)\right]\end{aligned} \tag{13}$$

$$\begin{aligned}M_x=\int_{-h/2}^{h/2}\sigma_x\left(1+\varphi\frac{z}{R}\right)zdz=-X_{11}\frac{\partial^2 w}{\partial x^2}+\frac{X_{12}}{R^2}\left(-\frac{\partial^2 w}{\partial \theta^2}+\psi\frac{\partial v}{\partial \theta}-\varphi w\right)\\ +X_{13}\left[-\frac{2}{R}\frac{\partial^2 w}{\partial x\partial \theta}-\frac{1}{R^2}\left(\frac{1}{2}\psi+\varphi\right)\frac{\partial u}{\partial \theta}+\frac{1}{R}\left(\frac{3}{2}\psi+\varphi\right)\frac{\partial v}{\partial x}\right]\\ +\varphi\left[\frac{X_{11}}{R}\frac{\partial u}{\partial x}+\frac{X_{12}}{R^2}\left(\frac{\partial v}{\partial \theta}+w\right)+\frac{X_{13}}{R}\left(\frac{\partial v}{\partial x}+\frac{1}{R}\frac{\partial u}{\partial \theta}\right)\right]\end{aligned} \tag{14}$$

$$\begin{aligned}M_\theta=\int_{-h/2}^{h/2}\sigma_\theta zdz=-X_{21}\frac{\partial^2 w}{\partial x^2}+\frac{X_{22}}{R^2}\left(-\frac{\partial^2 w}{\partial \theta^2}+\psi\frac{\partial v}{\partial \theta}-\varphi w\right)\\ +X_{23}\left[-\frac{2}{R}\frac{\partial^2 w}{\partial x\partial \theta}-\frac{1}{R^2}\left(\frac{1}{2}\psi+\varphi\right)\frac{\partial u}{\partial \theta}+\frac{1}{R}\left(\frac{3}{2}\psi+\varphi\right)\frac{\partial v}{\partial x}\right]\end{aligned} \tag{15}$$

$$\begin{aligned}M_{x\theta}=\int_{-h/2}^{h/2}\tau_{x\theta}\left(1+\varphi\frac{z}{R}\right)zdz=-X_{31}\frac{\partial^2 w}{\partial x^2}+\frac{X_{32}}{R^2}\left(-\frac{\partial^2 w}{\partial \theta^2}+\psi\frac{\partial v}{\partial \theta}-\varphi w\right)\\ +X_{33}\left[-\frac{2}{R}\frac{\partial^2 w}{\partial x\partial \theta}-\frac{1}{R^2}\left(\frac{1}{2}\psi+\varphi\right)\frac{\partial u}{\partial \theta}+\frac{1}{R}\left(\frac{3}{2}\psi+\varphi\right)\frac{\partial v}{\partial x}\right]\\ +\varphi\left[\frac{X_{31}}{R}\frac{\partial u}{\partial x}+\frac{X_{32}}{R^2}\left(\frac{\partial v}{\partial \theta}+w\right)+\frac{X_{33}}{R}\left(\frac{\partial v}{\partial x}+\frac{1}{R}\frac{\partial u}{\partial \theta}\right)\right]\end{aligned} \tag{16}$$

where $X_{ij}=Y_{ij}h^2/12$, with $i,j=1,2,3$.

From Equations (11)–(16) it is noted that the integrating functions of the circumferential force N_θ and moment M_θ resultants are the same for the three different shell theories. Conversely, the integrating functions of the other force and moment resultants are different, since in the Flügge shell theory, they also include the term $(1+z/R)$ (in this theory the thinness assumption is delayed), whereas this term is ignored in the other two thin shell theories. On the one hand, considering this additional term within the integrating functions of the resultants certainly makes the Flügge shell theory more refined (in fact, it is able to correctly model the vibrations of even relatively thick shells). On the other hand, this considerably increases the number of terms within the expressions of such resultants, and therefore, it strongly increases the computational effort of the numerical analyses. The main goal of this paper is therefore to verify whether a less refined theory, but with lower computational effort, such as the Donnell or Sanders shell theory, can provide sufficiently accurate results in terms of SWCNT natural frequencies compared to those provided by the Flügge shell theory, as investigated in [26] for cylindrical shells.

By substituting the expressions of the force and moment resultants (11)–(16) into the dynamic Equation (10), the equations of motion for the anisotropic elastic shell model are obtained in the form:

$$\begin{aligned}
&\left\{Y_{11}\frac{\partial^2}{\partial x^2}+\left(\frac{2Y_{13}}{R}-\varphi\frac{3X_{13}}{2R^3}\right)\frac{\partial^2}{\partial x\partial\theta}+\left[\frac{Y_{33}}{R^2}+(\psi-4\varphi)\frac{X_{33}}{4R^4}\right]\frac{\partial^2}{\partial\theta^2}\right\}u\\
&+\left\{\left(Y_{13}+\varphi\frac{X_{13}}{R^2}\right)\frac{\partial^2}{\partial x^2}+\left(\frac{Y_{12}+Y_{33}}{R}-\psi\frac{3X_{33}}{4R^3}\right)\frac{\partial^2}{\partial x\partial\theta}+\left[\frac{Y_{23}}{R^2}\right.\right.\\
&\left.\left.-(\psi+\varphi)\frac{X_{23}}{2R^4}\right]\frac{\partial^2}{\partial\theta^2}\right\}v+\left\{\left(\frac{Y_{12}}{R}-\varphi\frac{X_{12}}{R^3}\right)\frac{\partial}{\partial x}+\left(\frac{Y_{23}}{R^2}-\varphi\frac{X_{23}}{R^4}\right)\frac{\partial}{\partial\theta}\right.\\
&-\varphi\frac{X_{11}}{R}\frac{\partial^3}{\partial x^3}+(1-6\varphi)\frac{X_{13}}{2R^2}\frac{\partial^3}{\partial x^2\partial\theta}+\left[(1-2\varphi)\frac{X_{33}}{R^3}-\varphi\frac{X_{12}}{R^3}\right]\\
&\left.\frac{\partial^3}{\partial x\partial\theta^2}+(1-2\varphi)\frac{X_{23}}{2R^4}\frac{\partial^3}{\partial\theta^3}\right\}w=\rho h\frac{\partial^2 u}{\partial t^2}
\end{aligned}\tag{17}$$

$$\begin{aligned}
&\left\{\left(Y_{13}+\varphi\frac{3X_{13}}{2R^2}\right)\frac{\partial^2}{\partial x^2}+\left[\frac{Y_{12}+Y_{33}}{R}-(3\psi+4\varphi)\frac{X_{33}}{4R^3}\right]\frac{\partial^2}{\partial x\partial\theta}+\left[\frac{Y_{23}}{R^2}-\frac{X_{23}}{2R^4}(\psi+2\varphi)\right]\right.\\
&\left.\frac{\partial^2}{\partial\theta^2}\right\}u+\left\{\left[Y_{33}+(9\psi+16\varphi)\frac{X_{33}}{4R^2}\right]\frac{\partial^2}{\partial x^2}+\left[\frac{2Y_{23}}{R}+(6\psi+5\varphi)\frac{X_{23}}{2R^3}\right]\frac{\partial^2}{\partial x\partial\theta}+\left(\frac{Y_{22}}{R^2}\right.\right.\\
&\left.\left.+\psi\frac{X_{22}}{R^4}\right)\frac{\partial^2}{\partial\theta^2}\right\}v+\left\{\left(\frac{Y_{23}}{R}-\varphi\frac{X_{23}}{R^3}\right)\frac{\partial}{\partial x}+\left(\frac{Y_{22}}{R^2}-\varphi\frac{X_{22}}{R^4}\right)\frac{\partial}{\partial\theta}-(3+2\varphi)\frac{X_{13}}{2R}\frac{\partial^3}{\partial x^3}\right.\\
&\left.-\left[\frac{X_{12}}{R^2}+(3+2\varphi)\frac{X_{33}}{R^2}\right]\frac{\partial^3}{\partial x^2\partial\theta}-(7+2\varphi)\frac{X_{23}}{2R^3}\frac{\partial^3}{\partial x\partial\theta^2}-\frac{X_{22}}{R^4}\frac{\partial^3}{\partial\theta^3}\right\}w=\rho h\frac{\partial^2 v}{\partial t^2}
\end{aligned}\tag{18}$$

$$\begin{aligned}
&\left[-\frac{Y_{12}}{R}\frac{\partial}{\partial x}-\frac{Y_{23}}{R^2}\frac{\partial}{\partial\theta}+\varphi\frac{X_{11}}{R}\frac{\partial^3}{\partial x^3}+(-\psi+4\varphi)\frac{X_{13}}{2R^2}\frac{\partial^3}{\partial x^2\partial\theta}-\psi\frac{X_{33}}{R^3}\frac{\partial^3}{\partial x\partial\theta^2}\right.\\
&\left.-(\psi+2\varphi)\frac{X_{23}}{2R^4}\frac{\partial^3}{\partial\theta^3}\right]u+\left\{-\frac{Y_{23}}{R}\frac{\partial}{\partial x}-\frac{Y_{22}}{R^2}\frac{\partial}{\partial\theta}+(3\psi+4\varphi)\frac{X_{13}}{2R}\frac{\partial^3}{\partial x^3}\right.\\
&\left.+\left[(\psi+\varphi)\frac{X_{12}}{R^2}+(3\psi+4\varphi)\frac{X_{33}}{R^2}\right]\frac{\partial^3}{\partial x^2\partial\theta}+(7\psi+6\varphi)\frac{X_{23}}{2R^3}\frac{\partial^3}{\partial x\partial\theta^2}+\psi\frac{X_{22}}{R^4}\frac{\partial^3}{\partial\theta^3}\right\}v\\
&+\left[-\frac{Y_{22}}{R^2}-\varphi\frac{X_{22}}{R^4}\frac{\partial^2}{\partial\theta^2}-X_{11}\frac{\partial^4}{\partial x^4}-\frac{4X_{13}}{R}\frac{\partial^4}{\partial x^3\partial\theta}-\left(\frac{2X_{12}+4X_{33}}{R^2}\right)\right.\\
&\left.\frac{\partial^4}{\partial x^2\partial\theta^2}-\frac{4X_{23}}{R^3}\frac{\partial^4}{\partial x\partial\theta^3}-\frac{X_{22}}{R^4}\frac{\partial^4}{\partial\theta^4}\right]w=\rho h\frac{\partial^2 w}{\partial t^2}
\end{aligned}\tag{19}$$

5. Solution Method

In this paper, a complex method is considered to analytically solve the dynamic equations of motions (17)–(19) and to obtain the natural frequencies of SWCNTs. Specifically, a complex variable is used to solve the partial differential Equations (17)–(19) by setting the real and imaginary zero.

In the present work, simply supported boundary conditions are adopted. These boundary conditions, for the complex method, impose the conditions $\mathrm{Re}(v)=\mathrm{Re}(w)=0$

at both ends $x = (0, L)$ of the SWCNT. The displacement field that satisfies these boundary conditions can be written as [39]:

$$\begin{aligned} u(x,\theta,t) &= \bar{U}\exp(i\lambda_q x)\cos(s\theta)\cos(\omega t) \\ v(x,\theta,t) &= -i\bar{V}\exp(i\lambda_q x)\sin(s\theta)\cos(\omega t) \\ w(x,\theta,t) &= -i\bar{W}\exp(i\lambda_q x)\cos(s\theta)\cos(\omega t) \end{aligned} \tag{20}$$

where $\left(\bar{U}, \bar{V}, \bar{W}\right)$ denote the displacement amplitudes along the longitudinal u, circumferential v, and radial w directions, respectively; i is the imaginary unit; λ_q is the wavenumber along the longitudinal direction, with $\lambda_q = q\pi/L$, where q is the number of longitudinal half-waves and L is the length of the SWCNT; s is the number of circumferential waves; and ω is the circular frequency.

Substituting Equation (20) into Equations (17)–(19), a set of algebraic equations for the displacement amplitudes $\left(\bar{U}, \bar{V}, \bar{W}\right)$ is obtained, which can be written in the form [39]:

$$\mathbf{E}(\lambda_q, s, \omega)_{3\times 3}\begin{bmatrix}\bar{U}\\ \bar{V}\\ \bar{W}\end{bmatrix} = \begin{bmatrix}0\\0\\0\end{bmatrix} \tag{21}$$

where $\mathbf{E}$ is a non-symmetric matrix, whose elements are:

$$E_{11} = Y_{11}\lambda_q^2 + \left(\frac{2Y_{13}}{R} - \varphi\frac{3X_{13}}{2R^3}\right)\lambda_q s + \left[\frac{Y_{33}}{R^2} + (\psi - 4\varphi)\frac{X_{33}}{4R^4}\right]s^2 - \rho h\omega^2 \tag{22}$$

$$E_{12} = -\left(Y_{13} + \varphi\frac{X_{13}}{R^2}\right)\lambda_q^2 - \left(\frac{Y_{12} + Y_{33}}{R} - \psi\frac{3X_{33}}{4R^3}\right)\lambda_q s - \left[\frac{Y_{23}}{R^2} - (\psi + \varphi)\frac{X_{23}}{2R^4}\right]s^2 \tag{23}$$

$$\begin{aligned} E_{13} = &-\left(\tfrac{Y_{12}}{R} - \varphi\tfrac{X_{12}}{R^3}\right)\lambda_q - \left(\tfrac{Y_{23}}{R^2} - \varphi\tfrac{X_{23}}{R^4}\right)s - \varphi\tfrac{X_{11}}{R}\lambda_q^3 \\ &+(1 - 6\varphi)\tfrac{X_{13}}{2R^2}\lambda_q^2 s + \left[(1 - 2\varphi)\tfrac{X_{33}}{R^3} - \varphi\tfrac{X_{12}}{R^3}\right]\lambda_q s^2 + (1 - 2\varphi)\tfrac{X_{23}}{2R^4}s^3 \end{aligned} \tag{24}$$

$$E_{21} = -\left(Y_{13} + \varphi\tfrac{3X_{13}}{2R^2}\right)\lambda_q^2 - \left[\tfrac{Y_{12}+Y_{33}}{R} - (3\psi + 4\varphi)\tfrac{X_{33}}{4R^3}\right]\lambda_q s - \left[\tfrac{Y_{23}}{R^2} - (\psi + 2\varphi)\tfrac{X_{23}}{2R^4}\right]s^2 \tag{25}$$

$$E_{22} = \left[Y_{33} + (9\psi + 16\varphi)\tfrac{X_{33}}{4R^2}\right]\lambda_q^2 + \left[\tfrac{2Y_{23}}{R} + (6\psi + 5\varphi)\tfrac{X_{23}}{2R^3}\right]\lambda_q s + \left(\tfrac{Y_{22}}{R^2} + \psi\tfrac{X_{22}}{R^4}\right)s^2 - \rho h\omega^2 \tag{26}$$

$$\begin{aligned} E_{23} = &\left(\tfrac{Y_{23}}{R} - \varphi\tfrac{X_{23}}{R^3}\right)\lambda_q + \left(\tfrac{Y_{22}}{R^2} - \varphi\tfrac{X_{22}}{R^4}\right)s + (3 + 2\varphi)\tfrac{X_{13}}{2R}\lambda_q^3 \\ &+\left[\tfrac{X_{12}}{R^2} + (3 + 2\varphi)\tfrac{X_{33}}{R^2}\right]\lambda_q^2 s + (7 + 2\varphi)\tfrac{X_{23}}{2R^3}\lambda_q s^2 + \tfrac{X_{22}}{R^4}s^3 \end{aligned} \tag{27}$$

$$E_{31} = -\frac{Y_{12}}{R}\lambda_q + \frac{Y_{23}}{R^2}s - \varphi\frac{X_{11}}{R}\lambda_q^3 - (\psi - 4\varphi)\frac{X_{13}}{2R^2}\lambda_q^2 s + \psi\frac{X_{33}}{R^3}\lambda_q s^2 - (\psi + 2\varphi)\frac{X_{23}}{2R^4}s^3 \tag{28}$$

$$\begin{aligned} E_{32} = &-\tfrac{Y_{23}}{R}\lambda_q + \tfrac{Y_{22}}{R^2}s - (3\psi + 4\varphi)\tfrac{X_{13}}{2R}\lambda_q^3 + \Big[(\psi + \varphi)\tfrac{X_{12}}{R^2} \\ &+(3\psi + 4\varphi)\tfrac{X_{33}}{R^2}\Big]\lambda_q^2 s - (7\psi + 6\varphi)\tfrac{X_{23}}{2R^3}\lambda_q s^2 + \psi\tfrac{X_{22}}{R^4}s^3 \end{aligned} \tag{29}$$

$$E_{33} = \frac{Y_{22}}{R^2} + X_{11}\lambda_q^4 - \varphi\frac{X_{22}}{R^4}s^2 - \frac{4X_{13}}{R}\lambda_q^3 s + \left(\frac{2X_{12}+4X_{33}}{R^2}\right)\lambda_q^2 s^2 - \frac{4X_{23}}{R^3}\lambda_q s^3 + \frac{X_{22}}{R^4}s^4 - \rho h\omega^2 \quad (30)$$

At this point it is useful to remember that by imposing the parameters $(\psi = 0, \varphi = 0)$, we obtain the elements of matrix **E** for the Donnell shell theory; by imposing the parameters $(\psi = 1, \varphi = 0)$, we obtain the elements of matrix **E** for the Sanders shell theory; and by imposing the parameters $(\psi = 0, \varphi = 1)$, we obtain the elements of matrix **E** for the Flügge shell theory. See Expansion (2) for the middle surface change in the curvature k_θ and torsion $k_{x\theta}$ of the shell.

For the non-trivial solution, the determinant of the set of Equation (21) must be equal to zero [40]:

$$\det\mathbf{E}(\lambda_q, s, \omega)_{3\times3} = 0 \quad (31)$$

Solving Equation (31), we get a third-order algebraic equation in ω^2; this last equation provides three different eigenfrequencies for each number of waves (q, s) that give three different vibration modes (i.e., longitudinal, torsional and radial modes). Since the highest natural frequency corresponds to the radial vibration mode, only the radial natural frequencies were computed in the numeric results.

6. Numeric Results

In this paper, the natural frequencies of SWCNTs were obtained in the framework of the Donnell, Sanders, and Flügge shell theories. An anisotropic elastic shell model was used to take into account the chirality effects of CNTs. Simply supported boundary conditions were imposed. Vibration modes with different number of waves along the longitudinal and circumferential directions were considered. SWCNTs with different values of radius R and aspect ratio L/R were investigated.

As known from the literature, two relevant open issues related to the continuous modelling of carbon nanotubes are due to their intrinsic anisotropic character and their discrete configuration. To this end, it is very important to adopt parameters and models able to correctly describe the actual molecular structure of carbon nanotubes.

In Table 1, the values of carbon-carbon bond parameters (a, k_ρ, k_θ) and equivalent continuous parameters (h, ρ) retrieved from the literature are reported. In particular, parameters k_ρ and k_θ, which denote force constants correlated to the variance of carbon-carbon bond length a and angle θ, respectively, were adopted to express the anisotropic elastic constants of SWCNTs via the molecular mechanics-based "stick-spiral model" developed by Chang [35].

Table 1. Mechanical parameters adopted in the anisotropic elastic continuous shell model [16,35].

C-C bond length a (nm)	0.142
C-C bond elongation k_ρ (nN/nm)	742
C-C bond angle variance k_θ (nN·nm)	1.42
Shell thickness h (nm)	0.0665
Mass density ρ (kg/m^3)	11,700

Moreover, in order to study the dynamics of the actual discrete CNT via a continuous cylindrical shell, an equivalent thickness h, which is derived from MD simulations of CNT energy, and an equivalent mass density ρ, resulting from graphite surface density, were considered; see Ref. [16] for more details.

6.1. Comparison of the Shell Theories with the Results of Molecular Dynamics Simulations

In this section, the natural frequencies of the radial breathing mode $(q = 0, s = 0)$ of the SWCNT of Table 1 with an aspect ratio $L/R = 10$ obtained by considering Donnell,

Sanders, and Flügge shell theories are compared with the results of molecular dynamics simulations available in literature for different chirality indices (n, m); see Table 2.

Specifically, the results of the molecular dynamics simulations reported in Table 2 were retrieved from Ref. [10]. In that work, the free vibrations of armchair, zigzag, and chiral SWCNTs with different aspect ratios and diameters were studied via MM3 potential. This potential considers bond stretching, change in angles between adjacent bonds, torsion of the bond, van der Waals forces, and the coupling among stretching, bending, and torsional deformations. In particular, the energy due to the bond stretching has terms that are quadratic, cubic, and quartic in the bond length; thus, the strain energy due to the bond stretching is not an even function of the change in the bond length; see Ref. [10] for more details.

Table 2. Natural frequencies of the radial breathing mode $(q = 0, s = 0)$ of the SWCNT of Table 1 with aspect ratio. $L/R = 10$. Comparisons among Donnell, Sanders, Flügge shell theories and molecular dynamics simulations.

Natural Frequency ω_{RBM} (cm^{-1})		**Difference (%)**		
Chirality Indices (*n*,*m*)	**Molecular Dynamics Simulations [10]**	**Donnell Shell Theory**	**Sanders Shell Theory**	**Flügge Shell Theory**
(10, 0)	290.463	2.38	1.32	0.42
(9, 6)	221.496	3.47	1.93	0.43
(8, 8)	209.008	4.05	2.25	0.51
(9, 9)	185.896	4.03	2.24	0.49
(16, 0)	181.747	3.10	1.72	0.04
(10, 10)	167.377	4.01	2.23	0.48
(11, 11)	152.207	4.00	2.22	0.47
(20, 0)	145.363	3.35	1.86	0.11
(22, 7)	110.808	3.71	2.06	0.32
(25, 10)	93.153	3.56	1.98	0.21
(30, 5)	88.750	3.51	1.95	0.16
(36, 5)	75.118	3.47	1.93	0.17
(33, 16)	67.221	3.56	1.98	0.23

Obviously, the results of molecular dynamics simulations are the most correct since these simulations are able to correctly take into account the actual molecular structure of carbon nanotubes. However, the computational effort of molecular dynamics simulations is very high, in particular when dealing with very long and thin structures (such as carbon nanotubes), which have an extremely large number of atoms, and therefore, these results are available only for a reduced range of CNT geometries. To this end, it seems useful to investigate which shell theory is more accurate in the vibration modelling of carbon nanotubes considered as continuous homogeneous structures.

From Table 2, it can be observed that for all considered chirality indices, by assuming molecular dynamics results as reference, the percentage differences of the Flügge shell theory are the lowest, the percentage differences of the Donnell shell theory are the highest, and the Sanders shell theory gives an intermediate response.

These results could easily be predicted considering the strain–displacement Relationship (2), in which the Donnell shell theory has fewer terms than the other two (and therefore is the least accurate), and observing the force and moment Resultants (11)–(16), in which the Flügge shell theory includes more terms (and therefore is the most accurate).

Even if, among the three considered shell theories, Flügge is obtained to be the most accurate, it is preferable to not adopt this theory, because it has a relatively high computational effort, due to the large number of terms in the expansions of the force and moment resultants. This aspect is relevant not so much in the study of the linear vibrations (natural frequencies) but instead in the analysis of the non-linear vibrations (amplitude-frequency responses), where in the strain–displacement relationships the non-linear terms are also considered, and therefore the computational weight in solving the corresponding dynamic equations of motion becomes much greater.

To this aim, it could be useful to analyse whether the Donnell or Sanders shell theories are sufficiently accurate in the continuous modelling of SWCNT vibrations. Therefore, in the following, the natural frequencies of simply supported SWCNTs with different geometries and wavenumbers will be computed in the framework of the Donnell and Sanders shell theories, where the results of Flügge shell theory will be considered as reference.

6.2. Comparison of the Shell Theories for Different SWCNT Geometries and Wavenumbers

In the present section, the natural frequencies of the simply supported SWCNT of Table 1 obtained by considering the Donnell, Sanders, and Flügge shell theories were compared for different chirality indices (n, m), aspect ratios L/R, numbers of longitudinal half-waves q, and circumferential waves s.

By taking as reference the results of the Flügge shell theory (which are the closest to those obtained from molecular dynamics simulations, see Table 2), we attempted to investigate the fields of applicability and limitations of the Donnell and Sanders shell theories.

Figures 2–11 show the comparisons between Donnell and Flügge shell theories, and between Sanders and Flügge shell theories, carried out by adopting increasing values of chirality indices (SWCNT radius), specifically $(n = 5, m = 5)$ (i.e., $R = 0.34$ nm), $(n = 10, m = 10)$ (i.e., $R = 0.68$ nm), and $(n = 25, m = 25)$ (i.e., $R = 1.70$ nm), and increasing values of SWCNT aspect ratio $L/R = (10, 15, 20)$ are shown.

It must be underlined that this analysis was carried out within the range commonly assumed for the radius of CNTs. In fact, it is reported that, "a single-wall nanotube is defined by a cylindrical graphene sheet with a diameter of about 0.5–10.0 nm, though most of the observed single-wall nanotubes have diameters < 2 nm", see Ref. [9].

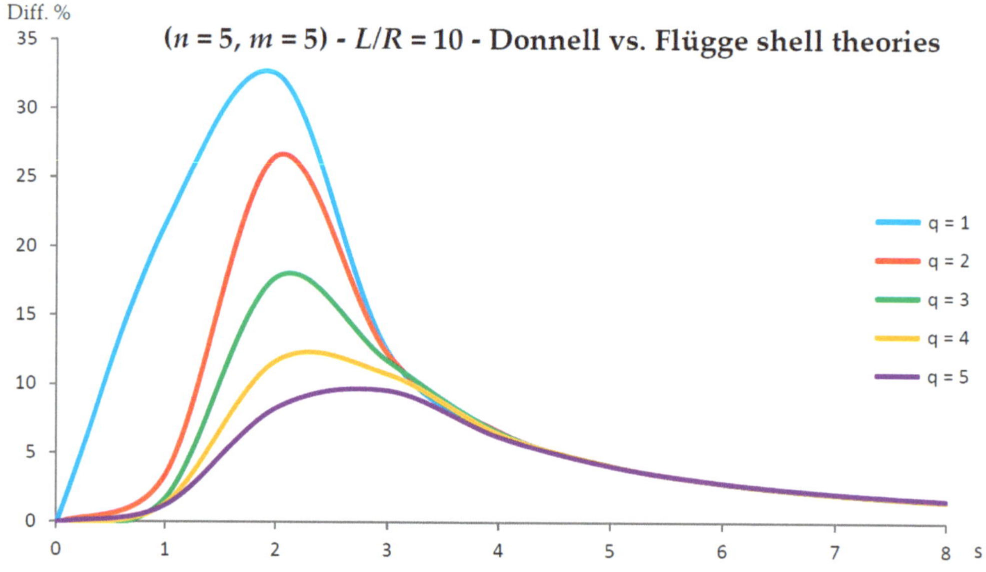

Figure 2. Percentage differences between the radial natural frequencies obtained via Donnell and Flügge shell theories (Flügge as the reference). Anisotropic elastic shell model. Simply supported SWCNT of Table 1 with chirality indices $(n = 5, m = 5)$ and aspect ratio $L/R = 10$. Number of longitudinal half-waves q. Number of circumferential waves s.

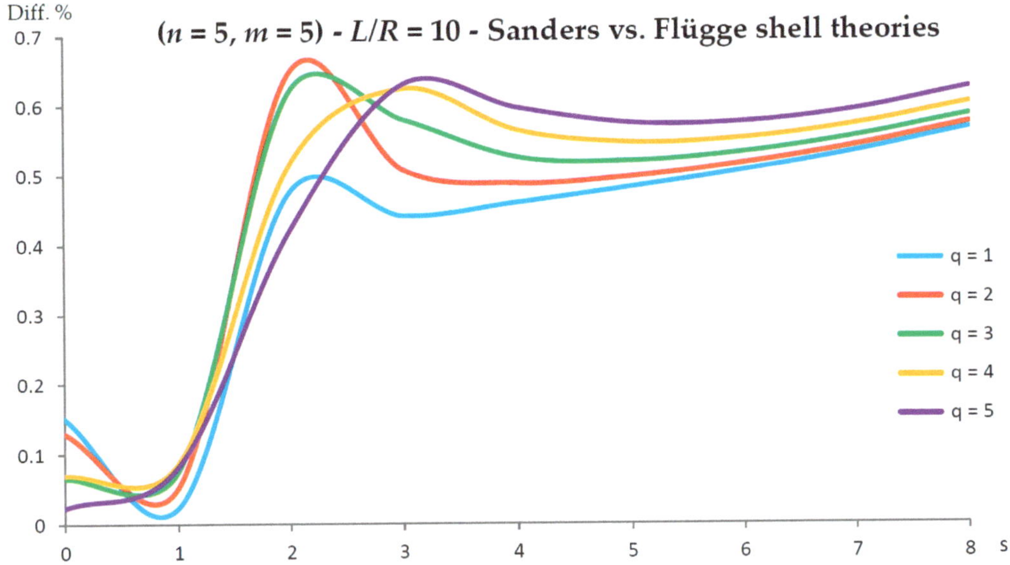

Figure 3. Percentage differences between the radial natural frequencies obtained via Sanders and Flügge shell theories (Flügge as the reference). Anisotropic elastic shell model. Simply supported SWCNT of Table 1 with chirality indices $(n = 5, m = 5)$ and aspect ratio $L/R = 10$. Number of longitudinal half-waves q. Number of circumferential waves s.

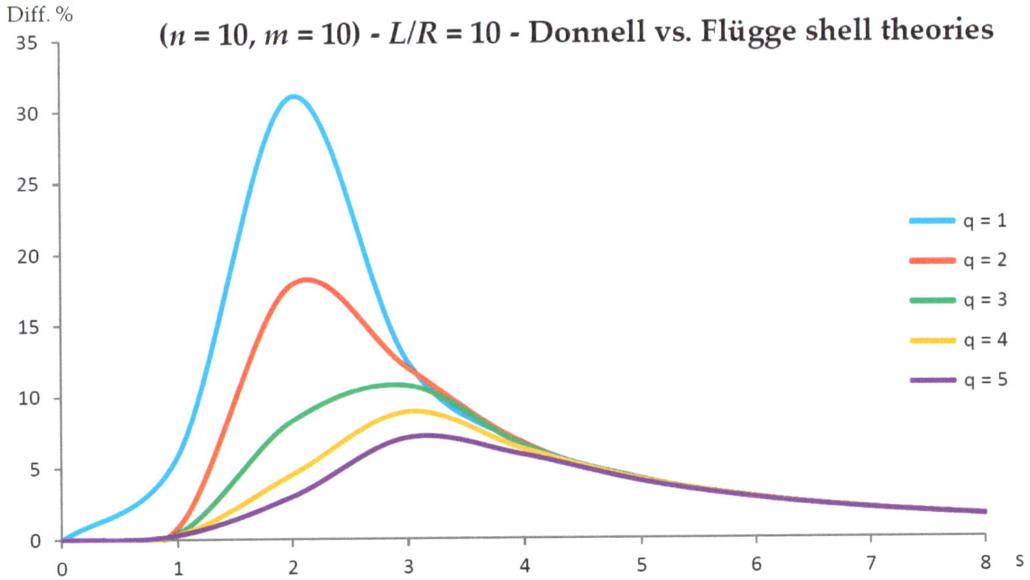

Figure 4. Percentage differences between the radial natural frequencies obtained via Donnell and Flügge shell theories (Flügge as the reference). Anisotropic elastic shell model. Simply supported SWCNT of Table 1 with chirality indices $(n = 10, m = 10)$ and aspect ratio $L/R = 10$. Number of longitudinal half-waves q. Number of circumferential waves s.

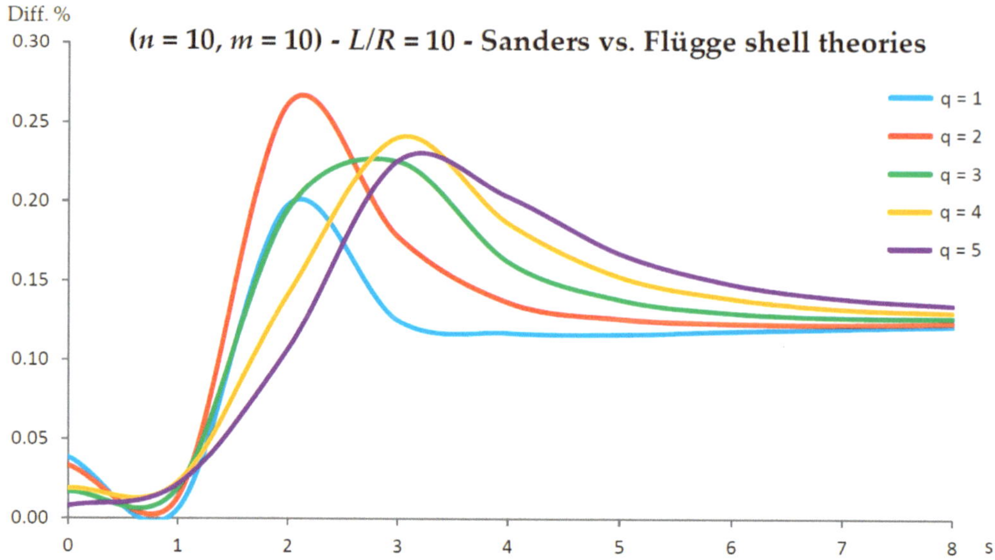

Figure 5. Percentage differences between the radial natural frequencies obtained via Sanders and Flügge shell theories (Flügge as the reference). Anisotropic elastic shell model. Simply supported SWCNT of Table 1 with chirality indices $(n = 10, m = 10)$ and aspect ratio $L/R = 10$. Number of longitudinal half-waves q. Number of circumferential waves s.

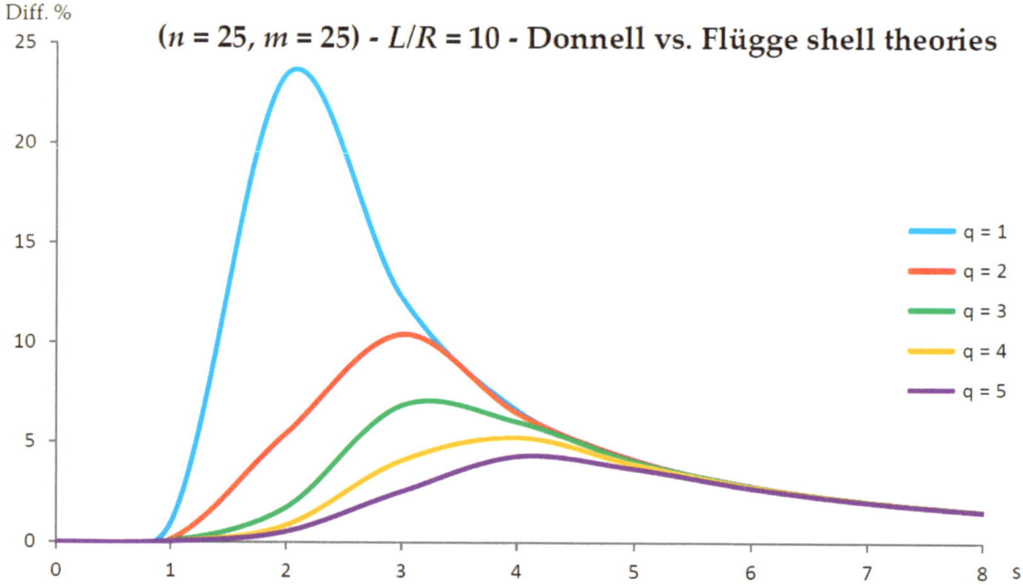

Figure 6. Percentage differences between the radial natural frequencies obtained via Donnell and Flügge shell theories (Flügge as the reference). Anisotropic elastic shell model. Simply supported SWCNT of Table 1 with chirality indices $(n = 25, m = 25)$ and aspect ratio $L/R = 10$. Number of longitudinal half-waves q. Number of circumferential waves s.

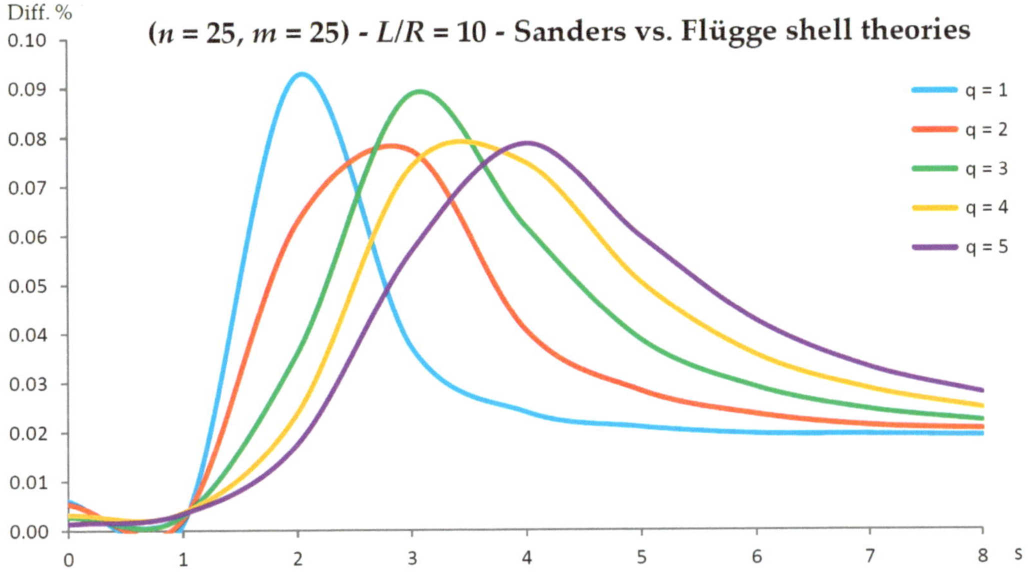

Figure 7. Percentage differences between the radial natural frequencies obtained via Sanders and Flügge shell theories (Flügge as the reference). Anisotropic elastic shell model. Simply supported SWCNT of Table 1 with chirality indices $(n = 25, m = 25)$ and aspect ratio $L/R = 10$. Number of longitudinal half-waves q. Number of circumferential waves s.

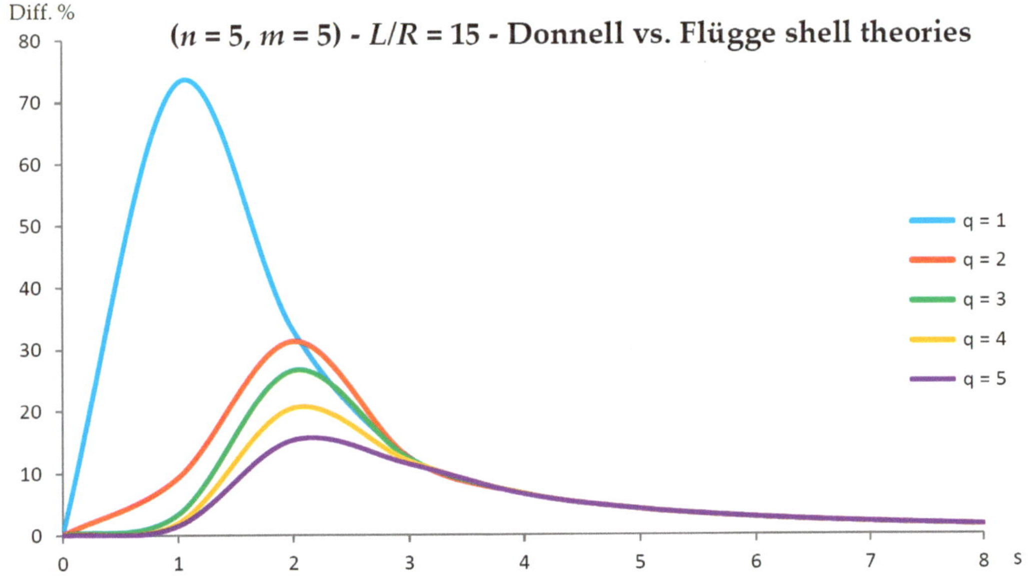

Figure 8. Percentage differences between the radial natural frequencies obtained via Donnell and Flügge shell theories (Flügge as the reference). Anisotropic elastic shell model. Simply supported SWCNT of Table 1 with chirality indices $(n = 5, m = 5)$ and aspect ratio $L/R = 15$. Number of longitudinal half-waves q. Number of circumferential waves s.

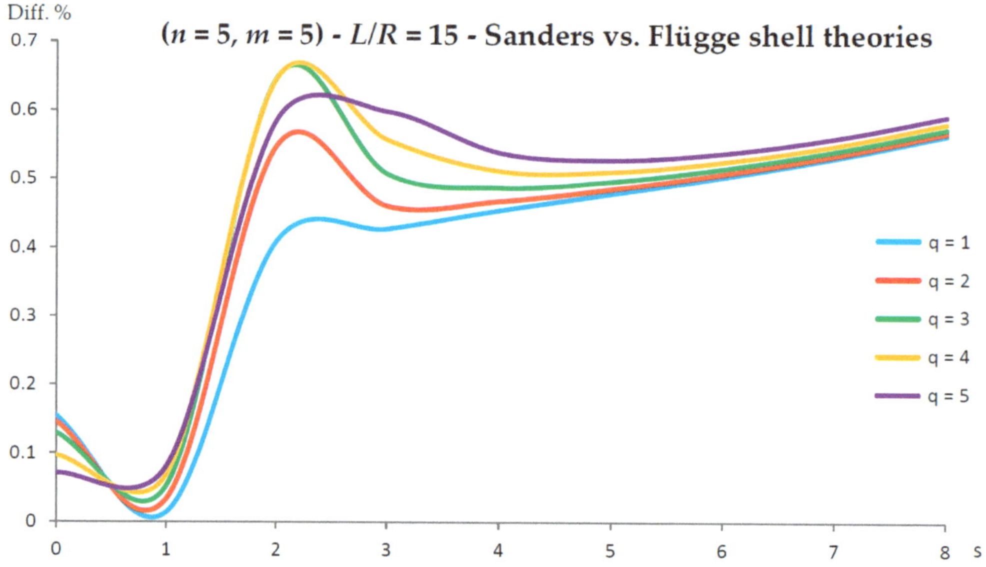

Figure 9. Percentage differences between the radial natural frequencies obtained via Sanders and Flügge shell theories (Flügge as the reference). Anisotropic elastic shell model. Simply supported SWCNT of Table 1 with chirality indices $(n = 5, m = 5)$ and aspect ratio $L/R = 15$. Number of longitudinal half-waves q. Number of circumferential waves s.

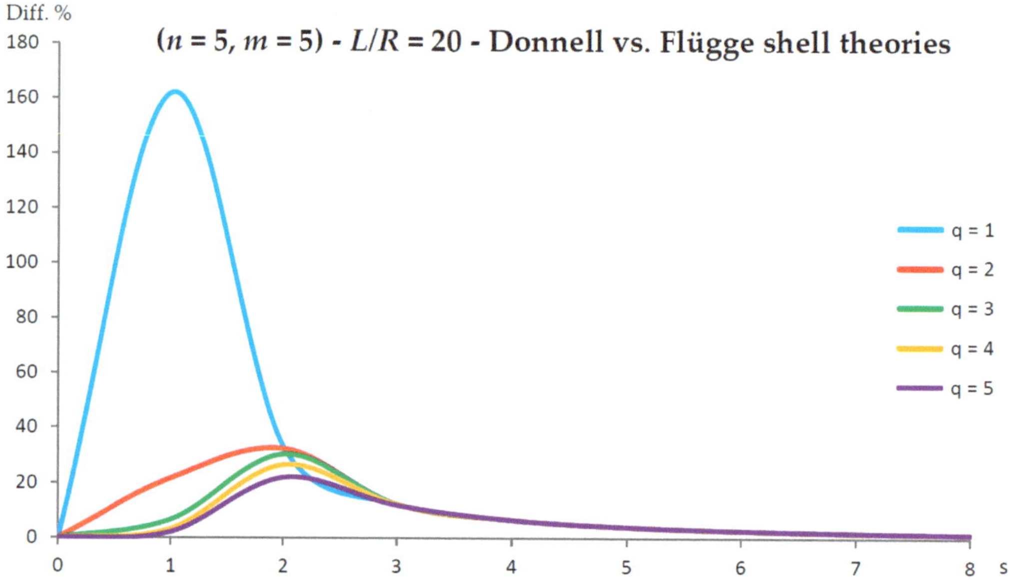

Figure 10. Percentage differences between the radial natural frequencies obtained via Donnell and Flügge shell theories (Flügge as the reference). Anisotropic elastic shell model. Simply supported SWCNT of Table 1 with chirality indices $(n = 5, m = 5)$ and aspect ratio $L/R = 20$. Number of longitudinal half-waves q. Number of circumferential waves s.

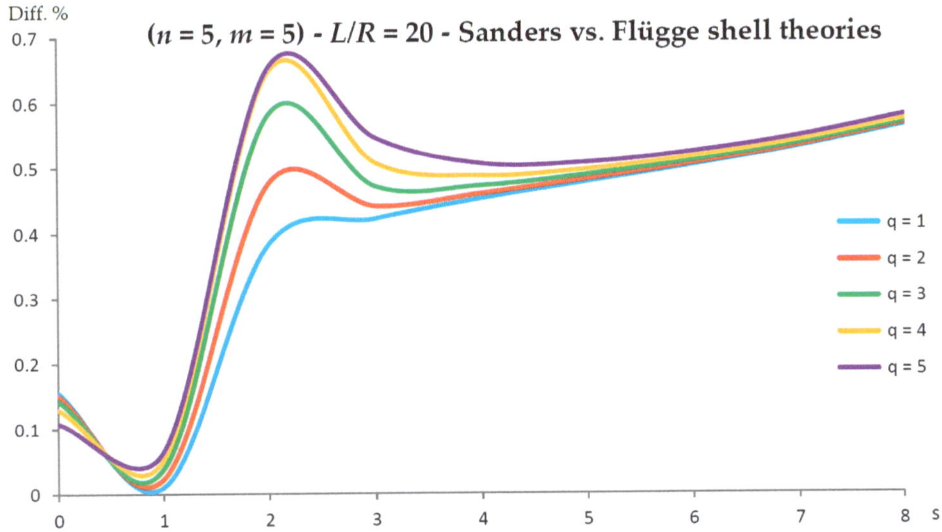

Figure 11. Percentage differences between the radial natural frequencies obtained via Sanders and Flügge shell theories (Flügge as the reference). Anisotropic elastic shell model. Simply supported SWCNT of Table 1 with chirality indices $(n = 5, m = 5)$ and aspect ratio $L/R = 20$. Number of longitudinal half-waves q. Number of circumferential waves s.

As for the Donnell shell theory, when increasing the radius R, the percentage difference with respect to the Flügge shell theory decreased for all longitudinal wavenumbers q, where the maximum value is obtained at $q = 1$ and decreased with increasing q; the maximum peak corresponding to $q = 1$ was always located at $s = 2$, whereas the peaks corresponding to $q = (2-5)$ moved to higher circumferential wavenumbers.

On the other hand, when increasing the aspect ratio L/R, the percentage difference with respect to Flügge shell theory increased for all longitudinal wavenumbers q, where the maximum value was found again at $q = 1$ and decreased with increasing q, but now the maximum peak corresponding to $q = 1$ moved from $s = 2$ to $s = 1$ (lower circumferential wavenumber), whereas the peaks corresponding to $q = (2-5)$ were always located at $s = 2$.

It should be underlined that the increase in the percentage difference at the longitudinal wavenumber $q = 1$ obtained by increasing the aspect ratio L/R is extremely higher than the corresponding decrease at the longitudinal wavenumber $q = 1$ obtained by increasing the radius R (i.e., the effect of aspect ratio is prevalent with respect to radius).

As for Sanders shell theory, increasing both the radius R and the aspect ratio L/R, it was always found that the percentage difference with respect to the Flügge shell theory was relatively low for every number of longitudinal and circumferential waves ($< 1\%$).

In particular, when increasing the radius R, the maximum percentage difference moved from $q = 2$ to $q = 1$, decreasing its value (from 0.7% to 0.1%); differently, when increasing the aspect ratio L/R, the maximum percentage difference moved from $q = 2$ to $q = 5$, preserving its value (0.7%) (i.e., no effect of both aspect ratio and radius).

Therefore, by considering the parametric analyses presented in Figures 2–11, it can be observed that the Donnell shell theory cannot be applied for the vibration modelling of SWCNTs with relatively low radius R and relatively high aspect ratio L/R, and for the vibration modelling of modes with relatively low numbers of longitudinal q and circumferential s waves. However, as previously reported, the effect of the aspect ratio is prevalent with respect to the radius (and also to the wavenumbers) in providing the very high percentage difference obtained between the Donnell and Flügge shell theories. This is due to the different expression of the middle surface change in curvature k_θ and torsion $k_{x\theta}$ of the shell for the two theories (see the strain–displacement Equation (2)), which present more terms in the Flügge than in the Donnell shell theory, and in particular to the middle surface torsion, which is very sensitive to the value of aspect ratio.

Specifically, for the relatively low value of aspect ratio $L/R = 10$, the maximum peak of percentage difference was located at the circumferential flexure mode $(q = 1, s = 2)$, where the effect of the middle surface change in the curvature k_θ was prevalent; see Figure 2. On the other hand, for the relatively high value of aspect ratio $L/R = 20$, the maximum peak of percentage difference was located at the beam-like mode $(q = 1, s = 1)$, where the effect of the middle surface torsion $k_{x\theta}$ prevailed; see Figure 10.

Conversely, by considering the parametric analyses presented in Figures 2–11, it can be observed that Sanders shell theory is able to model with very good accuracy the linear vibrations of SWCNTs for all considered geometries and wavenumbers, and therefore it can be adopted instead of the more complex Flügge shell theory to compute the natural frequencies of SWCNTs. This is due to the fact that, in the Sanders shell theory, the effect of the aspect ratio is not present (as the strain–displacement Equation (2) is very similar to that of the Flügge shell theory), but only the effect of the radius occurs.

In fact, from the numerical simulations carried out in the present work, it was found that for both the Donnell and Sanders shell theories, the difference in the natural frequencies with respect to the Flügge shell theory decreased with CNT radius, see as e.g., Figure 2 compared to Figure 6 (Donnell vs. Flügge), and Figure 3 compared to Figure 7 (Sanders vs. Flügge). This behaviour can be understood by observing once more the expressions of the middle surface change in curvature k_θ and torsion $k_{x\theta}$ of the shell; see Equation (2). Since in these two expressions, the radius R is located in the denominator, then, by increasing the radius R, their value (i.e., their influence) reduces, and therefore, the natural frequencies of the three theories tend to become closer.

Finally, in Figure 12, six different mode shapes of the simply supported SWCNT of Table 1 with chirality indices $(n = 5, m = 5)$ and aspect ratio $L/R = 10$ are shown, where the radial breathing mode $(q = 0, s = 0)$ corresponds to Rayleigh's inextensional symmetrical mode (i.e., uniform vibration). Moreover, it also shows the axisymmetric mode $(q = 1, s = 0)$ (no circumferential waves), beam-like mode $(q = 1, s = 1)$ (one circumferential wave, characteristic of structures with a very long aspect ratio, such as beams), and shell-like modes $(q = 1, s = 2-4)$ (two or more circumferential waves, characteristic of three-dimensional thin-walled structures, such as circular cylindrical shells). Such a graphical representation of the vibration modes could be useful for interpreting the previous results and comparisons.

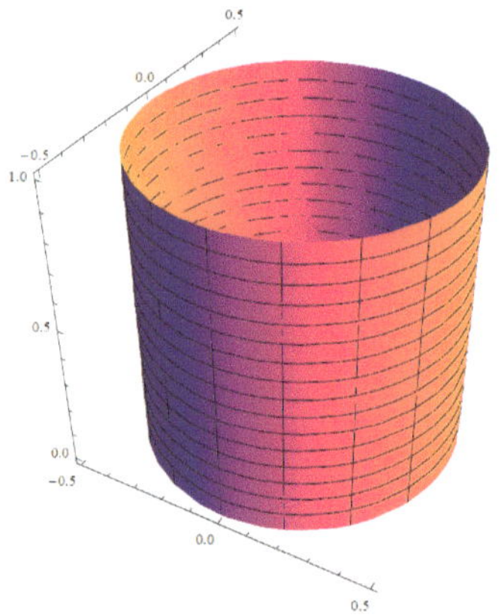

(a)

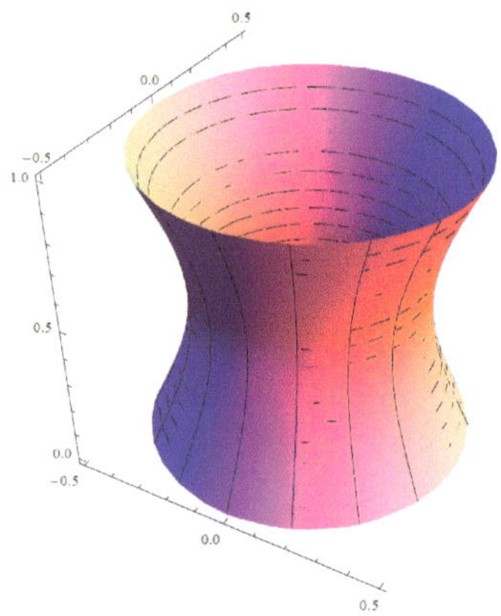

(b)

Figure 12. *Cont.*

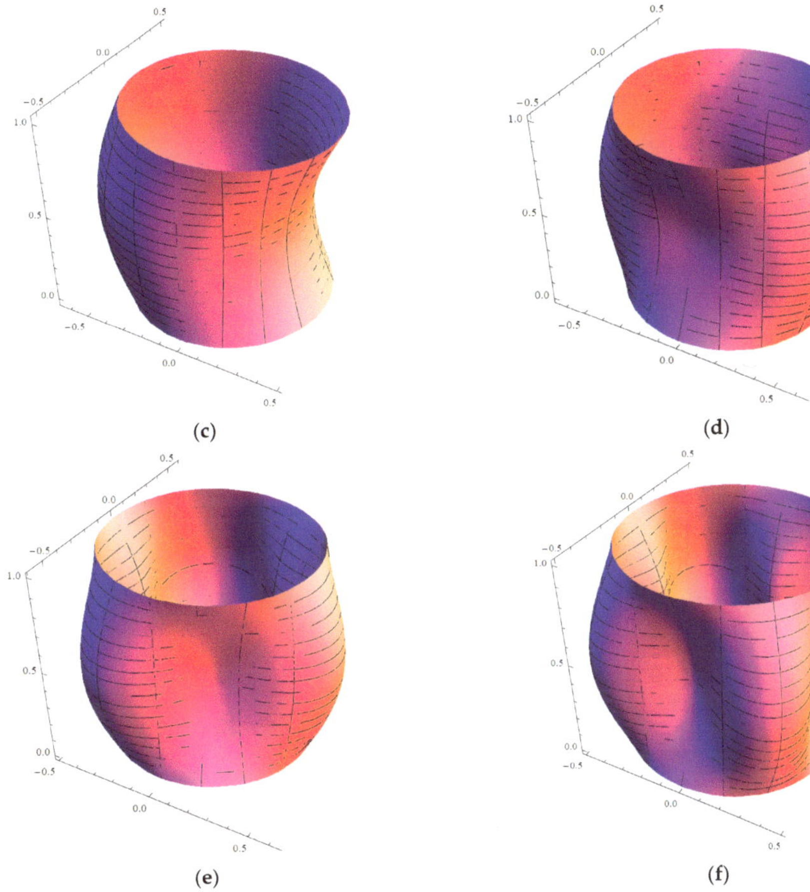

Figure 12. Mode shapes of the simply supported SWNT of Table 1 with chirality indices $(n = 5, m = 5)$ and aspect ratio $L/R = 10$: (**a**) Radial breathing mode $(q = 0, s = 0)$; (**b**) axisymmetric mode $(q = 1, s = 0)$; (**c**) beam-like mode $(q = 1, s = 1)$; (**d**) shell-like mode $(q = 1, s = 2)$; (**e**) shell-like mode $(q = 1, s = 3)$; and (**f**) shell-like mode $(q = 1, s = 4)$.

7. Conclusions

In this paper, the natural frequencies of SWCNTs obtained in the framework of the Donnell, Sanders, and Flügge shell theories were compared. An anisotropic elastic shell model was adopted to take into account the intrinsic chirality effects of CNTs. Simply supported boundary conditions were imposed. Vibration modes with different wavenumbers along the longitudinal and circumferential directions were studied. SWCNTs with different values of radius R and aspect ratio L/R were considered. The most important findings of the present paper are reported below.

- By means of comparisons with the results of molecular dynamic simulations reported in the literature, it was derived that the Flügge shell theory is the most accurate in the computation of the natural frequencies of SWCNTs.
- Since the Flügge shell theory requires a very high computational effort due to the large number of terms in the equations of motion, it was investigated whether a simpler shell theory is able to model with sufficient accuracy the linear vibrations of SWCNTs.

- It was found that the Donnell shell theory was not accurate for relatively low longitudinal and circumferential wavenumbers, for relatively low diameters, and for relatively high aspect ratios, and therefore, it is not able to properly model SWCNT vibrations.
- On the other hand, it was found that the Sanders shell theory was very accurate for all considered geometries and wavenumbers, and therefore, it can be adopted instead of the more complex Flügge shell theory to properly model SWCNT vibrations.

As first application of the results obtained in the present paper, the authors are planning to write a new manuscript on the effects of nonlocal elasticity and strain gradient on the linear vibrations of SWCNTs by considering an anisotropic elastic model in the framework of the Sanders shell theory.

A second relevant application of the findings of the present work regards the nonlinear vibrations of CNTs. It has been proven in the literature that, in the presence of a large number of carbon atoms, i.e., for relatively long or multi-walled carbon nanotubes, molecular dynamics simulations require higher computational effort than all equivalent continuous models, including the Flügge shell theory [3]. Moreover, by carrying out numerical simulations in linear field, it was found that all equivalent continuous models, among which are the Donnell, Sanders, and Flügge shell theories, are equally time costing [23]. The problem with the Flügge shell theory is due to the transition from linear to nonlinear analysis. On the one hand, the additional terms present in the expressions of forces and moments give the Flügge shell theory greater accuracy than that of Sanders and Donnell. On the other hand, these terms lead to very high computational effort in the numerical simulations in the nonlinear field, which are necessary to study the actual dynamic behaviour of carbon nanotubes and to investigate fluid–structure interactions. For this reason, especially for the nonlinear analyses, it is preferable to use the Sanders shell theory, since it is more accurate than the Donnell one (see the results of the present paper) and less computationally expensive than the Flügge one.

Author Contributions: Conceptualization: M.S. and I.E.E.; data curation: M.S. and M.B.; funding acquisition: M.C., R.R. and E.R.; investigation: M.S. and M.B.; methodology: M.S. and I.E.E.; project administration: M.S. and I.E.E.; supervision: M.C., R.R. and E.R.; visualization: M.S. and M.B.; writing—original draft: M.S., M.B. and I.E.E.; writing—review & editing: R.R., M.C. and E.R. All authors have read and agreed to the published version of the manuscript.

Funding: This research was funded by the Department of Sciences and Methods for Engineering, University of Modena and Reggio Emilia, Reggio Emilia, Italy grant number 020142_22_FRN_SOSTEGNO_RICERCA_DISMI.

Data Availability Statement: All data are available from the authors.

Acknowledgments: Authors M. Strozzi, E. Radi, M. Cocconcelli and R. Rubini are grateful to the Department of Sciences and Methods for Engineering, University of Modena and Reggio Emilia, Reggio Emilia, Italy (Grant 020142_22_FRN_SOSTEGNO_RICERCA_DISMI) for the financial support of this work. The present work is dedicated to the blessed memory of Leonid I. Manevitch, outstanding educator, respected professor, and admired scholar who died, after severe illness, on 20 August 2020, at the age of 82. The great passion and enthusiasm towards scientific research that Manevitch nurtured until the end of his life, in particular towards the much-loved carbon nanotubes, will remain forever in the heart of his collaborators, and will be for them a strong and ever-present motivation to progress in the knowledge of these wonderful structures.

Conflicts of Interest: The authors declare no conflict of interest.

References

1. Iijima, S. Helical microtubules of graphitic carbon. *Nature* **1991**, *354*, 56–58. [CrossRef]
2. Elishakoff, I.E.; Pentaras, D.; Dujat, K.; Versaci, C.; Muscolino, G.; Storch, J.; Bucas, S.; Challamel, N.; Natsuki, T.; Zhang, Y. *Carbon Nanotubes and Nanosensors: Vibration, Buckling and Ballistic Impact*; John Wiley & Sons: London, UK, 2012.
3. Jorio, A.; Dresselhaus, G.; Dresselhaus, M. Carbon Nanotubes. In *Advanced Topics in the Synthesis, Structure, Properties and Applications*; Springer: Berlin/Heidelberg, Germany, 2008.
4. Marulanda, J.M. *Carbon Nanotubes: Applications on Electron Devices*; InTech Open: Rijeka, Croatia, 2011.

5. Mahar, B.; Laslau, C.; Yip, R.; Sun, Y. Development of Carbon Nanotube-Based Sensors. A Review. *IEEE Sens. J.* **2007**, *7*, 266–284. [CrossRef]
6. Hierold, C.; Jungen, A.; Stampfer, C.; Helbling, T. Nano electromechanical sensors based on carbon nanotubes. *Sens. Actuators A* **2007**, *136*, 51–61. [CrossRef]
7. Rao, A.; Richter, E.; Bandow, S.; Chase, B.; Eklund, P.C.; Williams, K.; Fang, S.; Subbaswamy, K.R.; Menon, M.; Thess, A. Diameter-Selective Raman Scattering from Vibrational Modes in Carbon Nanotubes. *Science* **1997**, *275*, 187–191. [CrossRef] [PubMed]
8. Bandow, S.; Asaka, S.; Saito, Y.; Rao, A.; Grigorian, L.; Richter, E.; Eklund, P.C. Effect of the Growth Temperature on the Diameter Distribution and Chirality of Single-Wall Carbon Nanotubes. *Phys. Rev. Lett.* **1998**, *80*, 3779–3782. [CrossRef]
9. Dresselhaus, G.; Dresselhaus, M.S.; Hafner, J.H.; Hunter, M.; Jorio, A.; Lieber, C.M.; McClure, T.; Saito, R. Structural (n, m) Determination of Isolated Single-Wall Carbon Nanotubes by Resonant Raman Scattering. *Phys. Rev. Lett.* **2001**, *86*, 1118.
10. Gupta, S.S.; Bosco, F.G.; Batra, R.C. Wall thickness and elastic moduli of single-walled carbon nanotubes from frequencies of axial, torsional and inextensional modes of vibration. *Comput. Mater. Sci.* **2010**, *47*, 1049–1059. [CrossRef]
11. Cheng, H.C.; Liu, Y.L.; Wu, C.; Chen, W.H. On radial breathing vibration of carbon nanotubes. *Comput. Methods Appl. Mech. Eng.* **2010**, *199*, 2820–2827. [CrossRef]
12. Duan, W.; Wang, C.; Zhang, Y. Calibration of nonlocal scaling effect parameter for free vibration of carbon nanotubes by molecular dynamics. *J. Appl. Phys.* **2007**, *101*, 024305. [CrossRef]
13. Odegard, G.M.; Gates, T.S.; Nicholson, L.M.; Wise, K.E. Equivalent-Continuum Modeling of Nano-Structured Materials. *Compos. Sci. Technol.* **2002**, *62*, 1869–1880. [CrossRef]
14. Arroyo, M.; Belytschko, T. Continuum Mechanics Modeling and Simulation of Carbon Nanotubes. *Meccanica* **2005**, *40*, 455–469. [CrossRef]
15. Zhang, P.; Huang, Y.; Geubelle, P.H.; Hwang, K. On the continuum modeling of carbon nanotubes. *Acta Mech. Sin.* **2002**, *18*, 528–536.
16. Yakobson, B.I.; Brabec, C.J.; Bernholc, J. Nanomechanics of Carbon Tubes: Instabilities beyond Linear Response. *Phys. Rev. Lett.* **1996**, *76*, 2511–2514. [CrossRef]
17. Wang, C.; Ru, C.Q.; Mioduchowski, A. Applicability and Limitations of Simplified Elastic Shell Equations for Carbon Nanotubes. *J. Appl. Mech.* **2004**, *71*, 622–631. [CrossRef]
18. Silvestre, N.; Wang, C.; Zhang, Y.; Xiang, Y. Sanders shell model for buckling of single-walled carbon nanotubes with small aspect ratio. *Compos. Struct.* **2011**, *93*, 1683–1691. [CrossRef]
19. Silvestre, N. On the accuracy of shell models for torsional buckling of carbon nanotubes. *Eur. J. Mech.-A/Solids* **2012**, *32*, 103–108. [CrossRef]
20. Strozzi, M.; Pellicano, F. Nonlinear Resonance Interaction between Conjugate Circumferential Flexural Modes in Single-Walled Carbon Nanotubes. *Shock Vib.* **2019**, *2019*, 3241698. [CrossRef]
21. Leissa, A.W. *Vibration of Shells*; Acoustical Society of America: Columbus, OH, USA, 1993.
22. Yamaki, N. *Elastic Stability of Circular Cylindrical Shells*; Elsevier: Amsterdam, The Netherlands, 1984.
23. Amabili, M. *Nonlinear Vibrations and Stability of Shells and Plates*; Cambridge University Press: New York, NY, USA, 2008.
24. Soedel, W. *Vibrations of Shells and Plates*, 3rd ed.; Marcel Dekker: New York, NY, USA, 2004.
25. Ventsel, E.; Krauthammer, T. *Thin Plates and Shells: Theory, Analysis, and Applications*; The Pennsylvania State University, Marcel Dekker: New York, NY, USA, 2001.
26. Amabili, M. A comparison of shell theories for large-amplitude vibrations of circular cylindrical shells: Lagrangian approach. *J. Sound Vib.* **2003**, *264*, 1091–1125. [CrossRef]
27. Wang, C.M.; Tay, Z.Y.; Chowdhury, A.R.; Duan, W.H.; Zhang, Y.Y.; Silvestre, N. Examination of cylindrical shell theories for buckling of carbon nanotubes. *Int. J. Struct. Stab. Dyn.* **2011**, *11*, 1035–1058. [CrossRef]
28. Loy, C.T.; Lam, K.Y.; Reddy, J.N. Vibration of functionally graded cylindrical shells. *Int. J. Mech. Sci.* **1999**, *41*, 309–324. [CrossRef]
29. Pradhan, S.C.; Loy, C.T.; Lam, K.Y.; Reddy, J.N. Vibration characteristics of functionally graded cylindrical shells under various boundary conditions. *Appl. Acoust.* **2000**, *61*, 111–129. [CrossRef]
30. Arshad, S.H.; Naeem, M.N.; Sultana, N. Frequency analysis of functionally graded material cylindrical shells with various volume fractions laws. *J. Mech. Eng. Sci.* **2007**, *221*, 1483–1495. [CrossRef]
31. Carrera, E. Historical review of Zig-Zag theories for multi-layered plates and shells. *Appl. Mech. Rev.* **2003**, *56*, 287–308. [CrossRef]
32. Dumir, P.C.; Nath, J.K.; Kumari, P.; Kapuria, S. Improved Efficient Zigzag and Third Order Theories for Circular Cylindrical Shells Under Thermal Loading. *J. Therm. Stress.* **2008**, *31*, 343–367. [CrossRef]
33. Rahmani, O.; Khalili, S.M.R.; Malekzadeh, K. Free vibration response of composite sandwich cylindrical shell with flexible core. *Comp. Struct.* **2010**, *92*, 1269–1281. [CrossRef]
34. Lakis, A.; Selmane, A.; Toledano, A. Non-linear free vibration analysis of laminated orthotropic cylindrical shells. *Int. J. Mech. Sci.* **1998**, *40*, 27–49. [CrossRef]
35. Chang, T.; Geng, J.; Guo, X. Prediction of chirality- and size-dependent elastic properties of single-walled carbon nanotubes via a molecular mechanics model. *Proc. R. Soc. A Math. Phys. Eng. Sci.* **2006**, *462*, 2523–2540. [CrossRef]
36. Chang, T. A molecular based anisotropic shell model for single-walled carbon nanotubes. *J. Mech. Phys. Solids* **2010**, *58*, 1422–1433. [CrossRef]

37. Strozzi, M.; Elishakoff, I.E.; Manevitch, L.I.; Gendelman, O.V. Applicability and limitations of Donnell shell theory for vibration modelling of double-walled carbon nanotubes. *Thin-Walled Struct.* **2022**, *178*, 109532. [CrossRef]
38. Strozzi, M.; Smirnov, V.V.; Pellicano, F.; Kovaleva, M. Nonlocal anisotropic elastic shell model for vibrations of double-walled carbon nanotubes under nonlinear van der Waals interaction forces. *Int. J. Non-Linear Mech.* **2022**, *146*, 104172. [CrossRef]
39. Ghavanloo, E.; Fazelzadeh, S. Vibration characteristics of single-walled carbon nanotubes based on an anisotropic elastic shell model including chirality effect. *Appl. Math. Modell.* **2012**, *36*, 4988–5000. [CrossRef]
40. Fazelzadeh, S.; Ghavanloo, E. Nonlocal anisotropic elastic shell model for vibrations of single-walled carbon nanotubes with arbitrary chirality. *Comp. Struct.* **2012**, *94*, 1016–1022. [CrossRef]

nanomaterials MDPI

Article

Growth and Characterization of Carbon Nanofibers Grown on Vertically Aligned InAs Nanowires via Chemical Vapour Deposition

Muhammad Arshad [1,2,3], Lucia Sorba [4], Petra Rudolf [2] and Cinzia Cepek [1,*]

1 Istituto Officina dei Materiali—CNR, Laboratorio TASC Area Science Park—Basovizza, Edificio MM, Strada Statale 14, Km.163.5, I-34149 Trieste, Italy; muhammad.arshad@ncp.edu.pk
2 Zernike Institute for Advanced Materials, University of Groningen, Nijenborgh 4, NL-9747AG Groningen, The Netherlands; p.rudolf@rug.nl
3 Nanosciences and Technology Department, National Center for Physics, Quaid-i-Azam University Campus, Islamabad 2141, Pakistan
4 NEST, Istituto Nanoscienze-CNR and Scuola Normale Superiore, Piazza S. Silvestro 12, I-56127 Pisa, Italy; lucia.sorba@nano.cnr.it
* Correspondence: cepek@iom.cnr.it

Abstract: The integration of carbon nanostructures with semiconductor nanowires holds significant potential for energy-efficient integrated circuits. However, achieving precise control over the positioning and stability of these interconnections poses a major challenge. This study presents a method for the controlled growth of carbon nanofibers (CNFs) on vertically aligned indium arsenide (InAs) nanowires. The CNF/InAs hybrid structures, synthesized using chemical vapor deposition (CVD), were successfully produced without compromising the morphology of the pristine nanowires. Under optimized conditions, preferential growth of the carbon nanofibers in the direction perpendicular to the InAs nanowires was observed. Moreover, when the CVD process employed iron as a catalyst, an increased growth rate was achieved. With and without the presence of iron, carbon nanofibers nucleate preferentially on the top of the InAs nanowires, indicating a tip growth mechanism presumably catalysed by a gold-indium alloy that selectively forms in that region. These results represent a compelling example of controlled interconnections between adjacent InAs nanowires formed by carbon fibers.

Keywords: carbon nanofibers; indium arsenide nanowires; hybrid nanostructures; chemical vapor deposition; interconnects

Citation: Arshad, M.; Sorba, L.; Rudolf, P.; Cepek, C. Growth and Characterization of Carbon Nanofibers Grown on Vertically Aligned InAs Nanowires via Chemical Vapour Deposition. *Nanomaterials* **2023**, *13*, 3083. https://doi.org/10.3390/nano13243083

Academic Editors: Muralidharan Paramsothy and Jakob Birkedal Wagner

Received: 15 August 2023
Revised: 24 November 2023
Accepted: 27 November 2023
Published: 5 December 2023

1. Introduction

Carbon nanotubes (CNTs) and carbon nanofibers (CNFs) possess exceptional mechanical and electrical properties, making them promising materials for interconnecting wires in future very large-scale integration technology [1,2]. Simulations have indicated that the use of metallic CNT interconnects could lead to more energy-efficient and faster integrated circuits [3]. However, achieving precise control over the interconnection of individual nanowires (NWs) with CNFs/CNTs remains a significant challenge. One promising approach for integrating CNTs into real devices is the synthesis of controlled CNT/NW hybrid nanostructures. The possibility of obtaining well-defined CNT/metal or semiconducting NW hybrid structures has been widely explored in the literature [4] but controlled growth for practical electronic devices has not yet been realized. Semiconducting nanowires (NWs) can be grown in ordered and oriented arrays on semiconducting substrates using chemical beam epitaxy (CBE), and these precisely ordered one-dimensional nanostructures can serve as templates for synthesizing CNTs/CNFs and obtaining the desired hybrid structures [5]. Among the various techniques for CNT/CNF growth, catalytic chemical vapor deposition (CVD) holds the most promise due to its ability to control the location and diameter of the tubular structures, which is determined by the position and size of the catalyst nanoparticles. Moreover, the relatively low growth temperatures (400–900 °C) involved in CVD enable direct deposition onto electronic devices [6,7]. CVD, therefore, provides a cost-effective and straightforward means of synthesizing patterned building blocks for large-scale hybrid nanostructure fabrication.

In this study, we conducted CVD growth using C_2H_2 as the precursor gas on a network of vertically aligned InAs NWs, demonstrating the synthesis of CNFs without compromising the integrity of the pristine NW ensemble. Generally, the CVD growth of CNTs/CNFs requires transition metal (Ni, Fe, Co) catalysts to decompose the gaseous carbonaceous precursor [6]. Hence, we also investigated growth in the presence of Fe as a catalyst. The combination of an in situ X-ray photoelectron spectroscopy (XPS) study, ex situ characterization using scanning electron microscopy (SEM), and Raman spectroscopy allowed for a comprehensive analysis of the samples. These findings contribute to our understanding of how to realize controlled interconnections between adjacent nanowires with carbon fibers and hold great potential for the application of such hybrid nanostructures in energy-efficient integrated circuits and nanoscale devices.

2. Materials and Methods

2.1. *Growth of the Indium Arsenide Nanowire Arrays*

The vertical InAs NWs used in these experiments were fabricated at the Istituto Nanoscienze-CNR (Pisa, Italy). The InAs NWs have grown on an InAs(111)B substrate using chemical beam epitaxy (CBE) as detailed in ref. [8]. InAs substrates coated with a 0.1 nm thick Au film were introduced into the CBE chamber; pre-growth thermal treatments were performed in situ in the CBE chamber. The growth of all NWs followed the same protocol: (i) a temperature ramp was applied, starting from the standby temperature (~250 °C) and gradually increasing to the annealing temperature (520 $\pm$ 10 °C), while maintaining a tertiarybutylarsine (TBAs, Dockweiler Chemicals GmbH, Marburg, Germany, epigrade) line pressure of 5.3 mbar. The annealing step ensured the formation of Au nanoparticles through the thermal dewetting of the film and facilitated the desorption of surface oxide from the InAs substrate. (ii) Following the annealing step, the temperature was gradually decreased from the annealing temperature to a growth temperature of 430 $\pm$ 10 °C. (iii) The growth duration was 45 min, maintaining a line pressure of 0.4 mbar and 2.7 mbar for trimethylindium (TMIn, Dockweiler Chemicals GmbH, Marburg, Germany, optoelectronic grade) and TBAs, respectively. (iv) Subsequently, the temperature was lowered to ~250 °C without the presence of TMIn, while maintaining a linearly decreasing TBAs line pressure [8]. Transport to Trieste implied exposure to air.

2.2. *Chemical Vapor Deposition of the Carbon Nanofibers*

CVD growth and XPS analysis were performed in the INSPECT laboratory of the TASC-IOM-CNR institute, where all steps of the CVD process, including catalyst deposition, can be carried out in an ultra-high vacuum experimental setup (base pressure $< 1 \times 10^{-10}$ mbar). The typical CVD growth protocol comprised preliminary out-gassing of the InAs substrate at 300–400 °C using a home-made silicon heater, followed by H_2 (SIAD, Bergamo, Italy, grade 5) exposure to pre-treat and clean the surface (*vide infra*); the pressure in the growth system during this treatment was ~4×10^{-4} mbar. The temperature was measured using an infrared pyrometer. The typical duration of these pre-treatments was 15 min. Subsequently, the sample was annealed to the growth temperature (chosen in the range 490–550 °C) and CVD was performed with and without a catalyst (iron), typically for 25–40 min with C_2H_2 (SIAD, Bergamo, Italy, grade 2.5) and H_2 (SIAD, Bergamo, Italy, grade 5) as precursor gasses in the flux range 0.5–7.0 sccm; the pressure in the growth chamber during the process was in the range of 10^{-4} mbar. When a catalyst was used, Fe films were deposited using electron bombardment (Fe target from Sigma-Aldrich, St. Louis, MI, USA, 99.9% purity) onto the InAs substrate kept at room temperature at a growth rate of ~0.35 Å/min before starting CVD. The Fe deposition rate was obtained from the attenuation of the photoemission peaks of the In3*d* core level and confirmed using a home-made quartz microbalance. Catalyst deposition was followed by annealing to the chosen growth temperature.

2.3. Surface Characterization

X-ray photoelectron spectroscopy: The growth chamber is directly connected with the XPS analysis chamber, equipped with a conventional non-monochromatized MgKα X-ray source (hν = 1253.6 eV) and a 120° hemispherical electron energy analyser (RESOLVE 120, PSP Vacuum Technology, Macclesfield, UK). Thus, it was possible to control the chemical state of the substrate before and after the growth via XPS and to precisely monitor the influence of all CVD growth parameters. The XPS spectra were acquired before and after all CVD steps in normal emission geometry, with an energy resolution of 0.8 eV. Analysis of the data was conducted by performing a non-linear mean square fit, reproducing the photoemission intensity using Doniach-Sunjic line shapes superimposed onto a Shirley background. Binding energies were calibrated by fixing the C1*s* binding energy of adventitious carbon to 284.6 eV [9]. Binding energy (BE) positions are given with a precision of $\pm$0.1 eV.

Scanning Electron Microscopy: SEM observations were performed using a Field Emission In-lens SEM (JSM 890, JEOL, Basiglio, MI, Italy) at 10 kV accelerating voltage and a 1×10^{-12} A probe current, delivering a spatial resolution of 1 nm. Images were obtained with both secondary electron and back-scattered signals. The instrument is equipped with an X-Act detector (Oxford Instruments) for energy dispersive X-ray spectroscopy (EDS).

Raman spectroscopy: Raman scattering measurements were performed with a μ-Raman spectrometer (Renishaw 1000, Wotton-under-Edge, UK) using the 514.5 nm line of an Ar^+ laser as an excitation source with a spot diameter of about 1 μm. All measurements were performed at room temperature in backscattering geometry, with the wires in the plane of incidence of the incoming light.

3. Results and Discussion

3.1. Substrate Pre-Treatment

Scanning electron micrographs of Au-catalysed InAs NWs grown using chemical beam epitaxy on an InAs(111)B substrate are presented in Figure 1a,c and show vertically aligned nanowires of different diameters, which randomly cover the substrate surface. The length of the InAs nanowires seems different after H_2 pre-treatment because the SEM images are collected at different tilt angles. Energy-dispersive X-ray spectroscopy (not shown) performed together with scanning electron microscopy revealed that the Au nanoparticles, used as catalysts, reside at the tips of the InAs NWs. Note that the density of InAs NWs varies from sample to sample; when discussing the CNF growth, we shall, therefore, refer to the density of CNFs/InAs nanowire, expressed as a percentage. We tested the thermal stability of the InAs NWs to annealing in UHV at increasing temperatures using the same annealing time as typically used in the CVD process ($\approx$25 min). The SEM images in Figure 1 demonstrate that no melting or significant morphological changes occur up to ~520 °C when annealing without (Figure 1b) or with H_2 (Figure 1d), annealing at a temperature higher than 530 °C partially destroys the NWs (note that the InAs melting temperature in bulk form is 942 °C [10] and the InAs NW growth temperature is 420–440 °C). The H_2 pre-treatment is essential because when CVD growth (*vide infra*) was performed without this pre-treatment, we never observed the growth of tubular carbon nanostructures, and the SEM images were similar to those of the as-received samples (not shown). This is most probably due to the oxide layer formed on the whole surface of the substrate and of the NWs after air exposure, which does not support any CNF nucleation within the CVD parameter window we used. The optimized procedure before CVD was, therefore, the following: the InAs NWs were first annealed at ~430 °C for 10 min, then underwent the hydrogen pre-treatment at ~520 °C using a flux of 3 sccm for 15 min (pressure during the treatment: $\sim 4 \times 10^{-4}$ mbar).

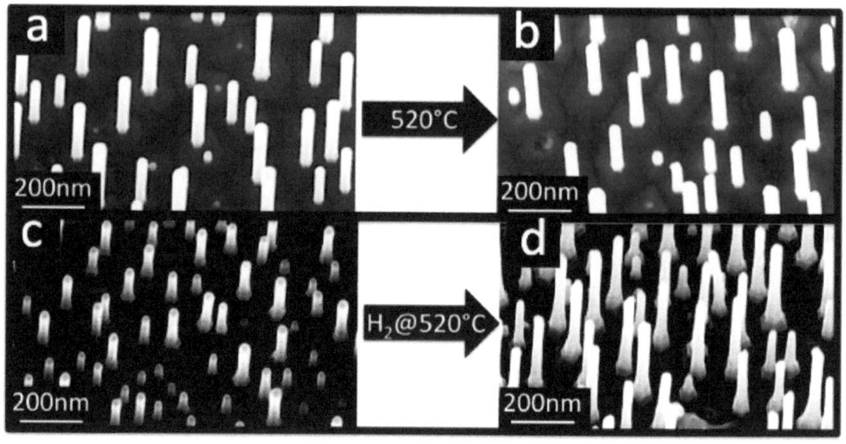

Figure 1. SEM images of as-grown InAs NWs (**a**,**c**) and after pre-treatment of the same sample at 520 °C without (**b**) or with (**c**) 3 sccm of H_2 (see text for more details). The images (**a**,**b**) were obtained at a tilt angle of 10.0° and the images (**c**,**d**) were acquired at tilt angles of 10.0° and 20.0°, respectively, which explains why the length of the InAs NWs seems different.

To evaluate whether the cleaning procedure was efficient in removing the oxide layer formed by exposure to air, the chemical effects of these treatments were studied using X-ray photoelectron spectroscopy. Figure 2 shows the As3*d* and In3*d* XPS spectra of the as-received sample, after annealing at 430 °C, 500 °C and after H_2 pre-treatment at 525 °C. Both the In3*d* and As3*d* core level spectra of the as-received sample (Figure 2a,b bottom) are fitted with two components; the more intense one corresponds to the In-As bond (In3$d_{5/2}$ peaked at a BE of ≈444.6 eV, As3*d* at ≈41.0 eV, marked in blue). The other component in each spectrum is associated with the native surface oxide (marked in green), composed of In_2O_3 (In3$d_{3/2}$ component peaked at a BE of ≈445.5 eV) and of a mixture of As_2O_3 and As_2O_5, giving rise to the As3*d* component peaked at a BE of ≈45.6 eV [11,12]. The weak peak at about 443 eV in the In3*d* spectrum (marked in orange) is due to the MgKα satellites.

Annealing at 430 °C leads to the partial desorption of In oxide and the complete desorption of the As oxide (Figure 2a,b second spectra from bottom), as well as causing partial desorption of adventitious carbon (data not shown). Annealing at 500 °C and H_2 pre-treatment at 525 °C result in a slight further desorption of carbon and also of oxygen. The desorption of the oxides (Figure 2a,b top spectra) goes hand in hand with a change in the ratio between the As3*d* and In3*d* intensities. While the as-received InAs NWs are stoichiometric within the experimental error, annealing at high temperatures causes the partial desorption of As, with the extent increasing as the temperature rises (Figure 2a, inset). It is worth noting that the photoelectron escape depth in our experimental conditions is approximately 25 Å for As and 18 Å for In [13]. Therefore, As desorption occurs at least within the topmost 2–3 nm but does not affect the core of the NWs, as evidenced by the Raman spectra discussed below. Arsenic desorption after annealing in UHV is a well-known phenomenon resulting from the partial decomposition of InAs. The temperature and rate of arsenic desorption depend on the structure of the InAs NWs [14].

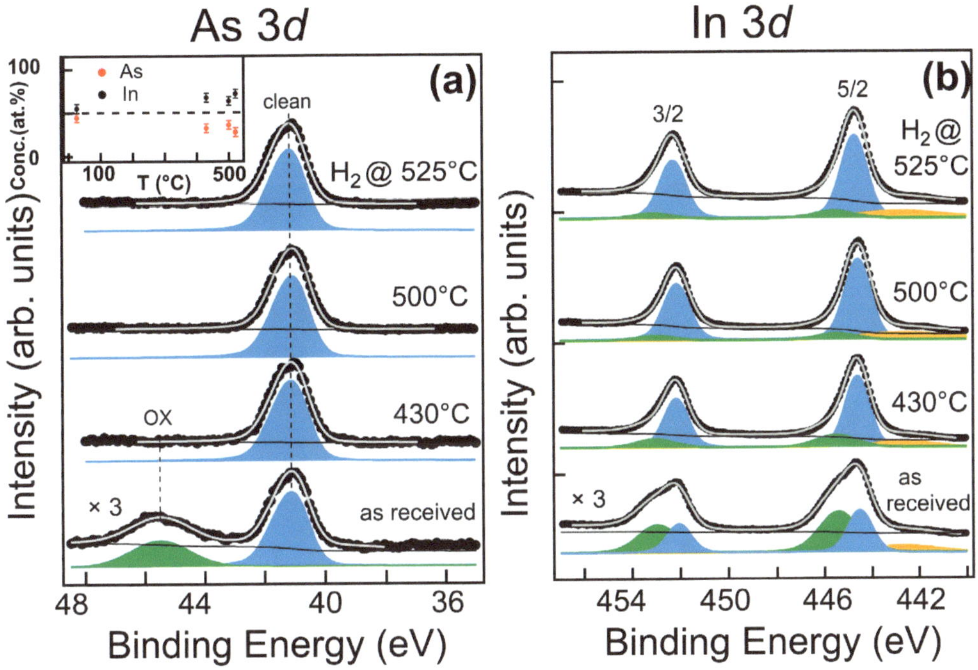

Figure 2. XPS spectra of the As3*d* (**a**) and In3*d* (**b**) core level regions of as-received InAs nanowires (bottom), after annealing at 430 °C, 500 °C and after H_2 pre-treatment at 525 °C (top) and corresponding fits. The inset shows the As (red) and In (black) concentrations after the same three pre-treatments. The dotted line corresponds to stoichiometric InAs (50%). The green and blue components refer to, respectively, oxidized and non-oxidized InAs-related components, while the yellow component is due to the MgKα satellites (see text for more details).

3.2. CVD Synthesis

3.2.1. Scanning Electron Microscopy Results

Synthesis of tubular carbon was successful only on substrates that underwent H_2 pre-treatment. CVD growth was carried out with and without the use of iron as a catalyst, employing various H_2 and C_2H_2 flow rates ranging from 0.5 sccm to 7.0 sccm. The growth temperatures were within the range of 470–550 °C, i.e., in a temperature window where the density, orientation, and length of the InAs NWs remained nearly unchanged after annealing in ultra-high vacuum (Figure 1) and beyond. When the growth was attempted at temperatures below approximately 490 °C, no significant changes compared to the as-grown InAs substrate were observed. This suggests that these temperatures are insufficient to decompose C_2H_2 on InAs NWs in our experimental conditions. The formation of new structures at the tips of the nanowires was observed within the temperature window of 500–530 °C. Figure 3a shows the SEM image acquired after CVD growth performed without catalyst at ~525 °C using a flux of ~7 sccm of C_2H_2 for 20 min (pressure in the preparation chamber during growth: ~8×10^{-4} mbar). In this case, a deposit of almost spherical nanoparticles is visible on both the InAs NWs and the InAs substrate, together with short tubular structures mainly located at the NW tips. As seen in the SEM micrograph shown in Figure 3b, the addition of H_2 to C_2H_2 during CVD helped to nearly eliminate the presence of the spherical carbon nanoparticles and to enhance the synthesis of CNFs. In this case, we used an acetylene flux of 3.6 sccm together with a hydrogen flux of 2 sccm at 525 °C (pressure during growth ~4.5×10^{-4} mbar). We observed that in the appropriate growth conditions, these tubular structures have diameters of a few nanometers, nucleate preferentially at the InAs NW tips, grow along the direction perpendicular to the NW axis, and sometimes connect two NWs. In detailed SEM micrographs of the sample of Figure 3b, which are shown in Figure 4a,b, one clearly distinguishes the nucleation of CNFs at the NW tips (a) and the establishment of a CNF bridge (b) between two InAs NWs. That these tubular structures are indeed carbon nanofibers is demonstrated by the Raman spectra discussed below.

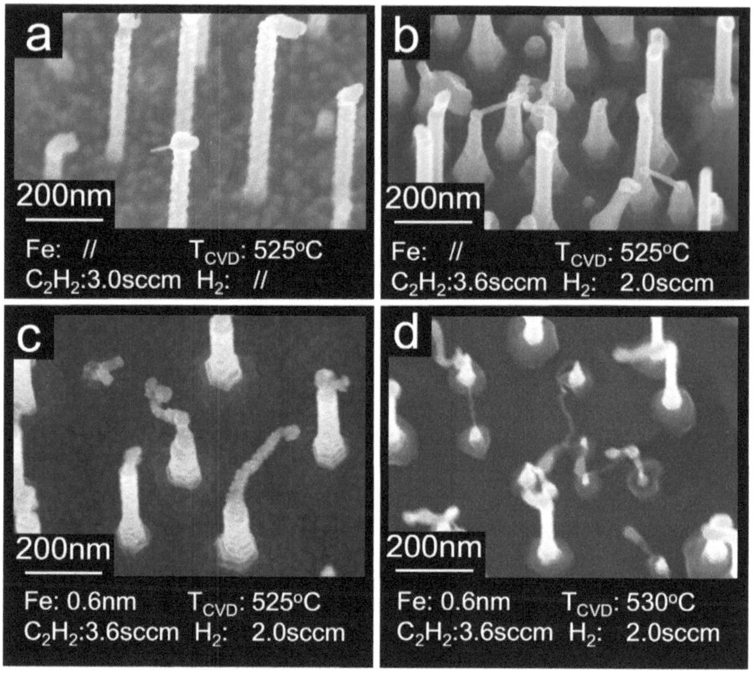

Figure 3. SEM images collected after different CVD processes: (**a**) without catalyst, ~8×10^{-4} mbar of C_2H_2 (3.0 sccm) for 20 min at ~525 °C; (**b**) without catalyst, ~4.5×10^{-4} mbar of C_2H_2 (3.6 sccm) and H_2 (2.0 sccm) for 40 min at 525 °C; (**c**) with catalyst (0.6 nm Fe), same CVD condition as (**b**); (**d**) same conditions as (**c**) but at higher temperature (530 °C), where it is clear that part of NWs are strongly distorted and/or etched (see text for more details). All the images were collected at a tilt angle of 20.0° (**a**,**b**,**d**) and 10.0° (**c**).

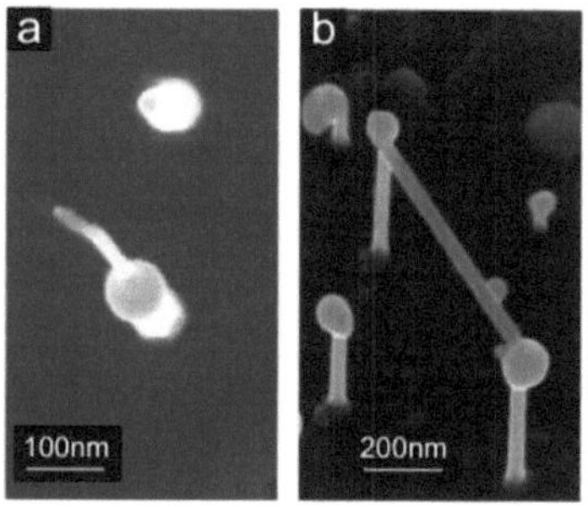

Figure 4. SEM images collected after different CVD processes: (**a**,**b**) are detailed views of the sample shown in Figure 3b grown without catalyst, with ~4.5×10^{-4} mbar of C_2H_2 (3.6 sccm) and H_2 (2.0 sccm) for 40 min at 525 °C. The images were collected at a tilt angle of 30.0° (**a**) and 0.0° (**b**).

In our attempt to optimize the growth parameters, we explored whether increasing the C_2H_2 flux, the H_2 flux, the growth time, and temperature could increase the density of CNFs produced. The results are shown in Figure 5. We found that the number of CNFs per InAs NW depends strongly on the growth conditions; in particular, first, we established, as illustrated in Figure 5a for a H_2 flux of 2 sccm, that a C_2H_2 flux of 3.6 sccm was optimal; we established that the maximum percentage of CNFs per InAs NW without Fe catalyst obtainable in our parameter window was around 15% (Figure 5a blue dots), while it reached 50% with the iron catalyst (Figure 5a red dots). The impact of growth temperature on the number of CNFs per InAs NW with and without the iron catalyst was explored as well and the results are summarized in Figure 5b. Notably, at temperatures below the melting point of the InAs NWs, the density of CNFs increases, reaching 50% when the catalyst is present (Figure 5b red dots), which is three times higher than that without the use of an iron catalyst (Figure 5b blue dots), while the tube diameter decreases. When attempting CVD growth at temperatures of 530 °C or higher, the SEM images (see for ex. Figure 3d) revealed partial melting of InAs nanowires, consistent with previous literature findings [15]. At temperatures exceeding 540 °C, no tubular structures were observed under any tested conditions, and the SEM images showed partially or completely melted InAs NWs (not shown).

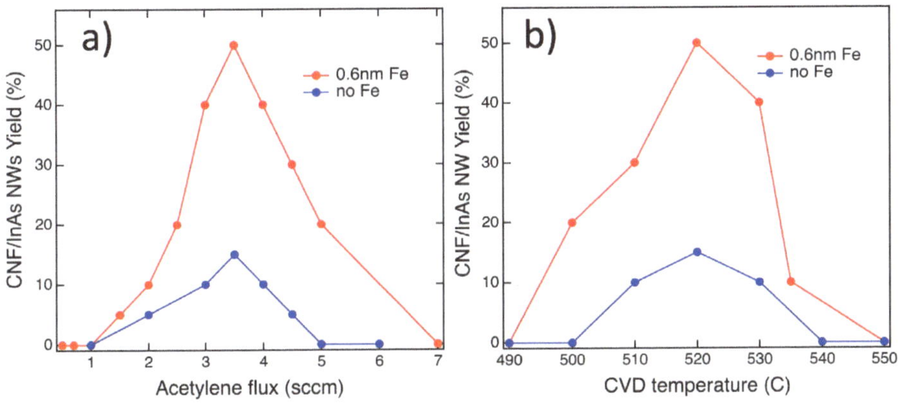

Figure 5. Yield of CNF growth: percentage of CNFs connecting InAs NWs when (**a**) the H_2 (2.0 sccm) flux is kept constant but the C_2H_2 flux is varied during CVD growth at 525 °C with (red dots) and without the iron catalyst (blue dots); (**b**) both the C_2H_2 (3.6 sccm) flux and the H_2 (2.0 sccm) flux are kept constant but the temperature is varied during CVD growth with (red dots) and without the iron catalyst (blue dots).

Figure 6 further illustrates CVD growth at the optimal conditions 525 °C, C_2H_2 flux 3.6 sccm and H_2 flux 2.0 sccm) with and without the iron catalyst. It is evident that the InAs NWs undergo shape changes, becoming approximately 15% to 20% shorter in length compared to before the growth, indicating partial melting. As seen in the SEM micrograph in Figure 6a, which refers to the sample shown in Figure 3c but on a larger scale, depositing a thin film of iron ($\approx$0.6 nm) before CVD growth causes a strong increase in the density of the tubular structures, which are also significantly longer and thinner than those obtained when growing without Fe (Figure 6b).

3.2.2. Spectroscopic Characterization

To gain a better understanding of the structures grown by CVD and the chemical changes induced by the growth process, Raman and XPS spectra were collected. While we have referred to the resulting structures as carbon nanofibers based on the CVD growth, the confirmation comes solely from the Raman data.

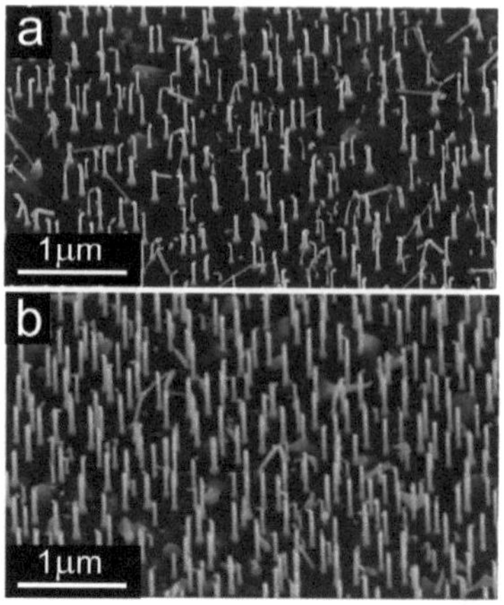

Figure 6. SEM image of (**a**) the sample shown in Figure 3c referring to CVD growth (~4.5×10^{-4} mbar of C_2H_2 (3.6 sccm) and H_2 (2.0 sccm) for 40 min) at 525 °C after deposition of 0.6 nm Fe on a larger scale; (**b**) the sample shown in Figure 3b referring to CVD growth under the same conditions but without the catalyst, also on a larger scale. These images were collected with a sample tilt of 20.0°.

Figure 7 displays the Raman spectra acquired on the InAs substrate covered with the as-received InAs NWs, after annealing at 430 °C, and after CVD growth as described in Section 2.2 above, namely approximately 4.5×10^{-4} mbar of C_2H_2 (3.6 sccm) and H_2 (2.0 sccm) for 40 min at 525 °C, without Fe catalyst. This sample is the one also shown in Figure 3b. The peaks at 212 cm^{-1} and 237 cm^{-1} (Figure 7, left panel) correspond to the typical tangential optical (TO) and longitudinal optical (LO) phonon modes of the InAs NWs [16]. These spectra show no significant differences before and after growth, indicating that the NWs remain largely unaffected by the various processes. No additional peaks appeared in this spectral region after CVD growth, ruling out the production of single-walled carbon nano- tubes (CNTs) under our growth conditions. In fact, if single-walled CNTs were formed, one would expect to observe the radial breathing mode (RBM) signature in this spectral region [17,18].

At higher wavenumbers, the spectrum acquired after CVD growth (Figure 7, right panel) exhibits a G peak at 1592 cm^{-1}, which is characteristic of sp^2 carbon, and a strong D peak at 1368 cm^{-1}, indicative of defects and disordered graphitic material [19–23], Both the D and G peaks are broad, suggesting the presence of disordered carbon and non-crystalline structures, such as carbon nanofibers and amorphous nanoparticles [24–26]. The intensity ratio I_D/I_G informs qualitatively on how ordered a carbon material is [27] and the high I_D/I_G ratio of 0.75, deduced from the spectrum in Figure 7, corresponds to a low degree of order typical for CNFs [28]. No significant changes in the Raman spectra were observed when the CVD growth was performed in the presence of the Fe catalyst (data not shown), indicating that in all cases, the carbon structures are similar and highly disordered.

In all instances (with and without H_2 during CVD growth, and with and without deposition of Fe), CNFs preferentially nucleate at the tips of the InAs NWs, where the Au nanoparticles used as catalysts for NW growth are located [29]. This suggests that the tips play a crucial role in CNF synthesis. It should be noted that if CNF growth were catalysed by In, As, and/or Fe (when used), one would expect to observe CNF nucleation on the substrate surface as well, along with potential chemical interactions between In or As and carbon, which could be detected using XPS. Additionally, the formation of iron compounds with indium and/or arsenide is unlikely at our growth temperatures [30,31].

In order to attempt to gain insights into the nucleation and growth process via the bonds formed, we acquired XPS spectra of the 3*d* core levels of In and As, as well as of the C1*s* core level region before and after CVD growth; the results are presented in Figure 8. The

C1*s* spectrum (Figure 8a) obtained after CVD growth reveals a sharp intensity decrease in the component at approximately 287.2 eV in binding energy (green component), primarily attributed to (C-O_x) contaminants that were apparently largely etched away during the CVD process. Furthermore, there is a decrease in the component at a BE of approximately 285.2 eV (white component), attributed to disordered and/or sp^3 carbon, and an increase in the component related to sp^2 bonds at approximately 284.4 eV BE (blue component) due to fiber nucleation. Meanwhile, the As3*d* spectrum after CVD growth (Figure 8b, top) shows the appearance of a component at a BE of 43.4 eV (green component), corresponding to As^{+1} [32]. This suggests the formation of As-C bonds [33,34]. It is already known that Au nanoparticles can catalyse the decomposition of InAs during high annealing in UHV [14,35]. Arsenic exhibits a low solubility in Au and at high temperatures, it sublimates in the form of As_x. Therefore, during the CVD process, it can react with C_2H_2 to form As carbide, as revealed by our XPS data. On the other hand, Indium has a high solubility in Au and readily forms Au-In alloys [36]. Unfortunately, due to the low Au concentration in our sample, we were not able to distinguish the formation of an In-Au alloy, and the In3*d* spectra acquired before and after the CVD process (Figure 8c) did not show any significant differences. Nevertheless, it is worth noting that the $AuIn_2$ alloy in the form of nanoparticles has previously been found to catalyze the growth of InAs nanotrees even at very low temperatures [29] and it may be responsible for the CNF synthesis in our case. The nanometric size of the Au-In wires in the samples studied here can further enhance their reactivity, as already observed in several nanostructured materials [37–40].

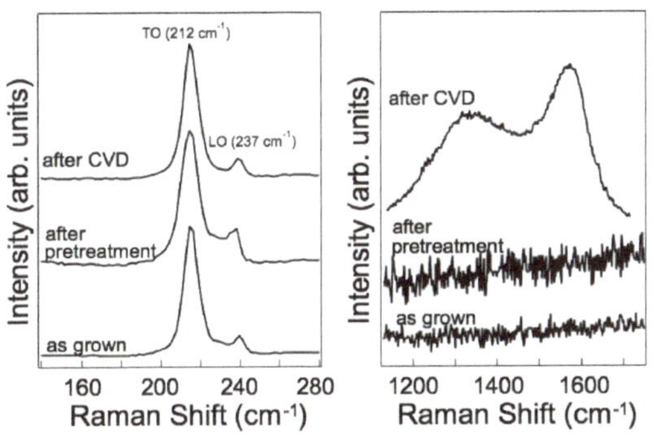

Figure 7. Raman spectra of as-grown InAs NWs, after annealing at 430 °C and after CVD growth resulting in the sample shown in Figure 3b. The left panel shows the low wavenumber region, while the right panel displays the high wavenumber region.

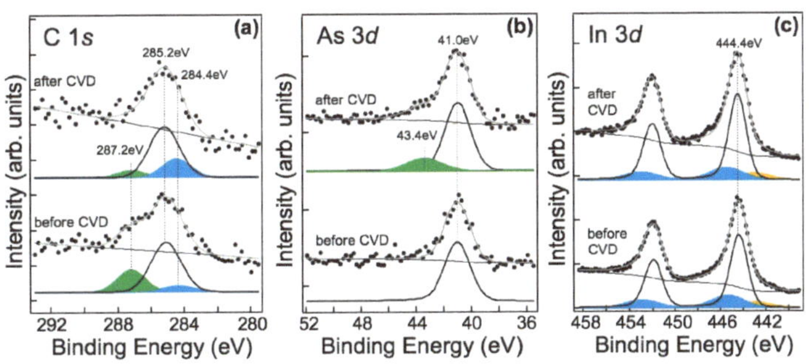

Figure 8. XPS spectra of the (**a**) C1*s*, (**b**) As3*d*, and (**c**) In3*d*, core level regions of recorded in situ on the InAs nanowire array before and after CVD and corresponding fits. The weak peak at about 443 eV in the In3*d* spectra (filled in yellow) is due to the Mg Kα satellites. The attribution of the various components in each spectrum is discussed in the text.

4. Conclusions

As summarized in the scheme in Figure 9, carbon nanofiber-InAs hybrid nanostructures were successfully synthesized by CVD at optimized growth conditions. SEM micrographs confirmed that the original ensemble of InAs nanowires is thermally stable (preserved) during annealing, H_2 pre-treatment, and CVD growth of carbon nanofibers.

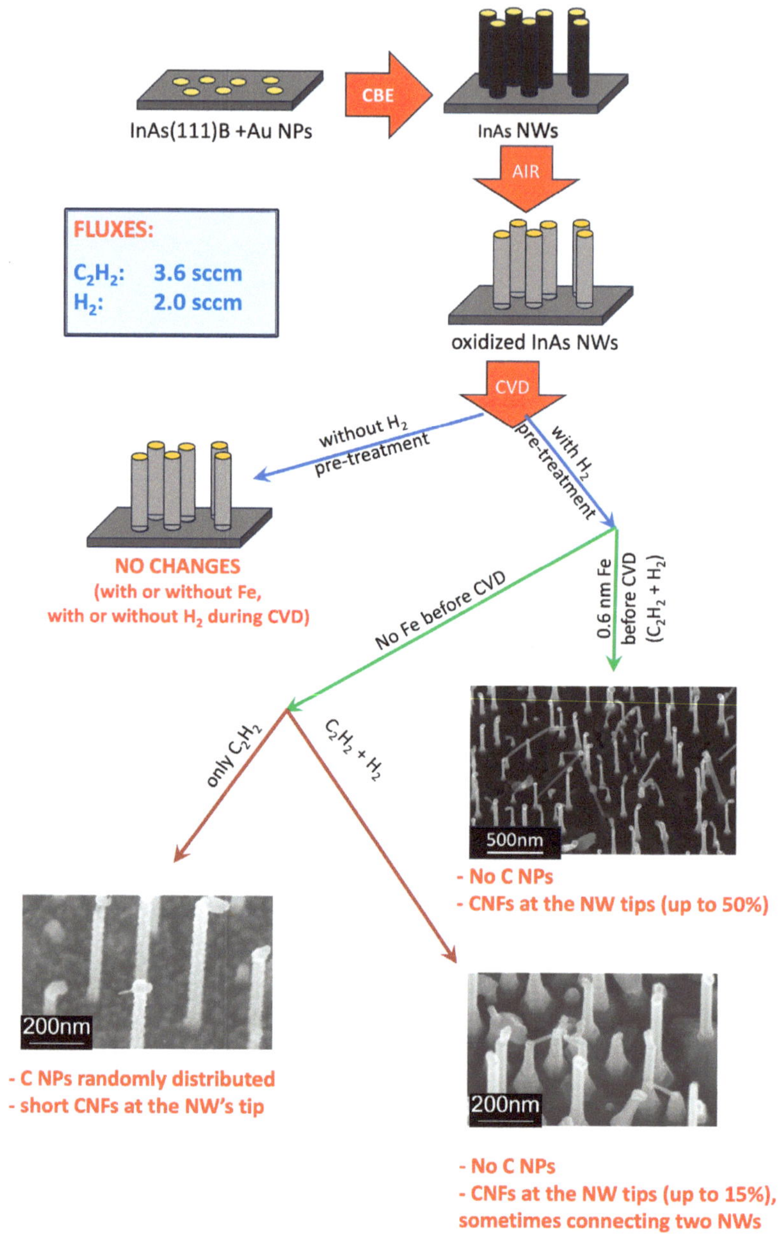

Figure 9. Scheme summarizing the procedures and findings.

When the growth was performed only with acetylene, carbon nanoparticles (C NPs) were found on the surface and on the InAs nanowires; C NPs were absent when hydrogen was added. In the latter conditions, Raman spectroscopy revealed the presence of graphitic-like carbon structures with a high number of defects, which points to carbon nanofibers. The carbon nanofibers preferentially nucleate at the tip of InAs nanowires, probably catalyzed by the gold-indium alloy, which can form only there. The number density of CNFs increased from 15% to 50% of decorated InAs nanowires when Fe was added as a catalyst in the CVD process. Our results demonstrate that interconnections between adjacent InAs NWs with carbon fibers can be obtained via CVD; however, to which of the neighboring NWs a connection is made, is random.

Author Contributions: Conceptualization, C.C. and L.S.; methodology M.A., C.C. and L.S.; validation, C.C., L.S. and P.R.; formal analysis, M.A. and C.C.; investigation, M.A. and C.C.; resources, C.C. and L.S.; data curation, M.A.; writing—original draft preparation, M.A. and C.C.; writing—review and editing, P.R. and L.S.; visualization, M.A. and C.C.; supervision, C.C. and P.R.; project administration, C.C.; funding acquisition, M.A. and C.C. All authors have read and agreed to the published version of the manuscript.

Funding: M. Arshad thanks the Higher Education Commission, Islamabad, Pakistan and Emerging Nations Science Foundation, Trieste, Italy for financial support during his stay abroad, as well as the ICTP for a TRIL grant. L.S. acknowledges the partial financial support by the FET-OPEN project And QC (H2020 Grant No. 828948).

Data Availability Statement: Data are contained within the article.

Conflicts of Interest: The authors declare no conflict of interest.

References

1. Naeemi, A.; Meindl, J.D. Carbon Nanotube Interconnects. *Annu. Rev. Mater. Res.* **2009**, *39*, 255–275. [CrossRef]
2. Zhao, W.-S.; Fu, K.; Wang, D.-W.; Li, M.; Wang, G.; Yin, W.-Y. Mini-Review: Modeling and Performance Analysis of Nanocarbon Interconnects. *Appl. Sci.* **2019**, *9*, 2174. [CrossRef]
3. Ko, H.; Zhang, Z.; Ho, J.C.; Takei, K.; Kapadia, R.; Chueh, Y.L.; Cao, W.; Cruden, B.A.; Javey, A.A. Flexible Carbon-Nanofiber Connectors with Anisotropic Adhesion Properties. *Small* **2010**, *6*, 22–26. [CrossRef]
4. Meng, G.; Han, F.; Zhao, X.; Chen, B.; Yang, D.; Liu, J.; Xu, Q.; Kong, M.; Zhu, X.; Jung, Y.J.; et al. A General Synthetic Approach to Interconnected Nanowire/Nanotube and Nanotube/Nanowire/Nanotube Heterojunctions with Branched Topology. *Angew. Chem. Int. Ed.* **2009**, *48*, 7166–7170. [CrossRef]
5. Liang, H.W.; Liu, S.; Yu, S.H. Controlled Synthesis of One-Dimensional Inorganic Nanostructures Using Pre-Existing One-Dimensional Nanostructures as Templates. *Adv. Mater.* **2010**, *22*, 3925–3937. [CrossRef]
6. Dai, H. Carbon Nanotubes: Synthesis, Integration, and Properties. *Acc. Chem. Res.* **2002**, *35*, 1035–1044. [CrossRef]
7. Vajtai, R.; Wei, B.; Ajayan, P. Controlled growth of carbon nanotubes. *Philos. Trans. R. Soc. A* **2004**, *362*, 2143–2160. [CrossRef]
8. Gomes, U.P.; Ercolani, D.; Zannier, V.; Beltram, F.; Sorba, L. Controlling the Diameter Distribution and Density of InAs Nanowires Grown by Au-assisted Methods. *Semicond. Sci. Technol.* **2015**, *30*, 115012. [CrossRef]
9. Briggs, D.; Seah, M.P. *Auger and X-ray Photoelectron Spectroscopy, Practical Surface Analysis 2nd ed*; John Wiley & Sons Ltd.: New York, NY, USA, 1990; Volume 1.
10. Milnes, A.G.; Polyakov, A.Y. Indium Arsenide: A Semiconductor for High Speed and Electro-optical Devices. *Mater. Sci. Eng. B* **1993**, *18*, 237–259. [CrossRef]
11. Mönch, W. *Semiconductor Surfaces and Interfaces*; Springer Series in Surface Sciences; Springer Science &. Business Media: Berlin, Heidelberg, Germany, 2013; Volume 26. [CrossRef]
12. Teodorescu, C.; Chevrier, F.; Brochier, R.; Richter, C.; Lakovac, V.; Heckmann, O.; De Padova, P.; Hricovini, K. Reactivity and Magnetism of Fe/InAs(100) Interfaces. *Eur. Phys. J. B* **2002**, *28*, 305–313. [CrossRef]
13. Tanuma, S.; Powell, C.J.; Penn, D.R. Calculations of Electron Inelastic Mean Free Paths. IX. Data for 41 Elemental Solids over the 50 eV to 30 keV ange. *Surf. Interface Anal.* **2011**, *43*, 689–713. [CrossRef]
14. Hertenberger, S.; Rudolph, D.; Becker, J.; Bichler, M.; Finley, J.J.; Abstreiter, G.; Koblmüller, G. Rate-limiting Mechanisms in High-temperature Growth of Catalyst-free InAs Nanowires with Large Thermal Stability. *Nanotechnology* **2012**, *23*, 235602. [CrossRef]
15. Park, H.; Gaillot, A.-C.; Prokes, S.; Cammarata, R. Observation of Size Dependent Liquidus Depression in the Growth of InAs Nanowires. *J. Cryst. Growth* **2006**, *296*, 159–164. [CrossRef]
16. Begum, N.; Piccin, M.; Jabeen, F.; Bais, G.; Rubini, S.; Martelli, F.; Bhatti, A.S. Structural Characterization of GaAs and InAs Nanowires by Means of Raman Spectroscopy. *J. Appl. Phys.* **2008**, *104*, 104311. [CrossRef]

17. Bachilo, S.M.; Strano, M.S.; Kittrell, C.; Hauge, R.H.; Smalley, R.E.; Weisman, R.B. Structure-Assigned Optical Spectra of Single-Walled Carbon Nanotubes. *Science* **2002**, *298*, 2361–2366. [CrossRef]
18. Dresselhaus, M.S.; Jorio, A.; Hofmann, M.; Dresselhaus, G.; Saito, R. Perspectives on Carbon Nanotubes and Graphene Raman Spectroscopy. *Nano Lett.* **2010**, *10*, 751–758. [CrossRef]
19. Jorio, A.; Fantini, C.; Dantas, M.S.S.; Pimenta, M.A.; Souza Filho, A.G.; Samsonidze, G.G.; Brar, V.W.; Dresselhaus, G.; Dresselhaus, M.S.; Swan, A.K.; et al. Linewidth of the Raman Features of Individual Single-wall Carbon Nanotubes. *Phys. Rev. B* **2002**, *66*, 115411. [CrossRef]
20. Lazzeri, M.; Piscanec, S.; Mauri, F.; Ferrari, A.C.; Robertson, J. Phonon Linewidths and Electron-phonon Coupling in Graphite and Nanotubes. *Phys. Rev. B* **2006**, *73*, 155426. [CrossRef]
21. Maultzsch, J.; Reich, S.; Schlecht, U.; Thomsen, C. High-Energy Phonon Branches of an Individual Metallic Carbon Nanotube. *Phys. Rev. Lett.* **2003**, *91*, 087402. [CrossRef]
22. Piscanec, S.; Lazzeri, M.; Robertson, J.; Ferrari, A.C.; Mauri, F. Optical Phonons in Carbon Nanotubes: Kohn Anomalies, Peierls Distortions, and Dynamic Effects. *Phys. Rev. B* **2007**, *75*, 035427. [CrossRef]
23. Telg, H.; Maultzsch, J.; Reich, S.; Hennrich, F.; Thomsen, C. Chirality Distribution and Transition Energies of Carbon Nanotubes. *Phys. Rev. Lett.* **2004**, *93*, 177401. [CrossRef] [PubMed]
24. Ferrari, A.C.; Meyer, J.; Scardaci, V.; Casiraghi, C.; Lazzeri, M.; Mauri, F.; Piscanec, S.; Jiang, D.; Novoselov, K.; Roth, S. Raman Spectrum of Graphene and Graphene Layers. *Phys. Rev. Lett.* **2006**, *97*, 187401. [CrossRef]
25. Wang, Y.; Serrano, S.; Santiago-Avilés, J.J. Raman Characterization of Carbon Nanofibers Prepared Using Electrospinning. *Synth. Met.* **2003**, *138*, 423–427. [CrossRef]
26. Zou, G.; Zhang, D.; Dong, C.; Li, H.; Xiong, K.; Fei, L.; Qian, Y. Carbon Nanofibers: Synthesis, Characterization, and Electrochemical Properties. *Carbon* **2006**, *44*, 828–832. [CrossRef]
27. Liu, Y.; Pan, C.; Wang, J. Raman Spectra of Carbon Nanotubes and Nanofibers Prepared by Ethanol Flames. *J. Mater. Sci.* **2004**, *39*, 1091–1094. [CrossRef]
28. Bhuvaneswari, M.S.; Bramnik, N.N.; Ensling, D.; Ehrenberg, H.; Jaegermann, W. Synthesis and Characterization of Carbon Nano Fiber/$LiFePO_4$ Composites for Li-ion Batteries. *J. Power Sources* **2008**, *180*, 553–560. [CrossRef]
29. Dick, K.A.; Geretovszky, Z.; Mikkelsen, A.; Karlsson, L.S.; Lundgren, E.; Malm, J.-O.; Andersen, J.N.; Samuelson, L.; Seifert, W.; Wacaser, B.A. Improving InAs Nanotree Growth with Composition-controlled Au–In Nanoparticles. *Nanotechnology* **2006**, *17*, 1344. [CrossRef]
30. Okamoto, H. The Co-In (Cobalt-Indium) System. *Bull. Alloy Phase Diagr.* **1990**, *11*, 137–139. [CrossRef]
31. Okamoto, H. The FeSe (Iron-Selenium) System. *JPE* **1991**, *12*, 383–389. [CrossRef]
32. Snigurenko, D.; Jakiela, R.; Guziewicz, E.; Przezdziecka, E.; Stachowicz, M.; Kopalko, K.; Barcz, A.; Lisowski, W.; Sobczak, J.J.; Krawczyk, M.; et al. XPS Study of Arsenic Doped ZnO Grown by Atomic Layer Deposition. *J. Alloys Compd.* **2014**, *582*, 594–597. [CrossRef]
33. Fukuda, Y.; Suzuki, Y.; Murata, J.; Sanada, N. Adsorption and Decomposition of Triethylindium on the Si(100) Surface Studied by X-ray and Ultra-violet Photoelectron Spectroscopy. *Appl. Surf. Sci.* **1993**, *65*, 593–597. [CrossRef]
34. Meharg, P.F.A.; Ogryzlo, E.A.; Bello, I.; Lau, W.M. Surface Damage and Deposition on Gallium Arsenide Resulting from Low Energy Carbon Ion Bombardment. *Surf. Sci.* **1992**, *271*, 468–476. [CrossRef]
35. Dick, K.A.; Deppert, K.; Karlsson, L.S.; Seifert, W.; Wallenberg, L.R.; Samuelson, L. Position-Controlled Interconnected InAs Nanowire Networks. *Nano Lett.* **2006**, *6*, 2842–2847. [CrossRef]
36. Dick, K.A.; Deppert, K.; Karlsson, L.S.; Wallenberg, L.R.; Samuelson, L.; Seifert, W. A New Understanding of Au-Assisted Growth of III–V Semiconductor Nanowires. *Adv. Funct. Mater.* **2005**, *15*, 1603–1610. [CrossRef]
37. Berchmans, S.; Thomas, P.J.; Rao, C.N.R. Novel Effects of Metal Ion Chelation on the Properties of Lipoic Acid-Capped Ag and Au Nanoparticles. *J. Phys. Chem. B* **2002**, *106*, 4647–4651. [CrossRef]
38. Moriarty, P. Nanostructured Materials. *Rep. Prog. Phys.* **2001**, *64*, 297. [CrossRef]
39. Suemori, K.; Yokoyama, M.; Hiramoto, M. Electrical Shorting of Organic Photovoltaic Films Resulting from Metal Migration. *J. Appl. Phys.* **2006**, *99*, 36109. [CrossRef]
40. Xu, Z.; Hou, Y.; Sun, S. Magnetic Core/Shell Fe_3O_4/Au and Fe_3O_4/Au/Ag Nanoparticles with Tunable Plasmonic Properties. *J. Am. Chem. Soc.* **2007**, *129*, 8698–8699. [CrossRef]

nanomaterials

MDPI

Article

Semi-Empirical Pseudopotential Method for Graphene and Graphene Nanoribbons

Raj Kumar Paudel [1,2,3], Chung-Yuan Ren [4] and Yia-Chung Chang [1,5,*]

1 Research Center for Applied Sciences, Academia Sinica, Taipei 11529, Taiwan; rajupdl6@gate.sinica.edu.tw
2 Molecular Science and Technology, Taiwan International Graduate Program, Academia Sinica, Taipei 11529, Taiwan
3 Department of Physics, National Central University, Chungli 320, Taiwan
4 Department of Physics, National Kaohsiung Normal University, Kaohsiung 824, Taiwan; cyren@nknu.edu.tw
5 Department of Physics, National Cheng-Kung University, Tainan 701, Taiwan
* Correspondence: yiachang@gate.sinica.edu.tw

Abstract: We implemented a semi-empirical pseudopotential (SEP) method for calculating the band structures of graphene and graphene nanoribbons. The basis functions adopted are two-dimensional plane waves multiplied by several B-spline functions along the perpendicular direction. The SEP includes both local and non-local terms, which were parametrized to fit relevant quantities obtained from the first-principles calculations based on the density-functional theory (DFT). With only a handful of parameters, we were able to reproduce the full band structure of graphene obtained by DFT with a negligible difference. Our method is simple to use and much more efficient than the DFT calculation. We then applied this SEP method to calculate the band structures of graphene nanoribbons. By adding a simple correction term to the local pseudopotentials on the edges of the nanoribbon (which mimics the effect caused by edge creation), we again obtained band structures of the armchair nanoribbon fairly close to the results obtained by DFT. Our approach allows the simulation of optical and transport properties of realistic nanodevices made of graphene nanoribbons with very little computation effort.

Keywords: semi-empirical pseudopotential (SEP); graphene; graphene nanoribbon; armchair graphene nanoribbon (AGNRs); density-functional theory (DFT); band structure; 2D materials; B-spline; mixed basis

Citation: Paudel, R.K.; Ren, C.-Y.; Chang, Y.-C. Semi-Empirical Pseudopotential Method for Graphene and Graphene Nanoribbons. *Nanomaterials* **2023**, *13*, 2066. https://doi.org/10.3390/nano13142066

Academic Editor: Fabrizio Pirri

Received: 15 June 2023
Revised: 6 July 2023
Accepted: 11 July 2023
Published: 13 July 2023

1. Introduction

The electronic properties of low-dimensional materials are unique due to unprecedented properties that are unparalleled by those in bulk counterparts [1,2]. A scientific breakthrough occurred in 2004 in the isolation of monolayer graphene by mechanical exfoliation. After the discovery of graphene, the research on two-dimensional (2D) materials stimulated a great deal of interest due to their promising optical, physical, and chemical properties. These materials are layered, weakly coupled materials that can exist in a few or single-layer forms. Many 2D materials can be easily fabricated due to recently developed cutting-edge technology.

Ab initio formalism such as density-functional theory (DFT) [3] has been used extensively in calculating the electronic structures [4] and thermoelectric [5] and optical properties of solids [6]. In most calculations, three-dimensional (3D) plane waves are used as the basis functions. Plane wave (PW) basis is easy to implement, and the convergence of the calculation can be checked systematically. DFT methods are still affected by the band-gap problem, which requires numerically expensive GW calculation [7] or hybrid functionals to correct the band gap and effective mass near the Fermi level. The other drawback of this basis is the requirement of a large number of PWs needed when the unit cell is large.

Li et al. [8–10] introduced a planar-basis method to utilize plane waves along periodic (x-y plane) and Gaussian functions along a non-periodic (z) direction for ab initio calculations. The planar basis has advantages over the conventional PW basis in that it resumes the layer-like local geometry that appears in surfaces or 2D materials. Moreover, one can calculate the total energy for an isolated slab instead of using a superlattice of alternating slab and vacuum regions. Thus, the work function can be easily obtained [8]. The planar-basis method was further improved by Ren et al. [11,12] via the use of B-spline functions instead of Gaussian functions along the non-periodic direction. B-splines are not associated with atomic positions, making the geometry optimization easy to use, and the relevant matrices are sparse. Another advantage of using a planar basis is the ease of handling charged 2D materials such as gated 2D materials. This can avoid the artificial long-range Coulomb interaction introduced by the supercell method [13].

To study bulk materials with one or few atoms per unit cell, the empirical pseudopotential method (EPM) [14,15] presents an efficient and accurate method. EPM is extremely powerful since very few parameters are needed in order to obtain information about many properties of the solid, such as band structure, optical properties, and dielectric properties. Unlike DFT, EPM does not involve a self-consistent solution of the Schrödinger equation for charge density, thus greatly reducing time and computational cost. Furthermore, this method has a great advantage when working with a large unit cell of many identical atoms since the full potential of each atom can be calculated and fitted with a smaller number of parameters. In EPM, the pseudopotential contains both local and non-local pseudopotentials. The latter depends upon the angular momentum and energy. The incorporation of both the non-local and local pseudopotentials of the system provides more precise band structures and wavefunctions [15,16].

DFT is a widely used computational method for studying the electronic structure of AGNRs [17,18]. DFT calculations provide a first-principles description of the system by solving the Kohn–Sham equations, taking into account the electron–electron interactions within an exchange–correlation functional. DFT calculations can accurately predict the band structure, electronic density of states, and other electronic properties of AGNRs. However, DFT calculations can be computationally demanding, particularly for larger AGNR systems, and certain approximations within the exchange–correlation functional can introduce errors in the result.

Empirical methods, on the other hand, offer an alternative approach to studying AGNRs [19]. These methods are based on empirical potentials. Empirical potentials capture the essential physics of the system by using simplified parameterized functions that reduce the computational cost compared to DFT. Empirical methods can provide insights into the electronic structure and transport properties of AGNRs with relatively lower computational requirements. They are particularly useful for studying larger systems and phenomena beyond the reach of DFT. For 2D materials, the conventional empirical pseudopotential, which depends on only a few form factors [14,15], does not work well since the contribution from self-consistent charge density and exchange–correlation potential is highly anisotropic for the 2D system. Therefore, the existing empirical potentials do not provide the accuracy desired to provide reliable simulation of the electronic or optical properties of graphene-related devices. Thus, the development of a more accurate SEP is warranted.

Graphene nanoribbons (GNRs) [17–25] are narrow strips or ribbons of graphene. They are characterized by their finite width, orientation, and well-defined edges. The width of GNRs can range from a few nanometers to tens of nanometers, and it strongly influences their electronic properties. GNRs can be classified into two main types based on their geometry along the edge: armchair-edge GNRs and zigzag-edge GNRs. In the current work, we solely focus on armchair graphene nanoribbons (AGNRs) to demonstrate the advantage of semi-empirical pseudopotential (SEP). AGNRs have edges formed by rows of carbon atoms that resemble the shape of an armchair. In this configuration, all carbon atoms at the edges are fully bonded, forming a stable structure. AGNRs are known for their semiconducting behavior [20], meaning they can exhibit a band gap in their electronic structure. The width of AGNRs determines the size of the band gap, with narrower ribbons having larger band gaps in general.

In electronics, AGNRs offer tunable bandgaps and unique edge states, making them potential candidates for high-performance transistors and logic devices [26]. AGNRs also exhibit attractive optical properties, with efficient light absorption and emission capabilities, making them suitable for optoelectronic applications such as photodetectors [27] and light-emitting diodes [28]. Furthermore, AGNRs have shown potential in topological physics [29–32], with the ability to host topological edge states and phenomena such as the quantum spin Hall effect. This opens up possibilities for the development of topological quantum devices and spintronics. Additionally, AGNRs can be tailored to exhibit magnetic properties [33,34] by introducing magnetic dopants or adatoms, enabling applications in spintronics, magnetic sensors, and information storage. AGNRs show promise in thermoelectric applications [35–38] due to their low thermal conductivity and tunable bandgap, enabling efficient energy conversion. The multifunctionality of armchair graphene nanoribbons makes them a highly versatile platform for advancing electronic, optical, topological, thermoelectric, and magnetic technologies, with ongoing research aimed at uncovering their full potential.

In this paper, we implement a more sophisticated semi-empirical pseudopotential method (SEPM) within the planar basis to study the atomic and electronic structure of 2D systems, using graphene and armchair graphene nanoribbons as examples. The applications of the currently developed SEP include optical and transport properties of graphene and related structures such as graphene nanoribbons and various junction devices.

2. Calculation Methods

2.1. B-Splines

To calculate the band structures of 2D materials in the semi-empirical pseudopotential (SEP) model, it is more efficient to adopt a planar basis. The basis consists of localized finite-element functions, i.e., B-splines [39] in the z coordinate multiplied by plane waves in the in-plane coordinates (x and y). B-splines of order κ consist of positive piecewise polynomials of z with degree $\kappa - 1$. These polynomials vanish everywhere outside the sub-inervals $t_i \le z < t_{i+k}$. The B-spline basis set with order κ defined by a knot sequence $\{t_i\}$ is generated by the following relationship:

$$B_{i\kappa}(z) = \left(\frac{z - t_i}{t_{i+\kappa-1} - t_i}\right) B_{i,\kappa-1}(z) + \left(\frac{t_{i+k} - z}{t_{i+k} - t_{i+1}}\right) B_{i+1,\kappa-1}(z) \tag{1}$$

with $B_{i1}(z) = \begin{cases} 1 & \text{if } t_i \le z < t_{i+1} \\ 0 & \text{otherwise} \end{cases}$, where $i = 1, 2, 3\ldots$ up to the number of knot sequence.

The first derivative of the B-spline of order κ is given by the following:

$$\frac{d}{dz} B_{i\kappa}(z) = \left(\frac{\kappa - 1}{t_{i+\kappa-1} - t_i}\right) B_{i,k-1}(z) - \left(\frac{\kappa - 1}{t_{i+\kappa} - t_{i+1}}\right) B_{i+1,\kappa-1}(z) \tag{2}$$

The derivative of B-splines of order κ is a linear combination of B-splines of order $\kappa - 1$, which is also a simple polynomial and is continuous across the knot sequence. Using the polynomial expansion

$$B_{i\kappa}(z) = \sum_{j=1}^{4} \sum_{n=0}^{\kappa-1} D_n^{i,j} z^n \ \text{ for} \in z\ (t_{i+j-1}, t_{i+j}), \tag{3}$$

we obtain the Fourier transform of $B_{i\kappa}(z)$ as the following:

$$\begin{aligned}\widetilde{B}_{i\kappa}(g) &= \frac{1}{\sqrt{L_c}}\sum_{j=1}^{4}\sum_{n=0}^{k} D_n^{i,j}\int_{t_{i+j-1}}^{t_{i+j}} dz e^{ig(z-\frac{L_c}{2})}z^n \\ &\equiv \frac{e^{ig(t_i-L_c/2)}}{\sqrt{L_c}}\sum_{j=1}^{4}\sum_{n=0}^{\kappa} D_n^{i,j} I_n^{i,j}(g).\end{aligned} \tag{4}$$

where $I_n^{i,j}(g)$ can obtain by the following recursion relationship:

$$I_n^{i,j}(g) = \int_{\tau_i}^{\tau_{i+j}} dz\, e^{igz}z^n = \left(\frac{z^n}{ig}e^{igz}\right)\Bigg|_0^{\tau_{i+j}-\tau_i} - \frac{n}{ig}I_j^{n-1}(g) \tag{5}$$

with $I_0^{i,j}(g) = \left(\frac{1}{ig}e^{igz}\right)\Big|_0^{\tau_{i+j}-\tau_i}$. Here, L_c is the period length used along z, and $I_n(g)$ can be obtained by the recursion relationship. The B-spline functions used for the current calculation are shown in in Figure 1.

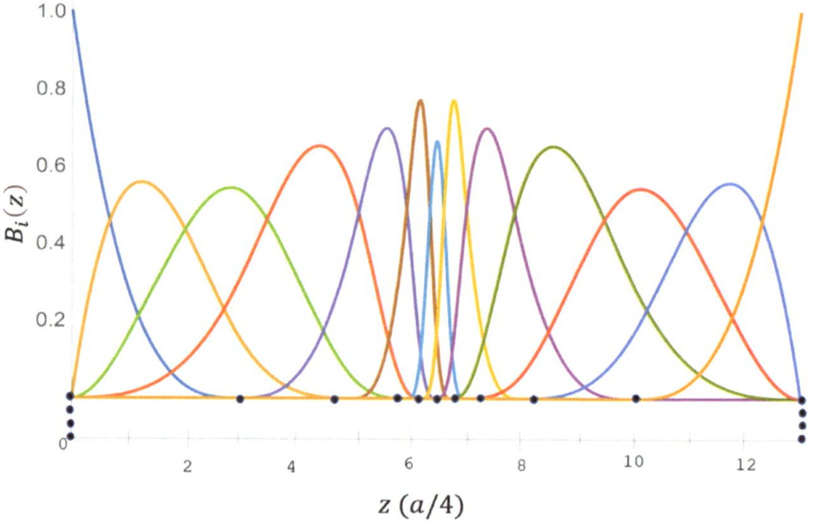

Figure 1. B-spline consists of an exponential-type of a 17-point knot sequence of order $\kappa = 4$. Here, (•) denotes the knot points in the sequence.

2.2. Kinetic and Overlap Matrix Elements within Planar Basis

Throughout this paper, we work in atomic units, where energy is in units of Ry, and distance is in units of Bohr. The Hamiltonian of the system in the ultrasoft pseudopotential approach [16] is given by the following:

$$\hat{H} = -\nabla^2 + V_{loc} + \hat{V}_{nl} \tag{6}$$

Here, V_{loc} and $\hat{V}_{nl}$ denote the local pseudopotential and non-local pseudopotential. Planar basis, used to expand $\sum_i$, is defined as the following:

$$\langle \mathbf{r}|\mathbf{k}+\mathbf{G};i\rangle = \frac{1}{\sqrt{A}}e^{i(\mathbf{k}+\mathbf{G})\cdot\boldsymbol{\rho}}B_{i\kappa}(z) \tag{7}$$

where $\mathbf{G}$ denotes the in-plane reciprocal lattice vector in 2D, and $\boldsymbol{\rho} = (x, y)$ is the projection of r in the x-y plane. $\mathbf{k}$ is the in-plane wave vector. A is the surface area of the sample. The overlap matrix elements between two planar basis functions are given by the following:

$$\langle \mathbf{k}+\mathbf{G}; i | \mathbf{k}+\mathbf{G}'; i' \rangle = \langle B_{i\kappa} | B_{i'} \rangle \delta_{\mathbf{G},\mathbf{G},} \quad (8)$$

where

$$\langle B_{i\kappa} | B_{i'\kappa} \rangle = \int dz B_{i\kappa}(z) B_{i'\kappa}(z) \equiv O_{ii,} \quad (9)$$

is the overlap integral between two B-spline functions. The kinetic energy matrix elements are given by the following:

$$T_{ii'}(\mathbf{k}+\mathbf{G}, \mathbf{k}+\mathbf{G}') = \left\langle \mathbf{k}+\mathbf{G}; i \middle| -\nabla^2 \middle| \mathbf{k}+\mathbf{G}; i' \right\rangle = \left[K_{ii'} + O_{ii'} |\mathbf{k}+\mathbf{G}|^2 \right] \delta_{\mathbf{G},\mathbf{G}'} \quad (10)$$

where

$$K_{ii'} = \int dz \frac{d}{dz} B_{i\kappa}(z) \frac{d}{dz} B_{i'\kappa}(z) \quad (11)$$

2.3. Implementation of the Semi-Empirical Local Pseudopotential for Graphene

The local pseudopotential of the crystal is given by the following:

$$V_{loc}(\mathbf{r}) = V_{ion}(\mathbf{r}) \ + V_H(\mathbf{r}) + V_{xc}(\mathbf{r}) \quad (12)$$

where the first term describes the ionic local potential with

$$V_{ion}(\mathbf{r}) = \sum_\sigma \sum_R V_L^\sigma(\mathbf{r} - \mathbf{R} - \boldsymbol{\tau}_\sigma) \quad (13)$$

in which $\mathbf{R}$ denotes a bulk lattice vector, and $\boldsymbol{\tau}_\sigma$ denotes the position of different atoms within the bulk unit cell. We note that $V_L^\sigma(\mathbf{r})$ consists of a long-range term that decays like Z^σ/r for large r. For charge-neutral systems, there is a counter long-range term in $V_H(\mathbf{r})$ due to the valence charges, and the sum of $V_{ion}(\mathbf{r}) + V_H(\mathbf{r})$ will be short-ranged.

For fitting empirical pseudopotentials of bulk materials, one adjusts the Fourier transform of the potential of $V_{loc}(\mathbf{r})$ (called "form factors" $\widetilde{V}_{loc}(\mathbf{G})$) at a small number of reciprocal lattice vectors until the band structure agrees with the experimental data or first-principal calculations [14]. To understand the nature of potential, a set of form factors for the first few shells at a large number of additional reciprocal lattice vectors is needed. A variety of algebraic forms have been used in the past for bulk materials [40–42]. We have not found an existing form that has sufficient flexibility to obtain the correct band structure for 2D materials. In our case, we try to mimic the full local potential by an analytic expression of the form.

$$V_{loc}(\mathbf{r}) = \sum_{\sigma,\mathbf{R}} V_0(\mathbf{r} - \boldsymbol{\tau}_\sigma - \mathbf{R}) + \sum_{\sigma,\mathbf{G}\neq 0} \Delta\widetilde{V}_{loc}(z, \mathbf{G}) e^{i\mathbf{G}\cdot(\boldsymbol{\rho}-\boldsymbol{\tau}_\sigma)} \quad (14)$$

where $\mathbf{G}$ denotes an in-plane 2D reciprocal lattice vector. The first term in Equation (14) denotes the main part of the local pseudopotential (denoted V_0), which is parametrized in terms of three spherical Gaussian functions, with C_s and α_s being fitted parameters. Namely, $V_0(\mathbf{r}) = \sum_{s=1}^3 C_s e^{-\alpha_s r^2}$. C_s and α_s are related to the spatial average of $V_{loc}(\mathbf{r})$ in the 2D plane as a function of z. We can rewrite $V_0(\mathbf{r})$ in terms of 2D plane waves as the following:

$$\sum_\mathbf{R} V_0(\mathbf{r} - \boldsymbol{\tau}_\sigma - \mathbf{R}) = \sum_{s=1}^3 C_s e^{-\alpha_s z^2} \frac{\pi}{A_c \alpha_s} \sum_{\sigma,\mathbf{G}} e^{-G^2/4\alpha_s} e^{i\mathbf{G}\cdot(\boldsymbol{\rho}-\boldsymbol{\tau}_\sigma)}. \quad (15)$$

Let $\widetilde{V}_{loc}(z, \mathbf{G})$ denote the Fourier transform of $V_{loc}(\mathbf{r})$ in the 2D reciprocal space. We have

$$\widetilde{V}_{loc}(z, 0) = \sum_{s=1}^3 C_s e^{-\alpha_s z^2} \frac{N_a \pi}{A_c \alpha_s} \equiv \sum_{s=1}^3 D_s e^{-\alpha_s z^2} \quad (16)$$

for the $\boldsymbol{\tau}_\sigma = \mathbf{0}$ term at $\mathbf{G} = \mathbf{0}$. Here, N_a denotes the number of atoms per unit cell, and A_c is the area of the 2D unit cell. We determine the fitting parameters D_s and α_s by fitting the corresponding $\widetilde{V}_{loc}(z,0)$, determined by a DFT calculation within the same mixed basis as defined in (7). The implementation of the DFT package within this basis for graphene and AGNRs was reported in [12]. The DFT results obtained by using the package developed in [12] were checked against results obtained with the VASP package [43], and the calculated results for graphene and related nanoribbons obtained by using both packages are essentially the same. The exchange–correlation functional used is deduced from the Monte Carlo results calculated by Ceperley and Alder [44] and parametrized by Perdew and Zunger [45]. The fitted coefficients for D_s can be directly converted to C_s through the relationship $C_s = D_s A_c \alpha_s/(N_a \pi)$. The best-fit parameters for graphene are given in the first row of Table 1, and the quality of the fit for $\widetilde{V}_{loc}(z,0)$ is shown in Figure 2a.

Table 1. Fitting parameters for the main part of local potential $V_0(\mathbf{r})$.

Exponents			Coefficients		
α_1	α_2	α_3	C_1	C_2	C_3
0.0396	1.4100	0.3461	−0.3682	−1.7360	−1.5710

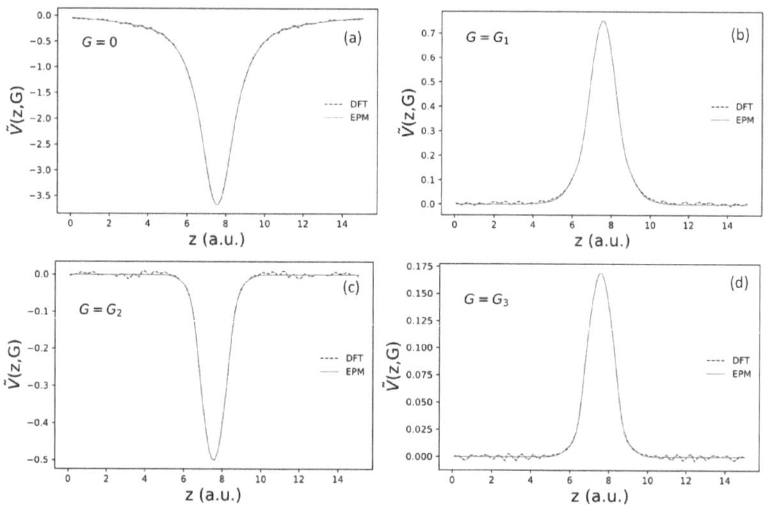

Figure 2. Fourier transform of effective local pseudopotential $\widetilde{V}_{loc}(z,\mathbf{G})$ for reciprocal lattice vectors ($\mathbf{G}$) in the first few shells. The magnitudes of $\mathbf{G}$ in subfigures are: (**a**) $G = 0$, (**b**) $G = G_1 = 1.56$ a.u., (**c**) $G = G_2 = 2.70$ a.u., and (**d**) $G = G_3 = 3.12$ a.u. for shells 0, 1, 2, and 3, respectively.

The second term of Equation (14) denotes the difference $V_{loc}(\mathbf{r}) - \sum_{\sigma,\mathbf{R}} V_0(\mathbf{r} - \boldsymbol{\tau}_\sigma - \mathbf{R})$, which is expressed in reciprocal space. Since $\mathbf{G} = 0$ is already well fitted by $\widetilde{V}_{loc}(z,0)$, we only have to consider the $\mathbf{G} \neq 0$ contribution. Due to point-group symmetry, the reciprocal lattice vector $\mathbf{G}$ can be sorted into many shells with the magnitude of $\mathbf{G}$ vectors being the same in each shell, and $\Delta\widetilde{V}_{loc}(z,\mathbf{G})$ is the same for all $\mathbf{G}$ vectors in the same shell. We found that for $\mathbf{G}$ vectors with magnitude $G > 4$ a.u., $\Delta\widetilde{V}_{loc}(z,\mathbf{G})$ can be well fitted by a short-range correction function of the form:

$$\Delta\widetilde{V}_S(z,\mathbf{G}) = f_S(z)\widetilde{D}_S(G)S_1(\mathbf{G})$$

with
$$\begin{cases} f_S(z) = \left(1 + C_1^S z^2 + C_2^S z^4 + C_3^S z^6 + C_4^S z^8\right) e^{-\alpha_0 z^2} \\ \widetilde{D}_S(G) = \left(P_1^S G^2 + P_2^S G^4 + P_3^S G^6\right) e^{-G^2/(4\alpha_2)} \end{cases} \tag{17}$$

$S_1(\mathbf{G}) = \frac{1}{A_c} \sum_\sigma e^{-i\mathbf{G}\cdot\boldsymbol{\tau}_\sigma}$ denotes the structure factor of graphene.

Note that the chosen form of $\Delta\widetilde{V}_S(z, \mathbf{G})$ goes to zero at $\mathbf{G} = 0$, so the fitting does not affect $\widetilde{V}_{loc}(z, 0)$. We can transform $\Delta\widetilde{V}_S(z, \mathbf{G})$ back to real space analytically and obtain the following:

$$\Delta V_S(\mathbf{r}) = \sum_{\sigma,\mathbf{G}} \Delta\widetilde{V}_S(z, \mathbf{G}) e^{i\mathbf{G}\cdot(\boldsymbol{\rho}-\boldsymbol{\tau}_\sigma)} = f_S(z) D_S(\boldsymbol{\rho}), \tag{18}$$

where $D_S(\boldsymbol{\rho}) = \sum_{\sigma,\mathbf{G}\neq 0} \widetilde{D}_S(G) e^{i\mathbf{G}\cdot(\boldsymbol{\rho}-\boldsymbol{\tau}_\sigma)}$.

The best-fit parameters for short-range correction functions defined above are listed in the first row of Table 2. Finally, for the three shells with $0 < G < 4$ a.u., we need to fit the difference $\Delta\widetilde{V}_{loc}(z, \mathbf{G}) - \Delta\widetilde{V}_S(z, \mathbf{G})$ by another long-range correction function of the form $\Delta\widetilde{V}_L(z, \mathbf{G}) = f_L(z) \widetilde{D}_L(G)$ with the following:

$$\begin{cases} f_L(z) = \left(1 + C_1^L z^2 + C_2^L z^4 + C_3^L z^6 + C_4^L z^8\right) e^{-\alpha_0 z^2} \\ \widetilde{D}_L(G) = \left(P_1^L G^2 + P_2^L G^4 + P_3^L G^6\right) e^{-G^2/(4\alpha_2)} \end{cases} \tag{19}$$

The best-fit parameters for fitting $f_L(z)$ and $\widetilde{D}_L(G)$ for the three shells are shown in the second row of Table 2. The Fourier transforms of the effective local pseudopotential, i.e., $\widetilde{V}_{\text{loc}}(z, \mathbf{G})$ for $\mathbf{G}$ shells with magnitude $G_1, G_2, G_3 = 1.56, 2.702$, and 3.12 a.u., respectively, are shown in Figure 2b–d. Similarly, we can transform $\Delta\widetilde{V}_L(z, \mathbf{G})$ back to real space analytically and obtain the following:

$$\Delta V_L(\mathbf{r}) = \sum_{\sigma,\mathbf{G}} \Delta\widetilde{V}_L(z, \mathbf{G}) e^{i\mathbf{G}\cdot(\boldsymbol{\rho}-\boldsymbol{\tau}_\sigma)} = f_L(z) D_L(\boldsymbol{\rho}) \tag{20}$$

with $D_L(\boldsymbol{\rho}) = \sum_{\sigma,\mathbf{G}} \widetilde{D}_L(G) e^{i\mathbf{G}\cdot(\boldsymbol{\rho}-\boldsymbol{\tau}_\sigma)}$. Then, we obtain the following relationship:

$$\widetilde{D}_\gamma(\mathbf{q}) = \frac{1}{A_C} \sum_{\sigma\in A_c} \int d\boldsymbol{\rho} e^{i\mathbf{q}\cdot\boldsymbol{\rho}} D_\gamma(\boldsymbol{\rho} - \boldsymbol{\tau}_\sigma) / S_1(\mathbf{q}).$$

where $\gamma = S$ or L labels the short-range or long-range term.

Table 2. Fitting parameters for the short-range and long-range shape function $f_S(z)$ and $f_L(z)$ for the correction terms to $\widetilde{V}_{loc}(z, \mathbf{G})$ used in this work.

Type	Exponent	Coefficients for $f_\gamma(z)$				Coefficients for $\widetilde{D}_\gamma(\mathrm{G})$		
γ	α_0	C_1^γ	C_2^γ	C_3^γ	C_4^γ	P_1^γ	P_2^γ	P_3^γ
Short Range (S)	2.07	2.0372	−16.164	13.912	−2.8969	0.04494	−0.00574	0.000224
Long Range (L)	2.07	2.6251	−5.6668	2.1280	1.0239	−0.1650	0.03132	−0.002615

Combining all above, the total local potential is given by the following:

$$V_{loc}(\mathbf{r}) = V_0(\mathbf{r}) + \Delta V_S(\mathbf{r}) + \Delta V_L(\mathbf{r}). \tag{21}$$

For convenience, $D_\gamma(\boldsymbol{\rho})$ can be expressed as a polynomial multiplied by a Gaussian function.

$$D_\gamma(\boldsymbol{\rho}) = \sum_{m=0}^{3} b_m^\gamma \rho^{2m} e^{-\alpha_2 \rho^2} \tag{22}$$

where b_m^γ can be determined by the following relationship:

$$\widetilde{D}_\gamma(G) = \frac{1}{A_C}\int d\mathbf{r}e^{i\mathbf{G}\cdot\mathbf{r}}D_\gamma(\boldsymbol{\rho}) = \frac{1}{A_C}\sum_{m=0}^{3} b_m^\gamma \int d\boldsymbol{\rho}e^{i\mathbf{G}\cdot\boldsymbol{\rho}}\rho^{2m}e^{-\alpha_2\rho^2} \quad (23)$$

which leads to the following relationship between b_m^γ and the set of parameters $\{P_1^\gamma, P_2^\gamma, P_3^\gamma\}$.

$$\begin{pmatrix} b_0^\gamma \\ b_1^\gamma \\ b_2^\gamma \\ b_3^\gamma \end{pmatrix} = \frac{A_C\alpha_2}{\pi}\begin{pmatrix} 1 & 4\alpha_2 & 32\alpha_2^2 & 384\alpha_2^3 \\ 0 & -4\alpha_2^2 & -64\alpha_2^3 & -1152\alpha_2^4 \\ 0 & 0 & 16\alpha_2^4 & 576\alpha_2^5 \\ 0 & 0 & 0 & -64\alpha_2^6 \end{pmatrix}\begin{pmatrix} 0 \\ P_1^\gamma \\ P_2^\gamma \\ P_3^\gamma \end{pmatrix} \quad (24)$$

To check if the real space form of the local pseudopotential $\boldsymbol{V_{loc}}(\mathbf{r})$ obtained by the current SEPs as given by Equation (21) can truthfully represent the DFT results, we compared the results for $V_{loc}(\mathbf{r})$ obtained by DFT and by the current method in Figure 3. In Figure 3a, we show the x-dependence of $V_{loc}(\mathbf{r})$ at $z = -0.5\Delta z$ and $y = a/\left(2\sqrt{3}\right)$ (a line passing through the center of a C atom). Here, $\Delta z = W/104$ is the grid size for the fast Fourier transform used in the DFT calculation, and W is the width of the domain along the z-axis used to define the B-spline basis.

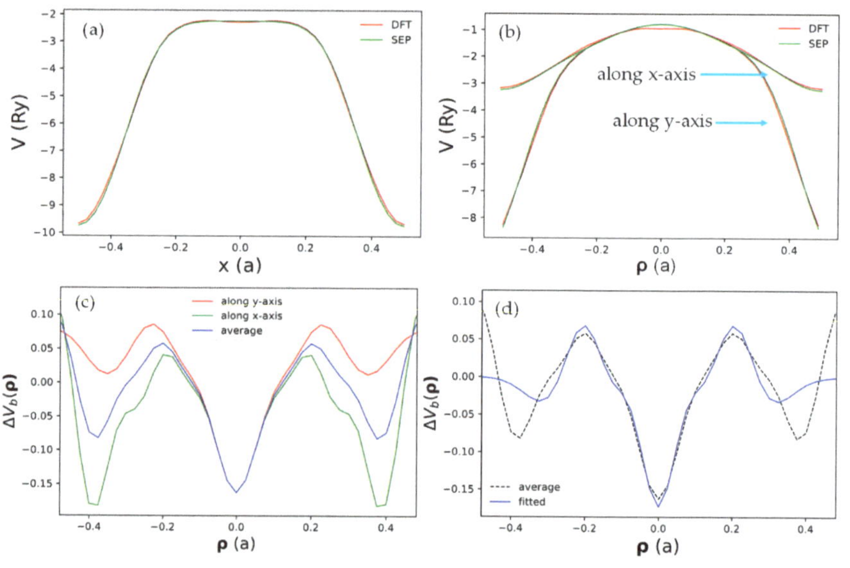

Figure 3. The net local pseudopotential in real space $V_{loc}(\boldsymbol{\rho}, z)$ along various lines in the plane with $z = -0.5\Delta z$ (half a grid from the center of the supercell). (**a**) Along the line at $y = \frac{a}{2\sqrt{3}}$, which passes through a row of C atoms in the graphene sheet. (**b**) Along two perpendicular lines (along the x-axis and y-axis) both passing through the origin, which is the center of the hexagon cell. (**c**) The difference of $V_{loc}(\boldsymbol{\rho}, z)$ obtained by DFT and SEP plotted along the x-axis (green) or y-axis (red). The blue curve indicates the average of the red and green curves. (**d**) The best-fit results of the average $\Delta V_b(\boldsymbol{\rho}, z)$.

It is seen that the result obtained by SEP matches the corresponding DFT results very well. In Figure 3b, we show both the x- and y-dependence of $V_{loc}(\mathbf{r})$ at $z \approx 0$ along lines in x- and y-directions, with both passing through the center of the hexagon cell. The agreement with the DFT results is still very good, except there is a small deviation near the center ($\rho \approx 0$). We also noticed that the x- and y-dependence almost coincide for $|\boldsymbol{\rho}| < 0.2a$, indicating the potential has nearly cylindrical symmetry near the center of the hexagon cell. We then took the difference of the local potential obtained by DFT and by SEP and plotted the difference function $\Delta V_b(\boldsymbol{\rho}, 0)$ in Figure 3c. It turns out that such a minor correction will still be important for the states derived from the σ-bonds, but it has a negligible effect on states derived from the π-bonds since the π-orbitals have negligible amplitude at the center of the hexagon cell. This minor correction is caused by the fact that all terms included in $V_{loc}(\mathbf{r})$ in Equation (21) are centered at atomic sites, while the effect of charge redistribution due to valence charges in the solid cannot be fully absorbed by the atom-centered terms. Thus, we should also consider contributions described by functions localized at the center of hexagon cells in the graphene lattice. We shall call such a term the "bond-charge contribution".

To determine this contribution, we fit the difference of the local potential obtained by DFT and by SEP, i.e., $\Delta V_b(\boldsymbol{\rho}, 0)$, in Figure 3c by the following analytic function:

$$\Delta V_b(\boldsymbol{\rho}, 0) = \sum_{n=0}^{4} C_n^b \rho^{2n} e^{-\alpha_b \rho^2} \equiv D_b(\boldsymbol{\rho}). \quad (25)$$

The best-fit parameters $\left\{C_n^b; n = 0, \ldots, 4\right\}$ and α_b are listed in Table 3. We approximate the net local potential near the center of the hexagon by a separable form:

$$V_{loc}(\boldsymbol{\rho}, z) = V_{loc}(\boldsymbol{\rho}, 0) f_b(z) \quad (26)$$

where $f_b(z) = V_{loc}(\mathbf{0}, z)/V_{loc}(\mathbf{0}, 0)$ describes the z-dependence of the bond-charge contribution. Here, $f_b(z)$ describes the variation of $V_{loc}(\boldsymbol{\rho}, z)/V_{loc}(\mathbf{0}, 0)$ along the z-axis at $\boldsymbol{\rho} = \mathbf{0}$, i.e., the center of the hexagon cell. By taking the difference between $V_{loc}(\mathbf{r})$ obtained by DFT and by SEP, we obtain the following:

$$\Delta V_b(\boldsymbol{\rho}, z) \approx D_b(\boldsymbol{\rho}) f_b(z), \quad (27)$$

where we assume that the z-dependent functions obtained by DFT and SEP are the same.

Table 3. Fitting parameters for the bond-charge contribution in pseudopotential localized at the center of the hexagon cell as described by Equations (25)–(27).

Exponents in Equations (25) and (28)			Coefficients in Equation (25)					
α_b	α_{h1}	α_{h2}	C_0^b	C_1^b	C_2^b	C_3^b	C_4^b	a_h
3.0053	0.3601	0.0383	−0.1727	1.5253	−6.4817	11.5249	−5.0681	0.6930

The net local potential of graphene, $V_{loc}(\boldsymbol{\rho}, z)$, evaluated at $\boldsymbol{\rho} = \mathbf{0}$ and obtained by DFT is shown in Figure 4. For convenience, we fit its normalized z-dependent function $f_b(z) = V_{loc}(\boldsymbol{\rho}, z)/V_{loc}(\mathbf{0}, z)$ by the following expression:

$$f_b(z) = a_h e^{-\alpha_{h1} z^2} + (1 - a_h) e^{-\alpha_{h2} z^2} \quad (28)$$

The best-fit parameters a_h, α_{h1}, and α_{h2} are given in Table 3. Adding this bond-charge contribution, the net local potential in our model becomes the following:

$$V_{loc}(\mathbf{r}) = V_0(\mathbf{r}) + \Delta V_S(\mathbf{r}) + \Delta V_L(\mathbf{r}) + \Delta V_b(\mathbf{r}). \quad (29)$$

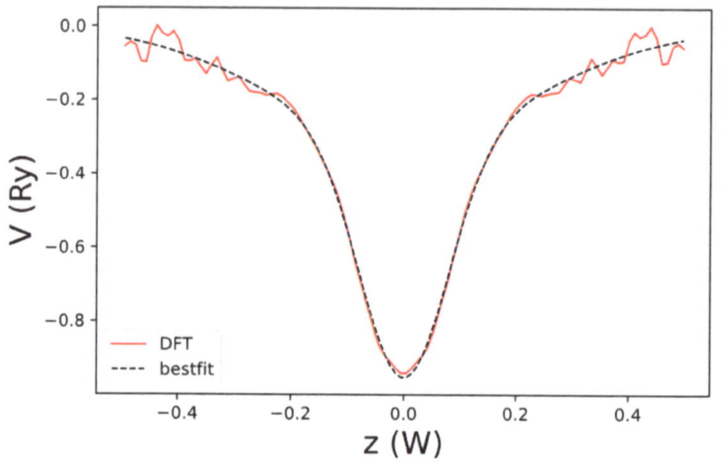

Figure 4. The net local potential of graphene, $V_{loc}(\mathbf{0}, z)$, as a function of z obtained by DFT (red curve) and the best-fit result to $V_{loc}(\mathbf{0}, z)$ (dashed black curve). Here, $W = 3.25a$ is the width of the domain along the z-axis used to define the B-spline basis.

2.4. Fitting of the Non-Local Pseudopotential for Graphene

The semi-empirical pseudopotential contains both local and non-local terms. The non-local pseudopotential contains angular momentum and energy-dependent terms [15]. Incorporating non-local in addition to local terms can provide a more accurate energy range for the valence band edge and best matches to the experimental data. The nonlocal potential is given by a separable form [16]:

$$\hat{V}_{nl} = \sum\nolimits_{\sigma lm,nn'} E_{lm}^{nn'} \left| \beta_{lm}^{n\sigma} \right\rangle \left\langle \beta_{lm}^{n'\sigma} \right| \tag{30}$$

where $\beta_{lm}^{n\sigma}(\mathbf{r})$ denotes the projector functions for an atom at position $\boldsymbol{\tau}_{\sigma}$. For each atom, the best-fit β functions take the following form:

$$\beta_{lm}^{n}(\mathbf{r}) = A_l^n(r) r^l Y_{lm}(\Omega). \tag{31}$$

for $l, m = 00, 10, 11, 20, 21, 22$. We fit the beta function for 2S and 2P orbitals for C atom from the Vanderbilt ultrasoft pseudopotential (USPP) [16]. The fitting potential used has the form for $r < R_s$, where R_s is the cut-off radius.

$$A_l^n(r) = \left(C_0 + C_1 r^2 + C_2 r^4 + C_3 r^6 + C_4 r^8\right) e^{-\alpha r^2} \tag{32}$$

For the second 2P orbital ($2P_2$), we break the β function into two segments with a dividing radius at $R_s = 0.9228$ (indicated by a vertical dashed line in Figure 5d). We fit the first segment (seg 1) for $r < 0.9228$ by expression (32). For the second segment (seg 2), we fit $A_l^n(r)$ with the following expression:

$$A_l^n(r) = \left(C_0 + C_1 r_s^2 + C_2 r_s^4 + C_3 r_s^6 + C_4 r_s^8\right) \text{ (for } 0.9228 < r < 1.2953) \tag{33}$$

where $r_s = r - 0.9228$. The bet-fit parameter obtained is given in Table 4.

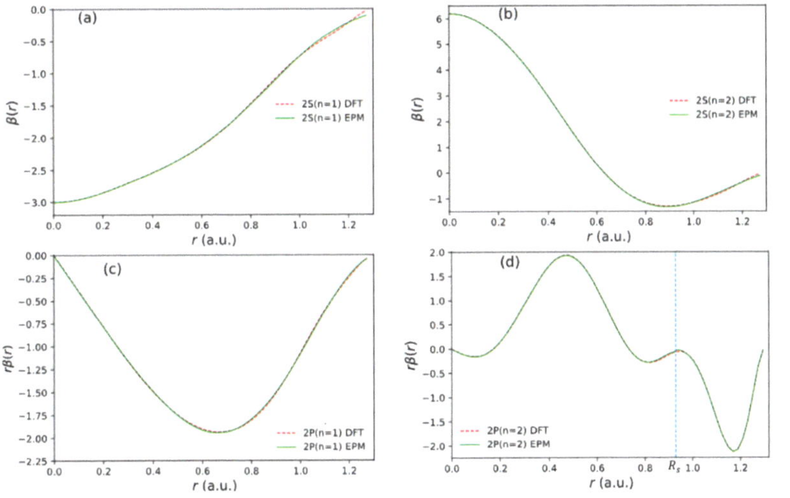

Figure 5. Fitting results of β functions used in the non-local pseudopotential of C atom. (**a**) $2S_1$ state. (**b**) $2S_2$ state. (**c**) $2P_1$ state. (**d**) $2P_2$ state.

Table 4. Fitting parameters (a.u.) for β functions used in the non-local pseudopotential of C atom.

Orbitals	α	R_s	C_0	C_1	C_2	C_3	C_4
$2S_1$	2.747	1.3174	−2.999	−4.209	−11.95	7.612	0
$2S_2$	2.171	1.3174	6.206	−9.434	−21.03	14.21	0
$2P_1$	0.5104	1.3174	−3.941	−1.411	5.485	−1.939	0
$2P_2$(seg1)	1.134	0.9228	−2.492	94.68	−365.4	480.9	−209.8
$2P_2$(seg2)	0.0	1.2953	−0.9771	−350.7	4210	−5711	−4377

The fitting result for each orbital is shown in Figure 5. The quality of fitting to the input of USPP [16] is excellent.

2.5. Matrix Elements of Local and Nonlocal Pseudopotential

The local pseudopotential $V_{loc}(\mathbf{r})$ is given by Equation (12). It consists of the atomic-like term $V_0(r)$, which is spherical in 3D space, and the correction term, which has a short-range part and a long-range part. We define the following:

$$I(\alpha_s; i,, i') = \int dz B_i(z) B_{i'}(z) e^{-\alpha_s z^2}. \tag{34}$$

$B_i(z)$ is the B-spline basis functions with the subscript κ dropped for brevity. Within the mixed basis, the matrix elements for the atomic-like term can be written as follows:

$$\langle \mathbf{G}; B_i | V_0 | \mathbf{G}'; B_{i'} \rangle = \sum_s C_s I(\alpha_s; i,, i') S_1(\Delta\mathbf{G}) \left(\frac{\pi}{\alpha_s}\right) e^{-\Delta\mathbf{G}^2/(4\alpha_s)} \tag{35}$$

within the B-spline basis. Here, $\Delta\mathbf{G} = \mathbf{G}' - \mathbf{G}$ and C_s denote the fitting parameters for a given C atom in the unit cell, as given in Table 1.

$$S_1(\Delta\mathbf{G}) = \frac{1}{A_c} \sum_\sigma e^{i\Delta\mathbf{G}\cdot\tau_\sigma} = \frac{2}{A_c} cos(\Delta\mathbf{G}\cdot\boldsymbol{\tau}_1) \tag{36}$$

denotes the structure factor for graphene. Due to the inversion symmetry of graphene, we choose the origin to be the center between two C atoms in the unit cell, and $\boldsymbol{\tau}_1$ is the position of one C atom in the unit cell. Thus, $S_1(\Delta G)$ is real.

Similarly, we obtain the matrix elements for the remaining terms for the local potential as follows:

$$\begin{aligned}\langle \mathbf{G}; B_i|\Delta V_{loc}|\mathbf{G}'; B_{i'}\rangle &= \int dz B_i(z)\Delta\widetilde{V}_{loc}(\Delta\mathbf{G}; z)B_{i'\kappa}(z)\\ = \int dz B_i(z)\Big[f_L(z)\widetilde{D}_L(\Delta\mathbf{G})S_1(\Delta\mathbf{G}) &+ f_S(z)\widetilde{D}_S(\Delta\mathbf{G})S_1(\Delta\mathbf{G}) + f_b(z)B_i(z)\Delta\widetilde{V}_b(\Delta\mathbf{G})\Big]B_{i'}(z)\end{aligned} \tag{37}$$

where

$$\Delta\widetilde{V}_b(\mathbf{q}) = \sum_{m=0}^{4} C_n^h\left(-\frac{\partial}{\partial\alpha_h}\right)^n \frac{\pi}{\alpha_h}e^{-q^2/4\alpha_h} \tag{38}$$

denotes the in-plane Fourier transform of the bond-charge contribution, $\Delta V_b(\boldsymbol{\rho}, 0)$, given in Equation (25).

2.6. Nonlocal Corrections in Overlap and Potential

The matrix elements for nonlocal potential read as follows [12]:

$$\begin{aligned}\langle \mathbf{K}; B_i|\hat{V}_{nl}|\mathbf{K}'; B_{i'}\rangle &= \sum_{\sigma lm,nn'} E_{lm}^{nn'}\langle \mathbf{K}; B_i|\beta_{lm}^{n0}\rangle\langle\beta_{lm}^{n'0}|\mathbf{K}'; B_{i'}\rangle e^{i(\mathbf{G}'-\mathbf{G})\cdot\boldsymbol{\tau}_\sigma}\\ &= \frac{1}{A_c}\sum_{\sigma lm}\sum_{nn'} E_{lm}^{nn'}P_{lm}^{in}(\mathbf{K})P_{lm}^{i'n'*}(\mathbf{K}')e^{i(\mathbf{G}'-\mathbf{G})\cdot\boldsymbol{\tau}_\sigma}.\end{aligned} \tag{39}$$

Here, $\mathbf{K} = \mathbf{k} + \mathbf{G}$; $\mathbf{K}' = \mathbf{k} + \mathbf{G}'$. $\boldsymbol{\tau}_\sigma = \pm\boldsymbol{\tau}_1$ for $\sigma = 1, 2$. $P_{lm}^{in}(\mathbf{K}) = i^l\sqrt{A_c}\langle \mathbf{K}; B_i|\beta_{lm}^{n0}\rangle$ denotes the projection of the β functions (with the center shifted to the origin) into our basis. A detailed description of the evaluation of $P_{lm}^{in}(\mathbf{K})$ is given in Appendix A. For calculating the band structure, we also need to evaluate the correction to the overlap matrix elements. The correction to the overlap matrix can be written as follows [12]:

$$\langle \mathbf{K}; B_i|\hat{S}|\mathbf{K}'; B_{i'}\rangle = \frac{1}{A_c}\sum_{\sigma lm}\sum_{nn'} q_{lm}^{nn'}P_{lm}^{in}(\mathbf{K})P_{lm}^{i'n'*}(\mathbf{K}')e^{i(\mathbf{G}'-\mathbf{G})\cdot\boldsymbol{\tau}_\sigma}. \tag{40}$$

This expression is identical to (39) except that the energy parameters $E_{lm}^{nn'}$ are replaced by the overlap parameters $q_{lm}^{nn'}$. Here, we adopt the same parameters for $E_l^{nn'}$ and $q_l^{nn'}$ as in the ab initio input data [12,16] to calculate the matrix elements of non-local pseudopotentials. The input parameter used for $E_l^{nn'}$ and $q_l^{nn'}$ are shown in Table 5.

Table 5. Parameters for the overlap and non-local pseudopotential terms.

n	n′	l	$E_l^{nn'}$	$q_l^{nn'}$
1	1	0	3.490422	−0.449056
1	2	0	0.207297	0.344889
2	2	0	−2.748230	−0.212785
3	3	1	2.474918	1.236379
3	4	1	−5.902130	−0.938122
4	4	1	9.289400	0.631727

The formulas derived above can be extended to other two-dimensional materials beyond graphene, such as transition-metal dichalcogenides. Due to the inversion symmetry in graphene, all matrix elements become real. Furthermore, there is a mirror symmetry for the z-axis. By adopting the B-spine basis along the z-axis and taking symmetric and antisymmetric combinations of these basis functions, we can decouple the eigenstates of the Hamiltonian within the symmetric and antisymmetric basis sets for any wave vector in the

2D plane. This significantly improves the speed of computation for the band structures. We found that using a direct solver for diagonalization becomes even faster than the iterative solver based on the conjugate-gradient (CG) approach [46] for this system with both inversion and mirror symmetry. The cut-off used for plane waves used in wavefunctions is 30 a.u. and that in pseudopotential is 180 a.u.

3. Results and Discussions

3.1. Band Structure of Graphene

The all-electron (AE) Bloch states of graphene are written as follows:

$$\Psi_{\nu,\mathbf{k}}(\boldsymbol{\rho},z)=\hat{O}\varphi_{\nu,\mathbf{k}}(\boldsymbol{\rho},z)=\hat{O}\sum\nolimits_{i,\mathbf{G}}Z_{i\mathbf{G}}^{\nu,\mathbf{k}}\widetilde{B}_i(z)\frac{1}{\sqrt{A}}e^{\mathrm{i}(\mathbf{k}+\mathbf{G})\cdot\boldsymbol{\rho}},\tag{41}$$

where $\varphi_{n,\mathbf{k}}(\boldsymbol{\rho},z)$ denotes the pseudo-wavefunction, $\widetilde{B}_i(z)$ denotes the orthogonalized B-spline function, and $\hat{O}$ is the overlap operator defined in (A5). The all-electron Kohn–Sham equation [47] reads as follows:

$$\hat{H}_{AE}\Psi_{\nu,\mathbf{k}}(\boldsymbol{\rho},z)=E_{\nu}(\mathbf{k})\Psi_{\nu,\mathbf{k}}(\boldsymbol{\rho},z)\tag{42}$$

where $\hat{H}_{AE}$ is the all-electron Hamiltonian operator, and $E_{\nu}(\mathbf{k})$ denotes the energy of the ν-th band at wavevector k. Substituting Equation (41) into Equation (42) gives rise to the following generalized eigenvalue problem:

$$\sum\nolimits_{i',\mathbf{G}'}\left\langle\mathbf{k}+\mathbf{G};\widetilde{B}_i\middle|\hat{H}\middle|\mathbf{k}+\mathbf{G}';\widetilde{B}_{i'}\right\rangle Z_{i'\mathbf{G}'}^{\nu,\mathbf{k}}=E_{\nu}(\mathbf{k})\sum\nolimits_{i'\mathbf{G}'}\left\langle\mathbf{k}+\mathbf{G};\widetilde{B}_i\middle|\hat{O}\middle|\mathbf{k}+\mathbf{G}';\widetilde{B}_{i'}\right\rangle Z_{i'\mathbf{G}'}^{\nu,\mathbf{k}}\tag{43}$$

where $\hat{H}$ is the Hamiltonian operator given in Equation (6).

We apply the current SEP to calculate the band structure of graphene. The real-space structure of graphene can be described by a 2D unit cell with primitive lattice vectors $\mathbf{a}_1=\left(1,-\sqrt{3}\right)\mathrm{a}/2$ and $\mathbf{a}_2=\left(1,\sqrt{3}\right)\mathrm{a}/2$. The positions of carbon atoms are $\left(1,-1/\sqrt{3}\right)\mathrm{a}/2$ and $\left(-1,1/\sqrt{3}\right)\mathrm{a}/2$. The basis vectors in reciprocal space are $\mathbf{b}_1=\frac{2\pi}{\sqrt{3}\mathrm{a}}\left(\sqrt{3},-1\right)$, $\mathbf{b}_2=\frac{2\pi}{\sqrt{3}\mathrm{a}}\left(\sqrt{3},1\right)$. Figure 6 shows the geometry of graphene in real space and reciprocal space.

In Figure 7, the band structure of graphene calculated by SEP with best-fitted parameters deduced in Section 2 is shown and compared with that obtained from a self-consistent calculation based on DFT [12]. The overall band structures obtained by the present method are in close agreement with the ab initio calculation. Since we solve the Bloch states with even and odd parities (with respect to the mirror symmetry about the z-axis) separately, the symmetry characteristics of the bands can be easily distinguished. Here, the even (odd) parity states are presented in green (red) color. It is noted that the red curves (with odd parity) are related to the p_z orbitals of the carbon atoms, and they form π-bonded bands, whereas the green curves (with even parity) are related to the s, p_x, andp_y orbitals.

As shown in Figure 7, the agreement between the SEP and DFT results obtained by using the method described in [12] for the lowest 20 bands is excellent. The advantage of the SEP is that it is easy to use and requires no self-consistent calculation to establish the charge density. The time needed to calculate the band structures via a direct solver on a laptop PC is less than 6 s, which is only a fraction (~1/5) of the time needed for the CG calculation used in the DFT package (after the self-consistent density is established). The saving is mainly due to the use of the inversion and mirror symmetry properties of graphene.

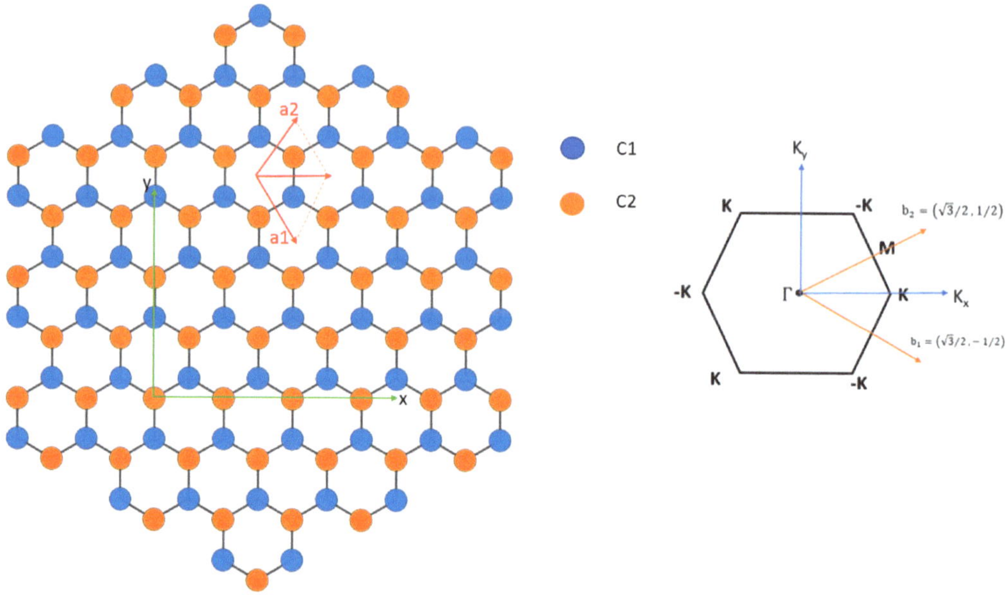

Figure 6. (**Left**) Primitive vectors and position of atoms for graphene. Atoms on sublattice A and B are colored blue and yellow, respectively. (**Right**) Hexagonal 2D Brillouin zone of graphene with main symmetry points. The primitive reciprocal lattice vectors shown are in units of $\frac{4\pi}{\sqrt{3}a}$.

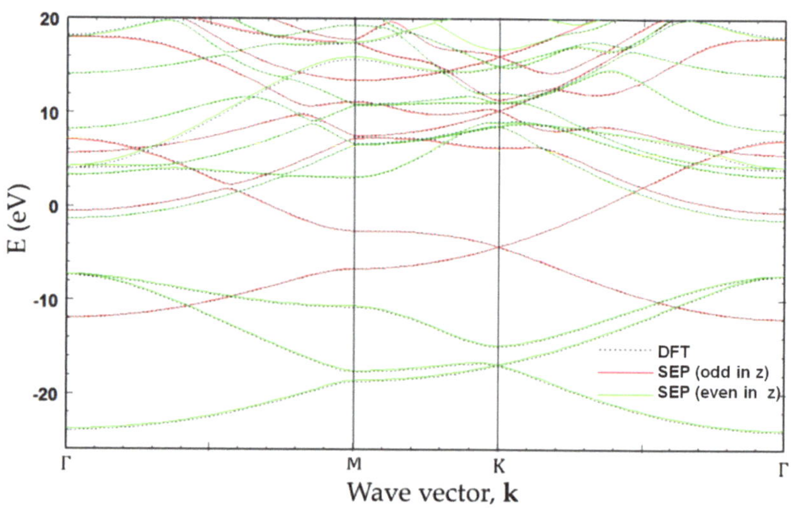

Figure 7. Band structure of graphene obtained by the present SEP with best-fitted parameters (solid curves). For comparison, the band structure obtained by self-consistent calculation based on DFT with Vanderbilt USPP is also included (dotted curves).

3.2. *Band Structure of Armchair Graphene Nanoribbon*

We consider a graphene nanoribbon with armchair edges with a width of Na, where a is the lattice constant of graphene. The supercell of the armchair graphene nanoribbon (AGNRs) contains $4N + 2$ atoms, as illustrated in Figure 8a. To adopt the 2D plane-wave basis, we introduce a vacuum region with the width of $(M - N)a$ inside the supercell such that the dimension of the rectangular supercell is $\sqrt{3}a \times Ma$. The reciprocal lattice vectors of a superlattice consisting of a 1D periodic of AGNRs along the x-axis can be written as $\mathbf{g}_{n_1 n_2} = n_1 \widetilde{\mathbf{b}}_1 + n_2 \widetilde{\mathbf{b}}_2$, where $\widetilde{\mathbf{b}}_1 = \frac{2\pi}{Ma}\hat{\mathbf{x}}$ and $\widetilde{\mathbf{b}}_2 = \frac{\pi}{a}\hat{\mathbf{y}}$ are the basis vectors in reciprocal space, and n_1 and n_2 are arbitrary integers. Let $\mathbf{g}_i$ $(i = 1, \ldots, 2M)$ denote those non-equivalent $\mathbf{g}$ vectors falling within the first Brillouin zone of graphene. (See Figure 8b.) Then, all $\mathbf{g}$ vectors of the AGNRs superlattice can be expressed as $\mathbf{g}_i + \mathbf{G}$, where $\mathbf{G}$ denotes the reciprocal lattice vectors of graphene.

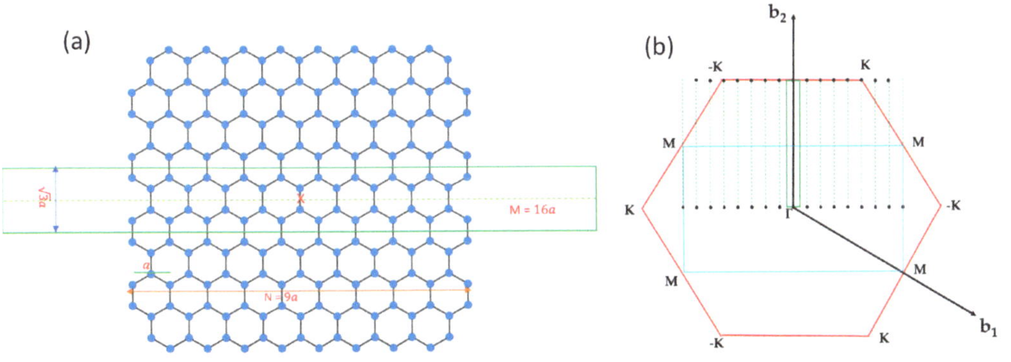

Figure 8. (**a**) Position of atoms of a 9×2 AGNR and the 16×2 supercell used in the calculation (enclosed within the green rectangular box). The red x marks the origin of the coordinate system to illustrate the inversion symmetry. (**b**) Rectangular 2D Brillouin zone (BZ) of the 16×2 supercell for the AGNRs are indicated with the green box, and the non-equivalent AGNRs reciprocal lattice vectors enclosed within the first BZ of graphene $\mathbf{g}_j$ $(j = 1, \ldots, 2M)$ are indicated by black dots. The blue rectangular box indicates the BZ of a 1×2 supercell for graphene. The black dots outside the red hexagon (BZ of graphene) can be shifted inside the box by adding a reciprocal lattice vector.

The Bloch states of the AGNRs superlattice can be written as linear combinations of basis functions that are products of 2D plane waves (labeled by wave vectors $\mathbf{k} + \mathbf{g}_j + \mathbf{G}$) and B-splines in z (i.e., the mixed-basis introduced in [11]). For wide AGNRs, this basis set can be quite large. The band structure of the AGNRs is expected to be quite close to the zone-folded band structure of graphene, with deviation mainly coming from the quantum confinement effect from the vacuum region and the effect due to the creation of edges. To describe such a change, it is computationally more efficient to start with the zone-folded band structures of graphene that can be calculated by using the graphene SEP code developed above at wave vectors of $\mathbf{k} + \mathbf{g}_j$ $(j = 1, \ldots, 2M)$, where $\mathbf{k}$ is within the first mini zone of the AGNR superlattice. A similar approach was applied to the Si 7×7 surface, and it was demonstrated that such a method works very well for large superstructures [48]. If the vacuum region introduced in the superlattice is thick enough, the coupling between adjacent AGNRs can be negligible, and $\mathbf{k}$ of interest will be along the long axis of the AGNR (taken to be parallel to $\hat{\mathbf{y}}$ here). The low-lying pseudo-Bloch states of graphene (with band labeled by ν) obtained at $\mathbf{k} + \mathbf{g}_j$ (denoted as $\varphi_{\nu,\mathbf{k}+\mathbf{g}_j}$) can then be used as a set of contracted basis functions for calculating the band structures of the AGNR. This set of contracted bases is a nearly complete basis for AGNR if we use a large number of graphene bands until the convergence is established for the calculated results.

Thus, we can write the Hamiltonian of the AGNR as $H = -\nabla^2 + U_1 + \hat{U}_2$, where

$$U_1(\mathbf{r}) = \left[V_0'(\mathbf{r}) + \Delta V_S'(\mathbf{r}) + \Delta V_L'(\mathbf{r}) + \Delta V_b'(\mathbf{r})\right] \quad (44)$$

denotes the sum of local potentials of C atoms defined in Equation (12) for atomic sites (indexed by σ) in the nanoribbon region A_{NR} [with $|x| \leq W/2$] inside the supercell used for AGNRs. $\hat{U}_2$ denotes the nonlocal pseudopotential of the nanoribbon. The AGNRs Hamiltonian matrix within the contracted basis is then given by the following:

$$\left\langle \varphi_{v,\mathbf{k}+\mathbf{g}_j} \middle| \hat{H}_{GNR} \middle| \varphi_{v',\mathbf{k}+\mathbf{g}_{j'}} \right\rangle = \left\langle \varphi_{v,\mathbf{k}+\mathbf{g}_j} \middle| -\nabla^2 \middle| \varphi_{v',\mathbf{k}+\mathbf{g}_{j'}} \right\rangle + \left\langle \varphi_{v,\mathbf{k}+\mathbf{g}_j} \middle| U_1 + \hat{U}_2 \middle| \varphi_{\prime,\mathbf{k}+\mathbf{g}_{j'}} \right\rangle \tag{45}$$

where the first term describes the kinetic energy, and the second term consists of the semi-empirical local (U_1) and nonlocal ($\hat{U}_2$) pseudopotentials for AGNR. The detailed expressions for these matrix elements are given in Appendix B.

In addition to the z-mirror symmetry, the AGNR also has x-mirror symmetry for any wave vector, i.e., **k**, along the y-axis. Thus, we can define symmetrized basis states with respect to the x-mirror as follows:

$$|\varphi^{s,\pm}_{v,\mathbf{k}+\mathbf{g}_j}\rangle = f_j\left(|\varphi^{s}_{v,\mathbf{k}+\mathbf{g}_j}\rangle \pm |\overline{\varphi}^{s}_{v,\mathbf{k}+\overline{\mathbf{g}}_j}\rangle \right)/\sqrt{2} \text{ for } g_{jx} \geq 0, \tag{46}$$

where $\overline{\mathbf{g}}_j = \left(-g_{jx}, g_{jy}\right)$, $\overline{\varphi}^{s}_{v,\mathbf{k}+\overline{\mathbf{g}}_j}(x,y,z) = \varphi^{s}_{v,\mathbf{k}+\mathbf{g}_j}(-x,y,z)$ (corresponding to flipping the sign of G_x in the plane-wave basis). $f_j = 1/\sqrt{2}$ for $g_{jx} = 0$, and $f_j = 1$ otherwise. The symbol "$\pm$" denotes even/odd parity states with respect to the x-mirror, while the superscript $s = e, o$ is the label for the even/odd parity states with respect to the z-mirror. The Hamiltonian matrix elements between symmetrized states of the same parity read as follows:

$$\left\langle \varphi^{s,\pm}_{v,\mathbf{k}+\mathbf{g}_j} \middle| \hat{H}_{GNR} \middle| \varphi^{s,\pm}_{v\prime,\mathbf{k}+\mathbf{g}_{j'}} \right\rangle = f_j f_{j'} \left[\left\langle \varphi^{s}_{v,\mathbf{k}+\mathbf{g}_j} \middle| \hat{H}_{GNR} \middle| \varphi^{s}_{v\prime,\mathbf{k}+\mathbf{g}_{j'}} \right\rangle \pm \left\langle \varphi^{s}_{v,\mathbf{k}+\mathbf{g}_j} \middle| \hat{H}_{GNR} \middle| \varphi^{s}_{v\prime,\mathbf{k}+\overline{\mathbf{g}}_{j'}} \right\rangle \right]. \tag{47}$$

The overlap matrix is given by the same expression above, with $\hat{H}_{GNR}$ replaced by $\hat{O}_{GNR}$.

Here, we divide the basis set into four different subsets according to different symmetry types labeled by $\{\pi_+, \pi_-, \sigma_+, \sigma_-\}$. π_+ and π_- states represent π-bonded (p_z-like) states with even and odd parity, respectively, under the x-mirror operation. σ_+ and σ_- states represent σ-bonded (p_x, p_y, or s-like) states with even and odd parity, respectively, under the x-mirror operation. Eigenstates of $\hat{H}_{GNR}$ with different symmetry types are decoupled based on group theory. Thus, we can block-diagonalize the Hamiltonian matrix into four diagonal sub-blocks, and each diagonal sub-block can be diagonalized separately. By taking advantage of these symmetry properties, the electronic structures of graphene nanoribbons can be solved very efficiently with the present SEP via a direct diagonalization method. Furthermore, the symmetry characteristics of different bands can be easily identified in the band structure by using different colors. This can help sort out the complex characteristics in AGNR electronic states and improve the understanding of the electronic properties of AGNR.

The graphene bulk basis adopted above can describe the band folding due to the quantum confinement effect in the nanoribbon well, but it requires a large number of basis states to fully capture the effect of localized edge states. The AGNRs band structures obtained by solving Equation (43) are shown in Figure 9a,b to compare with the DFT results. The number of graphene bands included in the calculation is 10 for the $\pi_+(\pi_-)$ states and 60 for the $\sigma_+(\sigma_-)$ states. Adding more bands produces no significant effect. For the odd states, the change of energy on the AGNR band structure is less than 0.005 eV within the energy window of interest. For the σ-bonded (even in z) states, most minibands remain nearly unchanged when the number of graphene bands included varies from 14 to 60, while the two pairs of minibands closest to the band-gap region change significantly and still do not reach the desired convergence level even with 60 graphene bands included in the basis. These edge-induced states correspond to the bonding and antibonding states of the dimers on two edges. Since there will be an edge-induced correction in the local potential, it is unnecessary to spend a significant effort to obtain fully convergent results for the preliminary investigation in this section.

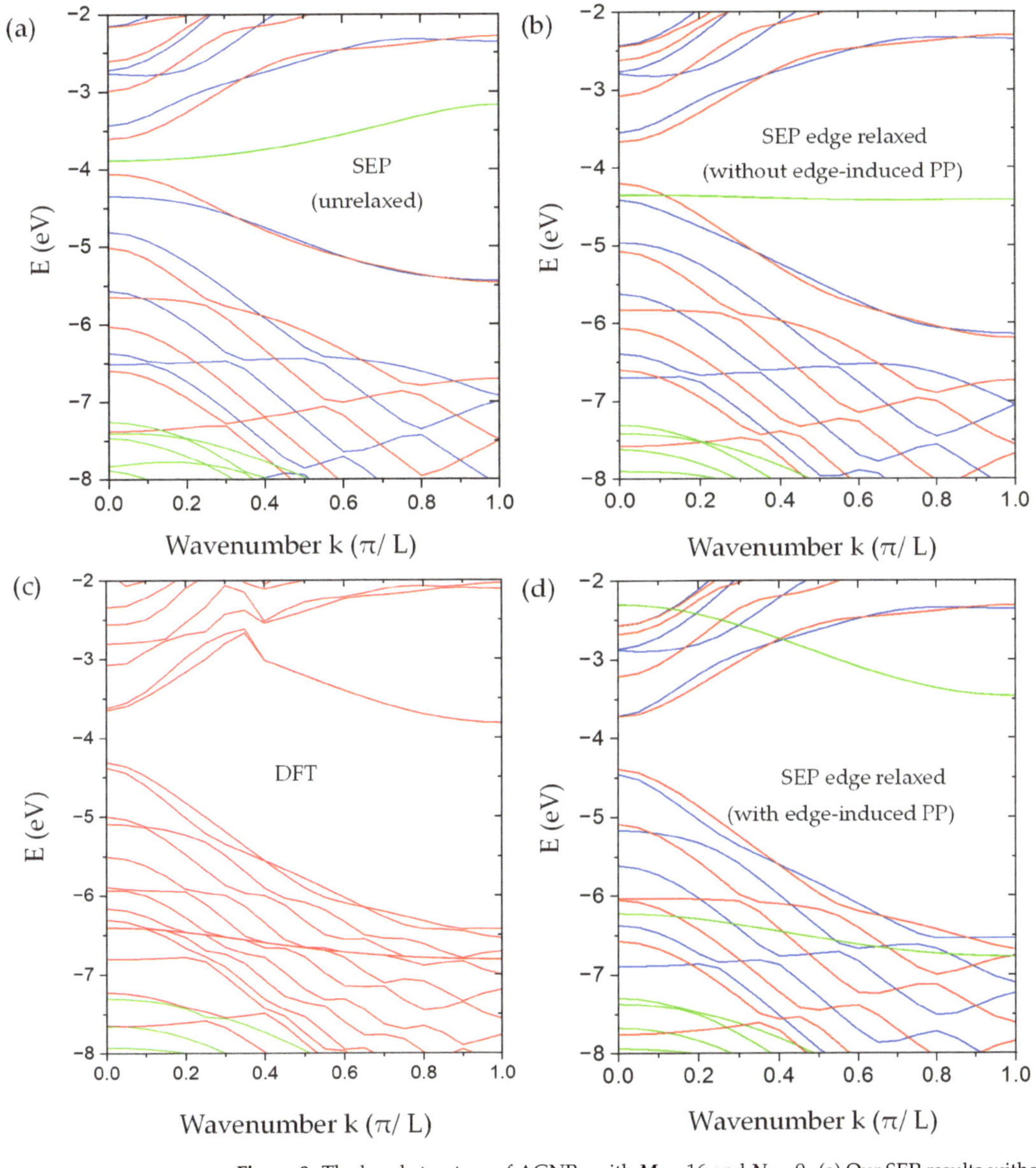

Figure 9. The band structure of AGNRs with $M = 16$ and $N = 9$. (**a**) Our SEP results without the relaxation of the edge atoms. (**b**) Our SEP results with edge relaxation but without modifying the pseudopotentials on the edge atoms. (**c**) DFT results were obtained by using the method described in [12] for the AGNRs with relaxed atomic positions for the edge atoms. (**d**) Our SEP results include the modification of pseudopotentials on the edge atoms. In SEP results (**a**,**b**,**d**), the bands in blue (**with** π_- **symmetry**) and red (with π_+ symmetry) are derived from π-bonded states (odd with respect to the z-mirror), while the bands in green are derived from the σ-bonded states (even with respect to the z-mirror). Here, we do not distinguish the σ_+ from σ_- states since the important states are edge states, and they are essentially degenerate.

We also performed DFT calculations of the same AGNR within the basis constructed by 2D plane waves multiplied by B-spline functions of z, as described in [12]. The edge atoms are allowed to relax to minimize the total energy of the AGNR. In the relaxed geometry, the displacement of the edge atom in the left and upper corner of the AGNR supercell (the green box) as shown in Figure 8a is described by $\Delta\boldsymbol{\tau} = (\Delta x, \Delta y)$, with $\Delta x = 0.2107\ a.u.$ and $\Delta y = 0.1788\ a.u.$ The displacement of three other edge atoms can be deduced by applying the x-mirror and y-mirror symmetry. The resulting band structures are shown in Figure 9c. To make sure that the relaxation of only the edge atoms is sufficient, we also performed the DFT calculation of the AGNR, in which the outermost four rows of atoms are allowed to relax, and we found that the major displacement occurs on the edge atoms, while the other inner rows only relax slightly, and the resulting band structure is quite similar to the one with relaxation only on the edge atoms. For comparison, we shifted the DFT band structure rigidly in energy, so the valence-band maximum is aligned with our SEP calculation.

We found that, as shown in Figure 9a, the band structure of the unrelaxed AGNR looks qualitatively similar to the corresponding DFT results (with edge relaxation) in Figure 9c. However, the band gap obtained is about 0.45 eV, which is substantially smaller than the DFT result of ~0.6 eV. Furthermore, we found a pair of mid-gap bands (in green) that are related to the σ-bonding states of the dimer atoms on the edge. There is also a pair of σ-antibonding states of the edge dimers with energy near 0 eV at the Γ point (not shown). These edge-dimer-related minibands are nearly doubly degenerate, corresponding to dimer states located at the left and right edges of the AGNR. Since the width of the ribbon is much larger than the spread of the wavefunctions localized on edges on opposite sides, these states form a closely spaced pair of minibands. In the DFT band structure, the corresponding σ-bonding states are lowered by ~2.5 eV at the Γ point, and there are no mid-gap states left. Since the DFT calculation did not separate the even and odd states, some of them even states look artificially entangled with the odd states.

The SEP band structures of the AGNR with the same relaxed atomic positions as determined by DFT but with the pseudopotential (PP) given by Equation (29) are shown in Figure 9b. With the relaxation, the band gap becomes enlarged to ~0.5 eV, and the overall band structure agrees better with the DFT results. The pair of σ-bonding states of edge-dimers (in green) is also lowered but only by ~0.5 eV at Γ point, and there are no states left in the mid gap. We note that for device applications, the π-bonded states (near the band gap region) play a dominant role. The band structure for these π-bonded states is not significantly changed due to the relaxation of atomic positions at the edges. However, there are still noticeable differences between the band structures obtained by SEP and by DFT, which are mainly caused by the deviation of local potential near the edge atoms as the graphene is cleaved to form a nanoribbon. We discuss the consequence of this effect and its remedy below.

3.3. Modification of Pseudopotential for Edge Atoms of Armchair Graphene Nanoribbon

Figure 10 shows the contour plot of the local potential of the 9 × 2 AGNR with edge relaxation. The ~5% shift of atomic positions near the edges is noticeable in this plot. To examine the deviation of local potential (V_{loc}) calculated by SEP from the DFT results, we plot the line cuts of $V_{loc}(x, y, z)$ as functions of x at the four different values of y ($y = -L/2, -3L/17, 0,$ and $6L/17$) and $z \approx 0$ in Figure 11. These lines are indicated by black lines in Figure 10. Here, $L = \sqrt{3}a$ is the length of the AGNR supercell along the y-axis.

As shown in Figure 11, we found that the local potentials obtained by SEP agree very well with the DFT results in the interior region of the AGNR, with noticeable deviation from the DFT results only near the two edges (with $|x| > 4.5a$). It implies that the SEP potential centered at an atom follows the position of the relaxed atom very well, while the cleavage of graphene to form an AGNR causes some charge redistribution that can lead to the change of the local potential near the edges.

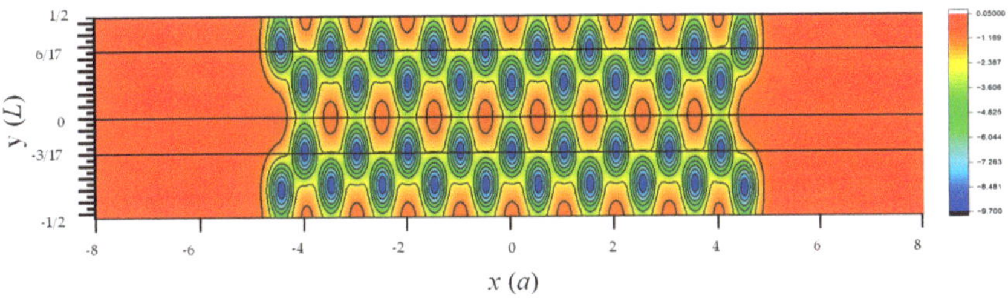

Figure 10. Contour plot of net local potential in one supercell of the AGNR obtained by DFT [11]. Here, a is the lattice constant of graphene, and $L = \sqrt{3}a$ is the the length of supercell along the y -axis.

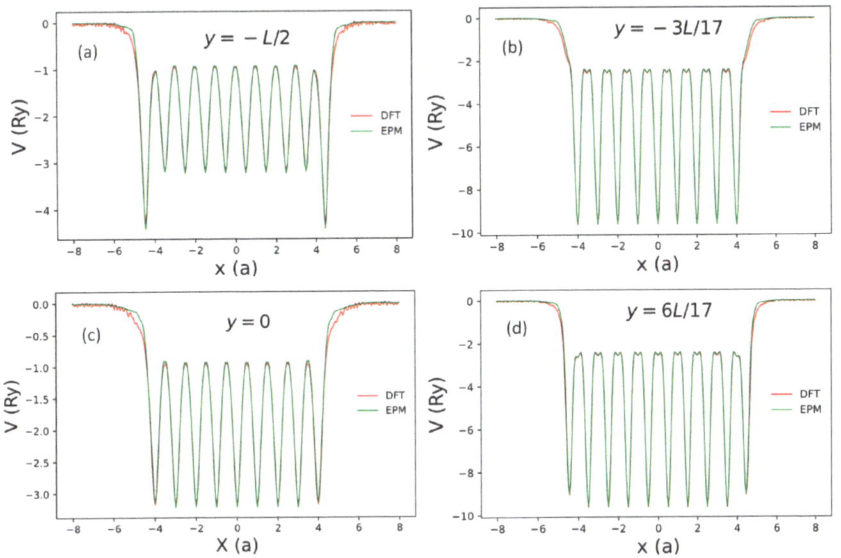

Figure 11. The net local pseudopotential in one supercell of the AGNRs obtained by the current SEP (green line) and by DFT (red line) along the lines with $z \approx 0$ (near the AGNRs plane) and some selected values of y. (**a**) $y = -L/2$ (along the bottom black line going through the centers of ten bonds in Figure 10), (**b**) $y = -3L/17$ (along the black line going through the centers of nine atoms in Figure 10), (**c**) $y = 0$ (along the middle black line going through the centers of nine bonds in Figure 10), and (**d**) $y = 6L/17$ (along the black line going through the centers of ten atoms in Figure 10).

To investigate the effect of local potential induced by the edge creation, we took the difference between the DFT (red curve) and SEP results (green curve) in Figure 11 for x between $3a$ and $6a$ at $y = -L/2, -3L/17, 0$, and $6L/17$. The results for the differences at the selected values of y are shown as dotted lines in Figure 12. These difference curves (for one of the two edges) can be reasonably fitted by the following functional form:

$$\Delta V_e(\boldsymbol{\rho}, 0) \approx \left[S_a e^{-\alpha_a(\boldsymbol{\rho}-\boldsymbol{\rho}_a)^2} + S_b e^{-\alpha_a(\boldsymbol{\rho}-\boldsymbol{\rho}_b)^2} + S_c e^{-\alpha_a(\boldsymbol{\rho}-\boldsymbol{\rho}_c)^2} + S_d e^{-\alpha_b(\boldsymbol{\rho}-\boldsymbol{\rho}_d)^2} \right], \quad (48)$$

where $\boldsymbol{\rho}_a = (x_a, -L/2)$, $\boldsymbol{\rho}_b = (x_b, -3L/17)$, $\boldsymbol{\rho}_c = (x_b, 0)$, and $\boldsymbol{\rho}_d = (x_b, y_d)$ denote the four locations of effective bond charges near the AGNR right edge. $y_d = y_e + \Delta y \approx 6L/17$ denotes the y-coordinate of the relaxation edge atom, including the relaxation Δy. The AGNR potential also has a y-mirror symmetry, so we duplicate the terms at y_b and y_d and add the same potentials at $-y_b$ and $-y_d$. We note that the potentials at $y_a = -L/2$ (equivalent to $y_a = L/2$ due to periodicity) and $y_c = 0$ will be mapped to themselves by the y-mirror. Here, $x_a = 5a$ denotes the bond-charge location at $y = -L/2$, and $x_b = 4.65a$ denotes the common bond-charge location near the edges for other y values. The positions of these bond charges are related to the minimum of the difference in the local potentials shown in Figure 12. S_a, S_b, S_c, and S_d, describing the variation of potential strength along the y-axis. Here, we choose $\alpha_a = 0.4$ and $\alpha_b = 1.1$, which approximately describe the width of the potential wells shown in Figure 12. The best-fit results for $\Delta V_e(\boldsymbol{\rho}, 0)$ are shown as green curves in Figure 12, with best-fit values of S_a, S_b, S_c, and S_d listed in Table 6.

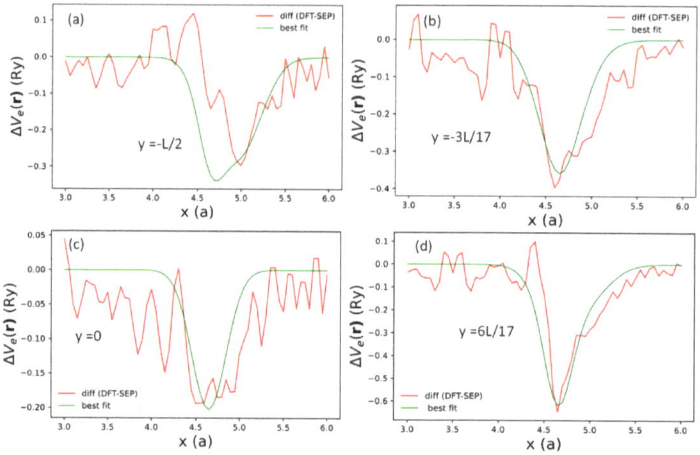

Figure 12. Difference in the local potentials between DFT and the current SEP results (red curves) and fitted by expression (48) (green curves) evaluated at (**a**) $y = -L/2$, (**b**) $y = -3L/17$, (**c**) $y = 0$, and (**d**) $y = 6L/17$, respectively.

Table 6. Fitting parameters S_a, S_b, and S_c (in a.u.) at four selected values of y.

S_a	S_b	S_c	S_d
−0.25	−0.34	0.10	−0.42

We approximate the net three-dimensional local potential, including the bond-charge redistribution near the edges by a separable form:

$$V_{loc}(\boldsymbol{\rho}, z) = V_{loc}(\boldsymbol{\rho}, 0) f_e(z) \tag{49}$$

where $f_e(z) = V_{loc}(\mathbf{0}, z)/V_{loc}(\mathbf{0}, 0)$ describes the z-dependence of the bond-charge contribution near the edges. $f_e(z)$ can be extracted from the net local potential of AGNR, $V_{loc}(x_b, 0, z)$, obtained by DFT with $x = x_b$ and $y = 0$. The result is shown in Figure 13, and it can be well fitted by the sum of two Gaussian functions. The normalized z-dependence of the local potential is given by the following:

$$f_e(z) = V_{loc}(x_b, 0, z)/V_{loc}(x_b, 0, 0) = C_1^e e^{-\alpha_{e1} z^2} + C_2^e e^{-\alpha_{e2} z^2} \tag{50}$$

where $C_2^e = 1 - C_1^e$ due to the normalization requirement similar to Equation (28). We obtain $C_1^e = 0.979$, $\alpha_{e1} = 1.0926$, and $\alpha_{e2} = 0.1026$. Thus, the correction to local pseudopotential due to the bond-charge redistribution near the edges can be approximately given by the following:

$$\Delta V_e(\boldsymbol{\rho}, z) \approx \Delta V_e(\boldsymbol{\rho}, 0) f_e(z) \tag{51}$$

where $\Delta V_e(\boldsymbol{\rho}, 0)$ is given in Equation (48) and $f_e(z)$ in Equation (50).

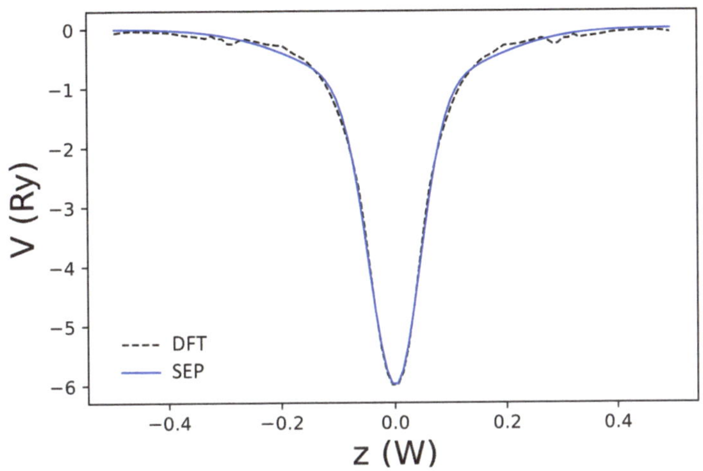

Figure 13. The net local potential of AGNR, $V_{loc}(x_b, 0, z)$, as a function of z obtained by DFT (dashed black curve) and the best-fit result to $V_{loc}(x_b, 0, z)$ (blue curve). Here, $W = 3.25a$ is the width of the domain along the z-axis used to define the B-spline basis.

Finally, we add the edge-induced correction in the local pseudopotential, $\Delta V_e(\boldsymbol{\rho}, z)$, to the Hamiltonian of the AGNR in our semi-empirical pseudopotential model. The matrix elements of $\Delta V_e(\boldsymbol{\rho}, z)$ within the contracted basis functions derived from the selected graphene eigenstates at special reciprocal lattice vectors for AGNR ($\mathbf{g}_j$) are given by the following:

$$\left\langle \varphi_{\nu,\mathbf{k}+\mathbf{g}_j} \middle| \Delta V_e \middle| \varphi_{\nu',\mathbf{k}+\mathbf{g}_{j'}} \right\rangle = \sum_{i\mathbf{G},i'\mathbf{G}'} Z_{i\mathbf{G}}^{\nu,\mathbf{k}+\mathbf{g}_j} Z_{i'\mathbf{G}}^{\nu',\mathbf{k}+\mathbf{g}_j} \sum_{m=1,2} C_m^e I(\alpha_{em}\, ; i, , i') \tilde{v}_e\left(\Delta \mathbf{g}_{jj'} + \Delta \mathbf{G}\right) \tag{52}$$

where $I(\alpha; i, , i')$ is defined in Equation (34), $\Delta \mathbf{g}_{jj'} + \Delta \mathbf{G} = \mathbf{g}_{j'} - \mathbf{g}_j + \mathbf{G}' - \mathbf{G}$, and

$$\tilde{v}_e(\mathbf{q}) = \frac{1}{A_{SC}} \int d\mathbf{r} e^{i\mathbf{q}\cdot\boldsymbol{\rho}} \Delta V_e(\boldsymbol{\rho}, 0) + c.c. \tag{53}$$

Here, $\Delta V_e(\boldsymbol{\rho}, 0)$ is the edge-induced correction in pseudopotential for the right edge, as given in Equation (48), and c.c. denotes the term contributed from the left edge, which is the complex conjugate of the previous term due to the inversion symmetry in the 2D plane of the AGNR. With the use of the analytic fitting functions introduced above, $\tilde{v}_e(\mathbf{q})$ can be written as follows:

$$\begin{gathered} \tilde{v}_e(\mathbf{q}) = \tfrac{2}{A_{SC}} \{ S_a I_0(\alpha_a, q) \cos(q_x x_a) \cos(q_x y_a) \\ + [2 S_b I_0(\alpha_a, q) \cos(q_x y_b) + S_c I_0(\alpha_a, q) + 2 S_d I_0(\alpha_b, q) \cos(q_x y_d)] \cos(q_x x_b) \} \end{gathered} \tag{54}$$

with $I_0(\alpha, q) = \frac{\pi}{\alpha} e^{-q^2/4\alpha}$. Due to the x-mirror symmetry, there are four corresponding bond charges at the left edge.

After adding the correction term $\Delta V_e(\boldsymbol{\rho}, z)$, the calculated band structure for the AGNR is shown in Figure 9d. To improve the convergence for the edge-dimer-related states, we included 90 graphene bands in our basis for the calculation of even states, while only 10 graphene bands are needed to achieve the desired convergence for odd states. We found that by adding more graphene bands to the calculation, the results remain nearly the same. Comparing with Figure 9c, where we did not include the edge-induced correction in pseudopotential, $\Delta V_e(\boldsymbol{\rho}, z)$, we found that the main effect of $\Delta V_e(\boldsymbol{\rho}, z)$ is to give a significant improvement of the miniband structures for the π-bonded states (in red or blue) of the AGNR, which agree very well with the DFT results as shown in Figure 9c, and the band gap obtained by SEP (~0.7 eV) also matches the DFT result well. The other important feature is the lowering of the σ-bonding states of edge dimers at the zone center from -4.3 eV in Figure 9b to -6.3 eV in Figure 9d, while the σ-antibonding states of edge dimers are also lowered by about 2 eV, and they appear in Figure 9d at -2.3 eV at the Γ point. The energy spacing between these edge-dimer bonding and anti-bonding states is about 4 eV at k = 0 (the zone center), and it reduces to about 3.3 eV at $k = \pi/L$ (the zone boundary). This feature also agrees reasonably well with the DFT result shown in Figure 9b. The energy position of these edge-dimer bonding states (near -6.75 eV at the zone boundary) is close to the DFT result (-6.8 eV), while the edge-dimer antibonding states sit around -3.46 eV, which is higher than the corresponding DFT results by about 0.3 eV. This discrepancy is likely due to other corrections in the pseudopotential that are not yet included by the current SEP. On the other hand, the band structure of other even states (in green) not related to the edge-dimer states are in very good agreement with the DFT results. Since the edge-dimer antibonding states have a minimum at the zone boundary, and their energies are higher than the conduction band minimum at the zone center, they may not play a significant role in certain device applications. In case they are needed, a "scissor operation" can be applied to shift these minibands rigidly to match the DFT results, and phenomena such as the Gunn effect and negative differential resistance [47,48] can be simulated with this simple approach, and it remains a reasonable approximation to use the wavefunctions obtained by SEP with the current set of basis to calculate the carrier-scattering process in the nano Gunn-diode application by using AGNRs. Since these edge-dimer states are very sensitive to the chemical modification of the AGNR edges, our model can be used to simulate the effect of edge modification on the I-V characteristics by imposing a model potential on the edge atoms.

3.4. Comparison with Experiments

Many experimental studies on graphene and AGNR-related devices have been reported in the last decade [49–61], which validate and complement the theoretical predictions, providing critical empirical evidence. Techniques such as scanning tunneling microscopy (STM) [51–57], electrical transport measurements [58–60], and angle-resolved photo emission spectroscopy (ARPES) [61] are commonly used in AGNR research. Experimental measurements verify the existence of energy bandgaps, electronic states, and other electronic and transport phenomena predicted by theory. The band gap of AGNR is determined by the energy difference between the conduction band and the highest valence band at the Γ point. We assume that the edge relaxation and the edge-induced correction in pseudopotential remain unchanged for AGNRs with $N = 3 \sim 12$. The corresponding numbers of dimer lines in these AGNRs are $N_d = 2N + 1 = 7 \sim 25$. The calculated band energies for the highest occupied molecular orbital (HOMO) and lowest unoccupied molecular orbital (LUMO) for AGNRs as functions of N_d are shown in Figure 14. Here, we only consider the cases with odd N_d so that the x-mirror symmetry holds. The computation time is less than 10 s for each case shown.

According to simple models that impose rigid boundary conditions on the edges of AGNRs, the band gap becomes zero when $N_d = 3m + 2$, where m is a positive integer [20,62]. Our calculations show that the band gaps indeed shrink to relatively small values at $N_d = 11, 17, 23$. However, at $N_d = 11$ and 17, the band gap is still appreciable (~0.14 eV and 0.13 eV). This is because the boundary conditions become not so rigid in the realistic situation. Our calculated value of ~0.14 eV is consistent with the small band gaps (~0.18 eV) observed in AGNRs of similar dimensions [51]. The band gap obtained by our calculation for the $N_d = 7$ case is 1.43 eV, which is close to the result of 1.47 eV obtained by a previous DFT calculation [63] and also consistent with the experimental result of ~1.4 eV based on Fourier-transformed scanning tunneling spectroscopy [52]. However, our result is ~0.9 eV, lower than that obtained by the dI/dV measurements for the 7-AGNR, which shows a band gap ~2.3 eV [52]. Since our calculation does not include the many-body effect for quasiparticle excitation (nor does the DFT calculation without GW correction), our result is expected to be significantly lower than the DFT-GW calculation [54], which predicts a band gap of ~3.94 eV when the many-body effect is included via the GW approximation [7]. However, most dI/dV measurements were taken for AGNR samples placed on Au substrate [51–57], and there is a strong screening effect on the many-body effect, which explains the large difference in band gap predicted by DFT-GW calculation and experimental results [63].

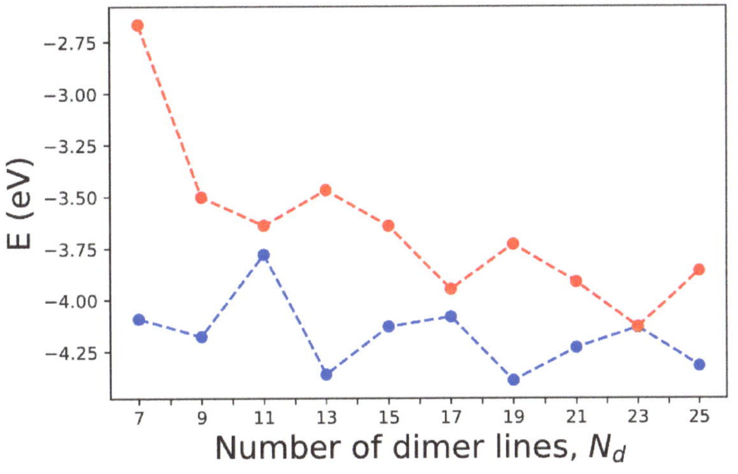

Figure 14. The HOMO (blue) and LUMO (red) levels of AGNRs with various numbers of dimer lines (N_d) calculated by SEPM.

For the case of 9-AGNR, the band gap obtained by the current SEPM is 0.68 eV, which is 0.7 eV lower than the 9-AGNR band gap measured by dI/dV measurements (~1.4 eV) [54] due to the many-body effect. For the 13-AGNR, our calculation predicts a band gap of 0.9 eV, and the LUMO level is ~0.8 eV lower than the 7-AGNR (See Figure 14). This energy difference in LUMO levels between 7- and 13-AGNRs is close to the result (~0.7 eV) obtained by the dI/dV measurements on a 7–13 AGNR heterojunction [22]. Since we adopted a B-spline basis along the *z*-axis, our calculated band energies are absolute values with respect to the vacuum level. Therefore, it is meaningful to make such a comparison. For the case of 15-AGNR, the band gap obtained by the current SEPM is 0.49 eV, which is ~0.5 eV lower than the dI/dV measurements of 1.03 eV [56]. For the case of 21-AGNR, the SEPM band gap is 0.32 eV, which is ~0.4 eV lower than the value of 0.7 eV obtained by dI/dV measurements [57]. Therefore, in general, the band gaps for AGNRs predicted by the current SEPM are fairly close to those obtained by other DFT calculations (without including the GW correction), and the values are consistently lower than the dI/dV measurements for AGNRs on Au substrate by an amount that varies from 0.4 eV at $N_d = 21$ to 0.9 eV at $N_d = 7$. Including the GW correction tends to obtain much larger band gaps in comparison to the experimental results [52], which can be attributed to the screening effect from the metal substrate on the GW correction. Thus, it is necessary to carry out DFT-GW calculations for AGNRs on Au substrate, as reported in [63], in order to determine the amount of band-gap correction due to the many-body effect.

4. Conclusions

We developed a semi-empirical pseudopotential (SEP) method that is easy to implement and capable of obtaining accurate band structures for graphene and armchair-edged graphene nanoribbons. For the π-bonded states (with odd symmetry with respect to the *z*-axis mirror), our SEP method can nearly reproduce all salient features of the DFT results with good accuracy. The time needed to compute the whole band structure associated with π-bonded states of graphene is only about a few seconds on a personal computer

(PC), and it takes around 100 s to compute the whole band structure for π-bonded states of an AGNR with supercell size of $16a \times \sqrt{3}a$. Thus, it will be a highly efficient method for modeling AGNR-related devices when the π-bonded states play the primary role. For the modeling of AGNR devices that involve the indirect valley minimum at the zone boundary, the edge-dimer antibonding states (which possess even symmetry with respect to the z-axis mirror) will be needed. For such a case, the current method becomes less efficient. It will take about one hour on a PC to obtain the nearly correct energy and dispersion of the localized σ-bonding and σ-antibonding states of the edge dimmers of the AGNR, while the corresponding DFT calculation will take more than one day.

The SEP method with the same edge-induced correction potential given in Equation (51) is also applicable for AGNRs of sizes different from the example used here since the dimer formation on AGNR edges should not be significantly affected by the width of the interior region. Thus, the current model is suitable for modeling any size of AGNR (as long as the width is at least $3a$) for device applications.

The current approach can still be improved by adding localized basis functions at the edges to make the convergence of σ-bonding and antibonding states at AGNR edges much faster. Furthermore, an SEP for the other popular graphene nanoribbons with zigzag edges will also be useful. The current approach can also be extended to develop similar SEPs for transition-metal dichalcogenides (TMDs) and related moiré superlattices. All these improvements are worthy topics for future research. For application to TMDs, some complications will occur due to the more complicated structure factors, which cannot be made real. Therefore, finding semi-empirical potentials to fit both the real and imaginary parts of the complex quasi-2D form factors, $\tilde{V}_{loc}(z, \mathbf{G})$, for two kinds of atoms will require more tedious procedures, although it can still be done. Once it is done, the more interesting application is to develop an SEPM for TMD moiré superlattices considering their high scientific impact. Here, we have demonstrated the success of using graphene eigenstates at different $\mathbf{g}_j$ points enclosed in the graphene Brillouin zone (BZ) (see Figure 8b) as a contracted basis set to calculate the AGNR band structures efficiently. The same idea can be applied to deal with twisted bilayer TMDs with a large supercell. Since there are no edge states to deal with in the moiré superlattices, we expect that only a small number of TMD bands need to be included in the contracted basis set. By using this contracted basis, the miniband structures of TMD moiré superlattices can be calculated very efficiently with good accuracy (similar to the π-bonded states in AGNR). The wavefunctions of AGNRs calculated by SEPM are very close to the DFT results. Therefore, they can be used to calculate the electron–phonon scattering with good accuracy. Furthermore, since all our AGNR wavefunctions (including edge states) are written as linear combinations of graphene Bloch states, we can relate the electron–phonon scattering matrix elements to those for graphene, and it will be convenient to model the transport properties in AGNR-related optoelectronic devices. The software developed here will be valuable for IC designs of 2D material-based nanoelectronics devices that will be of interest to the semiconductor industry.

The close interplay between theoretical and experimental studies fosters a deeper understanding of AGNRs. Moreover, experimental data provide feedback to refine and improve theoretical methods. The collaboration between theorists and experimentalists allows for the identification of new phenomena and the validation of theoretical models. Together, theoretical and experimental studies offer a comprehensive perspective on AGNRs' properties and pave the way for potential applications in nanoelectronics and beyond.

Author Contributions: Conceptualization, Y.-C.C.; methodology, Y.-C.C. and R.K.P.; software, R.K.P., Y.-C.C. and C.-Y.R.; validation, R.K.P., Y.-C.C. and C.-Y.R.; formal analysis, R.K.P. and Y.-C.C.; investigation, Y.-C.C. and C.-Y.R.; resources, Y.-C.C.; data curation, R.K.P. and C.-Y.R.; writing—original draft preparation, R.K.P. and Y.-C.C.; writing—R.K.P., Y.-C.C. and C.-Y.R.; supervision, Y.-C.C.; project administration, Y.-C.C.; funding acquisition, Y.-C.C. All authors have read and agreed to the published version of the manuscript.

Funding: This research was funded by the National Science and Technology Council, Taiwan, grant number MOST 111-2112-M-001-069.

Data Availability Statement: Theoretical methods and results are available from the authors.

Conflicts of Interest: The authors declare no conflict of interest.

Appendix A

Matrix Elements of Nonlocal Corrections in Overlap and Potential for Graphene

The nonlocal potential is given by $\hat{V}_{nl} = \sum_{\sigma lm,nn'} E_{lm}^{nn'} \left|\beta_{lm}^{n\sigma}\right\rangle\left\langle\beta_{lm}^{n'\sigma}\right|$. The matrix elements for nonlocal potential read as follows:

$$\begin{aligned}\left\langle \mathbf{K};B_i\right|\hat{V}_{nl}\left|\mathbf{K}';B_{i'}\right\rangle &= \sum_{\sigma lm,nn'} E_{lm}^{nn'}\left\langle \mathbf{K};B_i|\beta_{lm}^{n0}\right\rangle\left\langle \beta_{lm}^{n'0}\middle|\mathbf{K}';B_{i'}\right\rangle e^{\mathrm{i}(\mathbf{G}'-\mathbf{G})\cdot\boldsymbol{\tau}_\sigma} \\ &= \frac{1}{A_c}\sum_{\sigma lm}\sum_{nn'} E_{lm}^{nn'} P_{lm}^{in}(\mathbf{K})P_{lm}^{i'n'*}(\mathbf{K}')e^{\mathrm{i}(\mathbf{G}'-\mathbf{G})\cdot\boldsymbol{\tau}_\sigma}.\end{aligned} \tag{A1}$$

Here, $\mathbf{K} = \mathbf{k} + \mathbf{G}$, and $\mathbf{K}' = \mathbf{k} + \mathbf{G}'$. $\boldsymbol{\tau}_\sigma = \pm\boldsymbol{\tau}_1$ for $\sigma = 1,2$. In terms of 3D plane waves, indexed by $\mathbf{Q} = \mathbf{K} + g_z\hat{\mathbf{z}}$ (where g_z denotes the reciprocal lattice vectors along the z-axis for the supercell adopted), the projection of beta function in mixed basis (with respect to the position of a C atom) is given by the following:

$$P_{lm}^{in}(\mathbf{K}) = \mathrm{i}^l\sqrt{A_c}\left\langle \mathbf{K};B_i\middle|\beta_{lm}^{n0}\right\rangle = \frac{1}{\sqrt{L_c}}\sum_{\mathbf{g}_z}\widetilde{B}_i(g_z)I_l(Q)Y_{lm}(\hat{\mathbf{Q}}) \tag{A2}$$

where

$$I_l(Q) = 4\pi\int_0^{R_C} dr r^{l+2}A_l(r)j_l(Qr) = 4\pi\sum_i A_l(r_i)\left(r^{l+2}j_{l+1}(Qr)\right)\Big|_{r_{i+}}^{r_{i+}}$$
$$\int_{r1-}^{r1+} dr r^{l+2}j_l(Qr) = \sqrt{\frac{\pi}{2}}\int_{r1-}^{r1+} dr r^{l+2-1/2}j_{l+1/2}(Qr) = \left(r^{l+2}j_{l+1}(Qr)\right)\Big|_{r_{i+}}^{r_{i+}}/Q$$

Here, $\widetilde{B}_i(g_z)$ is the Fourier transform of $B_i(z)$ as given in Equation (4). For crystal structures with z-mirror symmetry, we obtain the following relationship (in non-orthogonal B-spline basis):

$$P_{lm}^{-in}(\mathbf{K}) = P_{lm}^{in}(\mathbf{K})(-1)^{l+m} \tag{A3}$$

where the superscript $-i$ labels the projection function into B-spline $B_{-i}(z)$, which is the mirror image of $B_i(z)$.

To calculate band structures, we can use either an iterative solver based on the conjugate gradient (CG) approach [46] or a direct solver to diagonalize the Hamiltonian matrix. We can rewrite Equation (A1) as follows:

$$\left\langle \mathbf{K};B_i\middle|V_{nl}\middle|\mathbf{K}';B_{i'}\right\rangle = \frac{1}{A_c}\sum_{\sigma l,m\geq 0}\sum_{n,n'} E_{lm}^{nn'} Re\left[P_{lm}^{in}(\mathbf{K})P_{lm}^{i'n'*}(\mathbf{K}')\right]d_m e^{\mathrm{i}(\mathbf{G}'-\mathbf{G})\cdot\boldsymbol{\tau}_\sigma}, \tag{A4}$$

where $d_0 = 1$, and $d_m = 2$ for $|m| > 0$.

For the conjugate gradient (CG) iterative solver, we need to calculate the product of the overlap $(\hat{O})$ and non-local potential operator $(\hat{V}_{nl})$ with the wavefunction (Ψ). Here,

$$\hat{O} = 1 + \sum_{\sigma lm,nn'} q_l^{nn'}\left|\beta_{lm}^{n\sigma}\right\rangle\left\langle\beta_{lm}^{n'\sigma}\right| \equiv 1 + \hat{S} \tag{A5}$$

and

$$\hat{V}_{nl} = \sum_{\sigma lm,nn'} E_l^{nn'}\left|\beta_{lm}^{n\sigma}\right\rangle\left\langle\beta_{lm}^{n'\sigma}\right|. \tag{A6}$$

We define $Q_{lm}^{in}(K) = \left|P_{lm}^{in}(\mathbf{K})\right|$ and obtain the following:

$$\langle\beta_{lm}^{n\sigma}|\Psi\rangle = \sum_{\mathbf{G};i}\langle\beta_{lm}^{n\sigma}|\mathbf{K};B_i\rangle\langle\mathbf{K};B_i|\Psi\rangle = \sum_{\mathbf{G};i}Q_{lm}^{in}(K)\left(\frac{K_x+iK_y}{K}\right)^m\langle\mathbf{K};B_i|\Psi\rangle e^{i\mathbf{K}\cdot\boldsymbol{\tau}_\sigma}. \quad (A7)$$

$$\langle\mathbf{K};B_i|\hat{S}\Psi\rangle = \sum_{\sigma lm}Q_{lm}^{in}\left(\frac{K_x-iK_y}{K}\right)^m e^{-i\mathbf{K}\cdot\boldsymbol{\tau}_\sigma}q_{lm}^{nn'}\langle\beta_{lm}^{n\sigma}|\Psi\rangle. \quad (A8)$$

Then, $\langle\mathbf{K};B_i|\hat{O}\Psi\rangle = \langle\mathbf{K};B_i|\Psi\rangle + \langle\mathbf{K};B_i|\hat{S}\Psi\rangle$ is the updated vector for $\hat{O}\Psi$, while $\langle\mathbf{K};B_i|\hat{V}_{nl}\Psi\rangle = \sum_{\sigma lm}Q_{lm}^{in}(K)\left(\frac{K_x-iK_y}{K}\right)^m e^{-i\mathbf{K}\cdot\boldsymbol{\tau}_\sigma}E_{lm}^{nn'}\langle\beta_{lm}^{n\sigma}|\Psi\rangle$ is the updated vector for $\hat{V}_{nl}\Psi$. For calculating the band structure, we need to evaluate the nonlocal correction to the overlap and pseudopotential matrix elements. The correction to the overlap matrix can be written as follows:

$$\langle\mathbf{K};B_i|\hat{S}|\mathbf{K}';B_{i'}\rangle = \sum_{n,n',l,|m|}q_l^{nn'}Q_{lm}^{in}(K)Q_{lm}^{i'n'}(K')\left(\frac{K_xK'_x+K_yK'_y}{KK'}\right)^{|m|}d_mS_1(\Delta\mathbf{G}) \quad (A9)$$

where $S_1(\Delta\mathbf{G})$ is the structure factor given in Equation (36). Similarly for the non-local pseudopotential, we obtain the following:

$$\langle\mathbf{K};B_i|\hat{V}_{nl}|\mathbf{K}';B_{i'}\rangle = \sum_{n,n',l,|m|}E_l^{nn'}Q_{lm}^{in}(K)Q_{lm}^{i'n'}(K')\left(\frac{K_xK'_x+K_yK'_y}{KK'}\right)^{|m|}d_mS_1(\Delta\mathbf{G}). \quad (A10)$$

Appendix B

Matrix Elements for the Hamiltonian of Armchair Graphene Nanoribbon

The matrix elements for the kinetic-energy term of armchair graphene nanoribbon (AGNR) in (46) of the main text are given by the following:

$$\left\langle\varphi_{\nu,\mathbf{k}+\mathbf{g}_j}\middle|-\nabla^2\middle|\varphi_{\nu',\mathbf{k}+\mathbf{g}_{j'}}\right\rangle = \sum_{i,i',\mathbf{G}}Z_{i\mathbf{G}}^{\nu,\mathbf{k}+\mathbf{g}_j}Z_{i'\mathbf{G}}^{\nu',\mathbf{k}+\mathbf{g}_j}T_{ii'}\left(\mathbf{k}+\mathbf{g}_j+\mathbf{G}\right)\delta_{jj'} \quad (A11)$$

with $T_{ii'}$ given by Equation (10).

The matrix elements for the local potential energy term (V_0') are as follows:

$$\left\langle\varphi_{\nu,\mathbf{k}+\mathbf{g}_j}\middle|V_0'\middle|\varphi_{\nu',\mathbf{k}+\mathbf{g}_{j'}}\right\rangle = \sum_{i\mathbf{G},i'\mathbf{G}'}Z_{i\mathbf{G}}^{\nu,\mathbf{k}+\mathbf{g}_j}Z_{i'\mathbf{G}}^{\nu',\mathbf{k}+\mathbf{g}_j}\sum_{s=1}^{3}C_sI(\alpha_s;i,,i')v_s\left(\Delta\mathbf{g}_{jj'}+\Delta\mathbf{G}\right) \quad (A12)$$

where $I(\alpha_s;i,,i')$ is defined in Equation (34), and $\Delta\mathbf{g}_{jj'}+\Delta\mathbf{G} = \mathbf{g}_{j'}-\mathbf{g}_j+\mathbf{G}'-\mathbf{G}$.

$$v_s(\mathbf{q}) = \frac{1}{A_{SC}}\sum_{\sigma\in A_{NR}}\int d\mathbf{r}e^{i\mathbf{q}\cdot\mathbf{r}}e^{-\alpha_s(\mathbf{r}-\boldsymbol{\tau}_\sigma)^2} = S'_{NR}(\mathbf{q})\frac{\pi}{\alpha_s}e^{-q^2/4\alpha_s}. \quad (A13)$$

$$S'_{NR}(\mathbf{q}) = \frac{1}{A_{SC}}\sum_{\sigma\in A_{NR}}e^{i\mathbf{q}\cdot\boldsymbol{\tau}_\sigma} \quad (A14)$$

Here, A_{SC} is the area of the whole AGNR supercell, and A_{NR} denotes the area covered by the AGNR only. $S'_{NR}(\mathbf{q})$ denotes the structure factor for the nanoribbon.

With the proper choice of the origin for the AGNR supercell, the inversion symmetry holds, and $S'_{NR}(\mathbf{q})$ becomes real.

$$\left\langle\varphi_{\nu,\mathbf{k}+\mathbf{g}_j}\middle|\Delta V'_\gamma\middle|\varphi_{\nu',\mathbf{k}+\mathbf{g}_{j'}}\right\rangle = \sum_{i\mathbf{G},i'\mathbf{G}'}Z_{i\mathbf{G}}^{\nu,\mathbf{k}+\mathbf{g}_j}Z_{i'\mathbf{G}}^{\nu',\mathbf{k}+\mathbf{g}_j}\int dzB_i(z)B_{i'}(z)f_\gamma(z)K_\gamma\left(\Delta\mathbf{g}_{jj'}+\Delta\mathbf{G}\right) \quad (A15)$$

where

$$K_\gamma(\mathbf{q}) = \frac{1}{A_{SC}}\sum_{\sigma\in A_{NR}}\int d\mathbf{r}e^{i\mathbf{q}\cdot\mathbf{r}}D_\gamma(\boldsymbol{\rho}-\boldsymbol{\tau}_\sigma) = S'_\gamma(\mathbf{q})\widetilde{D}_\gamma(\mathbf{q}) \quad (A16)$$

Here, $\gamma = S, L, b$ represent the short-range (S), long-range (L) range, and bond-charge (b) contributions to ΔV, and $D_\gamma(\boldsymbol{\rho})$ is related to $\tilde{D}_\gamma(\mathbf{q})$ by the Fourier transform. $S'_\gamma(\mathbf{q}) = S'_{NR}(\mathbf{q})$ for $\gamma = S$ or L, and the structure factor for the bond-charge contribution ($\gamma = b$) is as follows:

$$S'_b(\mathbf{q}) = \frac{1}{A_{SC}} \sum\nolimits_{h \in A_{NR}} e^{\mathrm{i}\mathbf{q}\cdot\boldsymbol{\tau}_h} \tag{A17}$$

where $\boldsymbol{\tau}_h$ denotes the center positions of hexagon cells inside the AGNR.

Similarly, the matrix elements of the nonlocal pseudopotential are as follows:

$$\begin{aligned} &\left\langle \varphi_{\nu,\mathbf{k}+\mathbf{g}_j} \middle| \hat{U}_2 \middle| \varphi_{\nu',\mathbf{k}+\mathbf{g}_{j'}} \right\rangle \\ &= \sum\nolimits_{i\mathbf{G},i'\mathbf{G}'} Z_{i\mathbf{G}}^{\nu,\mathbf{k}+\mathbf{g}_j} \left\langle \mathbf{k}+\mathbf{g}_j+\mathbf{G}; B_i \middle| \hat{V}'_{nl} \middle| \mathbf{k}+\mathbf{g}_{j'}+\mathbf{G}'; B_{i'} \right\rangle Z_{i'\mathbf{G}'}^{\nu\prime\mathbf{k}+\mathbf{g}_{j'}} \end{aligned} \tag{A18}$$

where $Z_{i\mathbf{G}}^{\nu,\mathbf{k}+\mathbf{g}_j}$ denotes the eigenvectors obtained by solving Equation (43) for the ν-th band at wavevector $\mathbf{k}+\mathbf{g}_j$. The kernel $\hat{V}'_{nl}$ in the above has a separable form, as given by Equation (A4), with $\mathbf{K} = \mathbf{k}+\mathbf{g}_j+\mathbf{G}$ and $\mathbf{K}' = \mathbf{k}+\mathbf{g}_{j'}+\mathbf{G}'$ and the index σ running through atoms in the nanoribbon region, A_{NR}. Thus, the sum can be evaluated efficiently. We obtain the following:

$$\begin{aligned} \left\langle \varphi_{\nu,\mathbf{k}+\mathbf{g}_j} \middle| \hat{U}_2 \middle| \varphi_{\nu',\mathbf{k}+\mathbf{g}_{j'}} \right\rangle &= \tfrac{1}{A_{SC}} \sum\nolimits_{\sigma\in A_{NR}} \sum\nolimits_{lm,nn'} E_{lm}^{nn'} \tilde{Z}_{\sigma lm}^{\nu n}\left(\mathbf{k}+\mathbf{g}_j\right) \tilde{Z}_{\sigma lm}^{\nu' n'*}(\mathbf{k}+\mathbf{g}_{j'}) \\ &= \sum\nolimits_{\sigma\in Gr} \sum\nolimits_{lm,nn'} E_{lm}^{nn'} \tilde{Z}_{\sigma lm}^{\nu n}\left(\mathbf{k}+\mathbf{g}_j\right) \tilde{Z}_{\sigma lm}^{\nu' n'*}(\mathbf{k}+\mathbf{g}_{j'}) \tilde{S}_{NR}\left(\mathbf{g}_{j'}-\mathbf{g}_j\right) \end{aligned} \tag{A19}$$

where $\sum_{\sigma\in Gr}$ indicates a sum over two graphene atoms closest to the origin.

$$\tilde{Z}_{\sigma lm}^{\nu n}\left(\mathbf{k}+\mathbf{g}_j\right) = \sum\nolimits_{i,\mathbf{G}} Z_{i\mathbf{G}}^{\nu,\mathbf{k}+\mathbf{g}_j} P_{lm}^{in}\left(\mathbf{k}+\mathbf{g}_j+\mathbf{G}\right) e^{-\mathrm{i}(\mathbf{k}+\mathbf{g}_j+\mathbf{G})\cdot\boldsymbol{\tau}_\sigma} \tag{A20}$$

and $P_{lm}^{in}\left(\mathbf{k}+\mathbf{g}_j+\mathbf{G}\right)$ is given in Equation (A2). $\tilde{S}_{NR}(\mathbf{q}) = \frac{1}{A_{SC}} \sum_{s\in A_{NR}} e^{\mathrm{i}\mathbf{q}\cdot\mathbf{R}_s}$, with the lattice vectors $\mathbf{R}_s$ running through all graphene unit cells within the area A_{NR}. Here, $\mathbf{R}_s = 0$ corresponds to the origin marked by a red cross in Figure 8a.

To find the eigenfunctions and eigenvalues of the AGNR, we express the eigenfunctions of the AGNR by the following:

$$\Phi_{\mathbf{k}}^{GNR} = \sum\nolimits_{\nu,j} Z_{\nu,j}^{GNR} \varphi_{\nu,\mathbf{k}+\mathbf{g}_j} \tag{A21}$$

Substituting Equation (A21) into the Kohn–Sham equation for the GNR gives the following:

$$\sum\nolimits_{\nu' j'} \left\langle \varphi_{n,\mathbf{k}+\mathbf{g}_j} \middle| \hat{H}_{GNR} \middle| \varphi_{\nu',\mathbf{k}+\mathbf{g}_{j'}} \right\rangle Z_{\nu' j'}^{GNR} = E(\mathbf{k}) \sum\nolimits_{\nu' j'} \left\langle \varphi_{\nu,\mathbf{k}+\mathbf{g}_j} \middle| \hat{O}_{GNR} \middle| \varphi_{\nu',\mathbf{k}+\mathbf{g}_{j'}} \right\rangle Z_{\nu' j'}^{GNR}, \tag{A22}$$

where

$$\left\langle \varphi_{\nu,\mathbf{k}+\mathbf{g}_j} \middle| \hat{O}_{GNR} \middle| \varphi_{\nu',\mathbf{k}+\mathbf{g}_{j'}} \right\rangle = \delta_{\nu,\nu'}\delta_{j,j'} + \left\langle \varphi_{\nu,\mathbf{k}+\mathbf{g}_j} \middle| \hat{S}_{GNR} \middle| \varphi_{\nu',\mathbf{k}+\mathbf{g}_{j'}} \right\rangle \tag{A23}$$

with

$$\left\langle \varphi_{\nu,\mathbf{k}+\mathbf{g}_j} \middle| \hat{S}_{GNR} \middle| \varphi_{\nu',\mathbf{k}+\mathbf{g}_{j'}} \right\rangle = \frac{1}{A_{SC}} \sum\nolimits_{\sigma\in A_{NR}} \sum_{lm,nn'} q_{lm}^{nn'} \tilde{Z}_{\sigma lm}^{\nu n}\left(\mathbf{k}+\mathbf{g}_j\right) \tilde{Z}_{\sigma lm}^{\nu' n'*}(\mathbf{k}+\mathbf{g}_{j'}). \tag{A24}$$

References

1. Novoselov, K.S.; Geim, A.K.; Morozov, S.V.; Jiang, D.; Zhang, Y.; Dubonos, S.V.; Grigorieva, I.V.; Firsov, A.A. Electric field effect in atomically thin carbon films. *Science* **2004**, *306*, 666–669. [CrossRef]
2. Mak, K.F.; Lee, C.G.; Hone, J.; Shan, J.; Heinz, T.F. Atomically thin MoS_2: A new direct-gap semiconductor. *Phys. Rev. Lett.* **2010**, *105*, 136805. [CrossRef]
3. Hohenberg, P.; Kohn, W. Inhomogeneous Electron Gas. *Phys. Rev.* **1964**, *136*, B864. [CrossRef]
4. Castro Neto, A.H.; Guinea, F.; Peres, N.M.R.; Novoselov, K.S.; Geim, A.K. The electronic properties of graphene. *Rev. Mod. Phys.* **2009**, *81*, 109–162. [CrossRef]
5. Xu, Y.; Li, Z.Y.; Duan, W.H. Thermal and thermoelectric properties of graphene. *Small* **2014**, *10*, 2182. [CrossRef]
6. Qiu, D.Y.; da Jordana, F.H.; Louie, S.G. Optical Spectrum of MoS_2: Many-Body Effects and Diversity of Exciton States. *Phys. Rev. Lett.* **2013**, *111*, 216805. [CrossRef]
7. Hybertsen, M.S.; Louie, S.G. Electron correlation in semiconductors and insulators: Band gaps and quasiparticle energies. *Phys. Rev. B* **1986**, *34*, 5390. [CrossRef]
8. Li, G.W.; Chang, Y.C. Planar-basis pseudopotential calculations of the Si(001)2$\times$1 surface with and without hydrogen passivation. *Phys. Rev. B* **1993**, *48*, 12032. [CrossRef]
9. Li, G.W.; Chang, Y.C. Electronic structure of As/Si(001)2$\times$1 and Sb/Si(001)2$\times$1 surfaces. *Phys. Rev. B* **1994**, *50*, 8675. [CrossRef]
10. Chang, Y.C.; Li, G.W. Planar-basis pseudopotential method and planar Wannier functions for surfaces and heterostructures. *Comput. Phys. Commun.* **1996**, *95*, 158. [CrossRef]
11. Ren, C.Y.; Hsue, C.S.; Chang, Y.C. A Mixed Basis Density Functional Approach for Low-Dimensional Systems with B-splines. *Comput. Phys. Commun.* **2015**, *188*, 94. [CrossRef]
12. Ren, C.Y.; Hsue, C.S.; Chang, Y.C. A Mixed Basis Density Functional Approach for One dimensional Systems with B-splines. *Comput. Phys. Commun.* **2016**, *202*, 188. [CrossRef]
13. Ren, C.Y.; Chang, Y.C. Density functional calculations of atomic structure, charging effect, and static dielectric constant of two-dimensional systems based on B-splines. *Phys. E* **2022**, *140*, 115203. [CrossRef]
14. Cohen, M.L.; Bergstresser, T.K. Band Structure and Pseudopotential Form Factors for Fourteen Semiconductors of the Diamond and Zinc-blende Structure. *Phys. Rev.* **1966**, *141*, 789. [CrossRef]
15. Chelikowsky, J.R.; Cohen, M.L. Nonlocal pseudopotential calculations for the electronic structure of eleven diamond and zinc-blende semiconductors. *Phys. Rev. B* **1976**, *14*, 556. [CrossRef]
16. Vanderbilt, D. Soft self-consistent pseudopotentials in a generalized eigenvalue formalism. *Phys. Rev. B* **1990**, *41*, 7892. [CrossRef]
17. Son, Y.W.; Cohen, M.L.; Louie, S.G. Energy Gaps in Graphene Nanoribbons. *Phys. Rev. Lett.* **2006**, *97*, 216803. [CrossRef]
18. Son, Y.-W.; Cohen, M.L.; Louie, S.G. Half-metallic graphene nanoribbons. *Nature* **2007**, *446*, 342. [CrossRef]
19. Fischetti, M.V.; Kim, J.; Narayanan, S.; Ong, Z.Y.; Sachs, C.; Ferry, D.K.; Aboud, S.J. Pseudopotential-based studies of electron transport in graphene and graphene nanoribbons. *J. Phys. Condens. Matter* **2013**, *25*, 473202. [CrossRef]
20. Nakada, K.; Fujita, M.; Dresselhaus, G.; Dresselhaus, M.S. Edge state in graphene ribbons: Nanometer size effect and edge shape dependence. *Phys. Rev. B* **1996**, *54*, 17954. [CrossRef]
21. Cai, J.; Ruffieux, P.; Jaafar, R.; Bieri, M.; Braun, T.; Blankenburg, S.; Muoth, M.; Seitsonen, A.P.; Saleh, M.; Feng, X.; et al. Atomically precise bottom-up fabrication of graphene nanoribbons. *Nature* **2010**, *466*, 470. [CrossRef] [PubMed]
22. Chen, Y.C.; Cao, T.; Chen, C.; Pedramrazi, Z.; Haberer, D.; de Oteyza, D.G.; Fischer, F.R.; Louie, S.G.; Crommie, M.F. Molecular bandgap engineering of bottom-up synthesized graphene nanoribbon heterojunctions. *Nat. Nanotechnol.* **2015**, *10*, 156–160. [CrossRef]
23. Yan, L.H.; Liljeroth, P. Engineered electronic states in atomically precise artificial lattices and graphene nanoribbons. *Adv. Phys. X* **2019**, *4*, 1651672. [CrossRef]
24. Rizzo, D.J.; Veber, G.; Jiang, J.W.; McCurdy, R.; Cao, T.; Bronner, C.; Chen, T.; Louie, S.G.; Fischer, F.R.; Crommie, M.F. Inducing metallicity in graphene nanoribbons via zero-mode superlattices. *Science* **2020**, *369*, 1597. [CrossRef] [PubMed]
25. Sevincli, H.; Topsakal, M.; Ciraci, S. Superlattice structures of graphene-based armchair nanoribbons. *Phys. Rev. B* **2008**, *78*, 245402. [CrossRef]
26. Llinas, J.P.; Fairbrother, A.; Borin Barin, G.; Shi, W.; Lee, K.; Wu, S.; Choi, B.Y.; Braganza, R.; Lear, J.; Kau, N.; et al. Short-channel field-effect transistors with 9-atom and 13-atom wide graphene nanoribbons. *Nat. Commun.* **2017**, *8*, 633. [CrossRef]
27. Wang, H.; Wang, H.S.; Ma, C.; Chen, L.; Jiang, C.; Chen, C.; Xie, X.; Li, A.P.; Wang, X. Graphene nanoribbons for quantum electronics. *Nat. Rev. Phys.* **2021**, *3*, 791–802. [CrossRef]
28. Xia, F.; Mueller, T.; Lin, Y.; Garcia, A.V.; Avouris, P. Ultrafast graphene photodetector. *Nat. Nanotech* **2009**, *4*, 839–843. [CrossRef]
29. Do, T.-N.; Gumbs, G.; Huang, D.; Hoi, B.D.; Shih, P.-H. Role Played by Edge-Defects in the Optical Properties of Armchair Graphene Nanoribbons. *Nanomaterials* **2021**, *11*, 3229. [CrossRef]
30. Sun, Q.; Yan, Y.; Yao, X.L.; Mullen, K.; Narita, A.; Fasel, R.; Ruffieux, P. Evolution of the topological energy band in graphene nanoribbons. *J. Phys. Chem. Lett.* **2021**, *12*, 8679. [CrossRef]

31. Rizzo, D.J.; Jiang, J.W.; Joshi, D.; Veber, G.; Bronner, C.; Durr, R.A.; Jacobse, P.H.; Cao, T.; Kalayjian, A.; Rodriguez, H.; et al. Rationally designed topological quantum dots in bottom-up graphene nanoribbons. *ACS Nano* **2021**, *15*, 20633. [CrossRef] [PubMed]
32. Groning, O.; Wang, S.; Yao, X.; Pignedoli, C.A.; Barin, G.B.; Daniels, C.; Cupo, A.; Meunier, V.; Feng, X.; Narita, A. Engineering of robust topological quantum phases in graphene nanoribbons. *Nature* **2018**, *560*, 209. [CrossRef] [PubMed]
33. Lin, K.S.; Chou, M.Y. Topological properties of gapped graphene nanoribbons with spatial symmetries. *Nano Lett.* **2018**, *8*, 7254. [CrossRef]
34. Wakabayashi, K.; Fujita, M.; Ajiki, H.; Sigrist, M. Electronic and magnetic properties of nanographite ribbons. *Phys. Rev. B* **1999**, *59*, 8271. [CrossRef]
35. Topsakal, M.; Sevincli, H.; Ciraci, S. Spin confinement in the superlattices of graphene ribbons. *Appl. Phys. Lett.* **2008**, *92*, 173118. [CrossRef]
36. Kuo, D.M.T.; Chang, Y.C. Contact effects on thermoelectric properties of textured graphene nanoribbons. *Nanomaterials* **2022**, *12*, 3357. [CrossRef]
37. Darancet, P.; Olevano, V.; Mayou, D. Coherent electronic transport through graphene constrictions: Subwavelength regime and optical analogy. *Phys. Rev. Lett.* **2009**, *102*, 136803. [CrossRef]
38. Zuev, Y.M.; Chang, W.; Kim, P. Thermoelectric and magnetothermoelectric transport measurements of graphene. *Phys. Rev. Lett.* **2009**, *102*, 096807. [CrossRef]
39. Carl, D.B. *A Practical Guide to Splines*; Springer: New York, NY, USA, 1987.
40. Wei, P.; Bao, W.Z.; Pu, Y.; Lau, C.N.; Shi, J. Anomalous thermoelectric transport of Dirac particles in graphene. *Phys. Rev. Lett.* **2009**, *102*, 166808. [CrossRef]
41. Wang, L.W.; Zunger, A. Pseudopotential calculations of nanoscale CdSe quantum dots. *Phys. Rev. B* **1996**, *53*, 9579. [CrossRef]
42. Sanchez, A.M.; Cristobal, A.G.; Bester, G. Semiempirical pseudopotential approach for nitride-based nanostructured and ab initio-based passivation of free surfaces. *Phys. Rev. B* **2012**, *86*, 205430. [CrossRef]
43. Kresse, G.; Furthmüller, J. Efficient iterative schemes for ab initio total-energy calculations using a plane-wave basis set. *Phys. Rev. B* **1996**, *54*, 11169. [CrossRef] [PubMed]
44. Ceperley, D.M.; Alder, B.J. Ground State of the Electron Gas by a Stochastic Method. *Phys. Rev. Lett.* **1980**, *45*, 566. [CrossRef]
45. Perdew, J.P.; Zunger, A. Self-interaction correction to density-functional approximations for many-electron systems. *Phys. Rev. B* **1981**, *23*, 5048. [CrossRef]
46. Shewchuk, J.R. *An Introduction to the Conjugate Gradient Method Without the Agonizing Pain*; Carnegie-Mellon University, Department of Computer Science: Pittsburgh, PA, USA, 1994.
47. Kohn, W.; Sham, L.J. Self-Consistent Equations Including Exchange and Correlation Effects. *Phys. Rev.* **1965**, *140*, A1133. [CrossRef]
48. Chang, Y.C. Theory of Electronic Properties of Si(111) 7×7 Surface. *J. Vac. Sci. Technol.* **1983**, *1*, 709. [CrossRef]
49. Bonilla, L.L.; Teitsworth, S.W. *Nonlinear Wave Methods for Charge Transport*; Wiley-VCH: Weinheim, Germany, 2010.
50. Carlstrom, J.E.; Plambeck, R.L.; Thornton, D.D. A Continuously Tunable 65-115 GHz Gunn Oscillator. *IEEE* **1985**, *33*, 610–619.
51. Tapasztó, L.; Dobrik, G.; Lambin, P.; Biro, L.P. Tailoring the atomic structure of graphene nanoribbons by scanning tunnelling microscope lithography. *Nat. Nanotechnol.* **2008**, *3*, 397–401. [CrossRef]
52. Ma, C.; Xiao, Z.; Puretzky, A.A.; Wang, H.; Mohsin, A.; Huang, J.; Liang, L.; Luo, Y.; Lawrie, B.J.; Gu, G.; et al. Engineering edge states of graphene nanoribbons for narrow-band photoluminescence. *ACS Nano* **2020**, *14*, 5090–5098. [CrossRef]
53. Talirz, L.; Söde, H.; Dumslaff, T.; Wang, S.; Sanchez-Valencia, J.R.; Liu, J.; Shinde, P.; Pignedoli, C.A.; Liang, L.; Meunier, V.; et al. On-Surface Synthesis and Characterization of 9-Atom Wide Armchair Graphene Nanoribbons. *ACS Nano* **2017**, *11*, 1380–1388. [CrossRef]
54. Rizzo, D.J.; Veber, G.; Cao, T.; Bronner, C.; Chen, T.; Zhao, F.; Rodriguez, H.; Louie, S.G.; Crommie, M.F.; Fischer, F.R. Topological band engineering of graphene nanoribbons. *Nature* **2018**, *560*, 204–208. [CrossRef]
55. Wang, S.; Kharche, N.; Costa Girão, E.; Feng, X.; Mullen, K.; Meunier, V.; Fasel, R.; Ruffieux, P. Quantum dots in graphene nanoribbons. *Nano Lett.* **2017**, *17*, 4277–4283. [CrossRef]
56. Merino-Díez, N.; Garcia-Lekue, A.; Carbonell-Sanromà, E.; Li, J.; Corso, M.; Colazzo, L.; Sedona, F.; Sánchez-Portal, D.; Pascual, J.I.; de Oteyza, D.G. Width-Dependent Band Gap in Armchair Graphene Nanoribbons Reveals Fermi Level Pinning on Au(111). *ACS Nano* **2017**, *11*, 11661–11668. [CrossRef]
57. Ma, C.; Liang, L.; Xiao, Z.; Puretzky, A.A.; Hong, K.; Lu, W.; Meunier, V.; Bernholc, J.; Li, A.P. Seamless staircase electrical contact to semiconducting graphene nanoribbons. *Nano Lett.* **2017**, *17*, 6241–6247. [CrossRef]
58. Franc, G.; Gourdon, A. Covalent networks through on-surface chemistry in ultra-high vacuum: State-of-the-art and recent developments. *Phys. Chem. Chem. Phys.* **2011**, *13*, 14283–14292. [CrossRef]
59. Kimouche, A.; Ervasti, M.M.; Drost, R.; Halonen, S.; Harju, A.; Joensuu, P.M.; Sainio, J.; Liljeroth, P. Ultra-narrow metallic armchair graphene nanoribbons. *Nat. Commun.* **2015**, *6*, 10177. [CrossRef]
60. Han, P.; Akagi, K.; Canova, F.; Shimizu, R.; Oguchi, H.; Shiraki, S.; Weiss, P.S.; Asao, N.; Hitosugi, T. Self-assembly strategy for fabricating connected graphene nanoribbons. *ACS Nano* **2015**, *9*, 12035–12044. [CrossRef]

61. Senkovskiy, B.V.; Usachov, D.Y.; Fedorov, A.V.; Haberer, D.; Ehlen, N.; Fischer, F.R.; Grüneis, A. Finding the hidden valence band of N= 7 armchair graphene nanoribbons with angle-resolved photoemission spectroscopy. *2D Mater.* **2018**, *5*, 035007. [CrossRef]
62. Zheng, H.; Wang, Z.F.; Luo, T.; Shi, Q.W.; Chen, J. Analytical study of electronic structure in armchair graphene nanoribbons. *Phys. Rev. B* **2007**, *75*, 165414. [CrossRef]
63. Liang, L.; Meunier, V. Electronic structure of assembled graphene nanoribbons: Substrate and many-body effects. *Phys. Rev. B* **2012**, *86*, 195404. [CrossRef]

nanomaterials

MDPI

Article

Spectromicroscopy Study of Induced Defects in Ion-Bombarded Highly Aligned Carbon Nanotubes

Sammar Tayyab [1,2,*], Alice Apponi [3], Maria Grazia Betti [1,2], Elena Blundo [1], Gianluca Cavoto [1,2], Riccardo Frisenda [1], Nuria Jiménez-Arévalo [1], Carlo Mariani [1,2,*], Francesco Pandolfi [2], Antonio Polimeni [1], Ilaria Rago [1,2], Alessandro Ruocco [3], Marco Sbroscia [1] and Ravi Prakash Yadav [1,2]

1 Dipartimento di Fisica, Sapienza Università di Roma, Piazzale Aldo Moro 2, 00185 Rome, Italy
2 Istituto Nazionale di Fisica Nucleare Sezione di Roma, Piazzale Aldo Moro 2, 00185 Rome, Italy
3 Dipartimento di Scienze, Università Degli Studi Roma Tre and Istituto Nazionale di Fisica Nucleare Sezione di Roma Tre, Via della Vasca Navale 84, 00146 Rome, Italy
* Correspondence: sammar.tayyab@uniroma1.it (S.T.); carlo.mariani@uniroma1.it (C.M.); Tel.: +39-06-49914281 (S.T.); +39-06-49914393 (C.M.)

Abstract: Highly aligned multi-wall carbon nanotubes were investigated with scanning electron microscopy (SEM), Raman spectroscopy and X-ray photoelectron spectroscopy (XPS) before and after bombardment performed using noble gas ions of different masses (argon, neon and helium), in an ultra-high-vacuum (UHV) environment. Ion irradiation leads to change in morphology, deformation of the carbon (C) honeycomb lattice and different structural defects in multi-wall carbon nanotubes. One of the major effects is the production of bond distortions, as determined by micro-Raman and micro-X-ray photoelectron spectroscopy. We observe an increase in sp^3 distorted bonds at higher binding energy with respect to the expected sp^2 associated signal of the carbon 1s core level, and increase in dangling bonds. Furthermore, the surface damage as determined by the X-ray photoelectron spectroscopy carbon 1s core level is equivalent upon bombarding with ions of different masses, while the impact and density of defects in the lattice of the MWCNTs as determined by micro-Raman are dependent on the bombarding ion mass; heavier for helium ions, lighter for argon ions. These results on the controlled increase in sp^3 distorted bonds, as created on the multi-wall carbon nanotubes, open new functionalization prospects to improve and increase atomic hydrogen uptake on ion-bombarded multi-wall carbon nanotubes.

Keywords: carbon nanotubes; ion bombardment; SEM; XPS; Raman

Citation: Tayyab, S.; Apponi, A.; Betti, M.G.; Blundo, E.; Cavoto, G.; Frisenda, R.; Jiménez-Arévalo, N.; Mariani, C.; Pandolfi, F.; Polimeni, A.; et al. Spectromicroscopy Study of Induced Defects in Ion-Bombarded Highly Aligned Carbon Nanotubes. *Nanomaterials* **2024**, *14*, 77. https://doi.org/10.3390/nano14010077

Academic Editor: Muralidharan Paramsothy

Received: 6 December 2023
Revised: 20 December 2023
Accepted: 23 December 2023
Published: 27 December 2023

1. Introduction

Since the pioneering work by Iijima on carbon nanotubes [1], multi-wall carbon nanotubes (MWCNTs) have generated enormous interest among scientists and engineers in the field of materials science [2]. These structures feature a set of unique properties due to small dimensions, closed topology and lattice helicity. The presence of atomic-scale defects in MWCNTs is responsible for changing or altering their mechanical and electronic properties [3]. Thus, the development of methods capable of controlling the amount and type of defects in MWCNTs is highly desirable, both for addressing the defect origin and for exploiting their properties to engineer the system characteristics. Among such methods, the use of heavy ions cause surface roughening and the removal of carbon atoms, leading to changes in the surface structure and properties of the nanotubes [4]. On the other hand, implanting lighter ions can affect the chemical composition and electronic properties of the nanotubes [5]. Both heavy and light ion bombardment can introduce vacancies and defects in the MWCNTs structure. These defects can act as active sites for chemical reactions or serve as traps for gas molecules, making them useful for various applications. We underline that functionalized MWCNTs can be applied in the fields of sensors [6,7] and biological sensors [8], composites [9], devices [10], energy storage systems [11,12], engineering [13], field effect transistors [14,15], power electronic devices [16], supercapacitor [17], etc.

Recent theoretical and experimental studies on the irradiation of carbon nanotubes with energetic particles have revealed a broad range of new interesting phenomena. Different ions like nitrogen [18], argon [19], helium [20], carbon and silver [21], with high [22] and low doses [23], and with varying energies, were used to bombard CNTs. These works aimed to induce defects, creating molecular junctions between the nanotubes [24], to form multiple species at the surface, to find a saturation damage, and to evaluate the stability and the evolution of the bonding. Effects on CNTs upon ion irradiation can be compared with analogous studies on its basic constituent, graphene. Carbon [25], helium [4], hydrogen [26] and deuterium were irradiated on graphene to functionalize it and to observe the charge carrier transport phenomena, to produce defects that lead to engineering process [27] and to unveil the band gap opening [26], useful for optoelectronic properties.

However, a quantification of the changes in hybridization and structure of MWCNTs that will eventually lead to a change in their electronic and physical properties, obtained by bombarding them with a sequence of non-interacting gases, remains an open question. To this purpose, we have investigated the influence of noble ion beams with different masses, at a fixed energy and given flux of ions. The choice of ion species, energy, and dosage, as well as the conditions under which the bombardment is conducted, can significantly influence the outcomes.

The main objective of the present study is to produce controlled defects on MWCNTs, in order to understand whether the different noble ion masses influence the quality and density of effects. Thus, we use heavy and light ions of noble gases (Ar^+, Ne^+ and He^+) to check the suitability for applications requiring controlled modifications and functionalization of MWCNTs. In particular, we present a careful characterization of produced defects in ultra-high-vacuum (UHV) conditions, offering an ultra-clean environment. The achievement of a controlled density of defects mainly associated with sp^3 hybridized levels in the C-atomic mesh of the CNTs would constitute active sites for CNTs functionalization. The increase in sp^3 bonds is an important prerequisite for perspective usage, like improving alkali metal adsorption [28] for charge accumulation applications, or favoring hydrogen uptake [26] towards a potential solid-state material for hydrogen storage.

2. Materials and Methods

2.1. CNT Growth by Chemical Vapor Deposition

Vertically aligned MWCNTs were grown by the chemical vapor deposition (CVD) method with a home-made thermal CVD reactor in UHV, at the TITAN laboratory in Sapienza University of Rome, as described elsewhere [29,30]. Silicon was used as growth substrate, with a 2 μm thick buffer layer of SiO_2. Electron beam evaporation was exploited over the substrate to deposit a 3 nm-thick layer of Fe catalyst. Samples were mounted inside a high vacuum reaction chamber with a base pressure in the low 10^{-7} mbar range and a two stage CNT synthesis was performed. First, we annealed the substrate for 4 min at 720 °C in a H_2 atmosphere, necessary to remove the oxide components of catalyst that possibly stick to the surface of the $Fe/SiO_2/Si$ substrate. Annealing also helps to activate the catalyst layer (dewetting of catalyst) and the nucleation of iron-based nanoparticles. Next step leads to the growth of carbon nanotubes, in which the reaction temperature was increased to 740 °C and a carbon precursor (acetylene) was introduced inside the CVD chamber at a 300 sccm flow rate, without any carrier gas, up to a partial pressure of about 50 mbar. The growth time is limited to 10–12 min, in which reaction between carbon precursor and iron nanoparticles occurs and nanotubes grow vertically on the substrate by lifting the catalyst upward. After cooling, the vertically aligned (VA) MWCNTs grown on the Si substrate are extracted from the UHV chamber.

2.2. High Resolution Imaging

Scanning Electron Microscopy (SEM) measurements were carried out at CNIS Laboratory of Sapienza University Rome, with a field emission Zeiss Auriga 405 instrument, with a resolution of 1 nm at maximum magnification, by using a beam energy of 19 keV with a working distance in the 1.5–3.5 mm range.

2.3. X-ray Photoelectron Spectroscopy

The VA-MWCNTs samples were transferred to the integrated X-ray Photoelectron Spectroscopy (XPS)/micro-Raman apparatus at the SMART Laboratory [31] of the Department of Physics of Sapienza University, to carry out high-resolution XPS measurements, in UHV with a base pressure in the low 10^{-11} mbar range. Pristine MWCNTs samples ($\sim$1 cm^2 size) were mounted on the sample holders in such a way to be able to measure both the top and lateral faces of the nanotubes (see Figure S5 in the Supplementary Information, SI), and introduced into the preparation chamber.

For the XPS measurements, core-level photoelectrons were excited by a monochromatized X-ray Al K_α (1486.8 eV) photon source (model SPECS XR50 MF) with focused beam, and analyzed by a PHOIBOS 150 analyzer (SPECS group, Berlin, Germany) working with 0.4 eV energy resolution, analyzed in constant pass energy (PE) mode set at 20 eV. Best spatial resolution of this setup was 30 μm. The electron binding energy (BE) scale was calibrated by using a gold foil that was in electrical contact with the sample, by acquiring the Au $4f_{7/2}$ core-level at 84.0 eV. The C 1s core level was taken in the (280–296) eV binding energy range.

2.4. Raman Spectroscopy

Micro-Raman (μ-Raman) measurements were performed at ambient conditions, with the sample being mounted on piezoelectric motors (by Attocube). The excitation laser was provided by a single frequency Nd:YVO_4 laser (DPSS series by Lasos) emitting at 532.2 nm. The laser light was focused on the sample by a 50× long-working-distance objective with numerical aperture NA = 0.5 (by Olympus), resulting in a spot of about 1 μm. The same objective allowed us to collect the scattered light, in a backscattering configuration. The laser light was filtered out by a very sharp long-pass Razor edge filter (by Semrock). The Raman signal was spectrally dispersed by a 75 cm focal length Acton monochromator (by Princeton Instruments) equipped with a 300 grooves/mm grating, and was detected by a back-illuminated N_2-cooled Si CCD camera (100BRX by Princeton Instruments). For depth-dependent Raman measurements, the laser was focused on the sample side. The laser spot was initially positioned at the very top of the carbon nanotubes (depth = 0) and the sample was then moved relative to the spot on-demand in order to have the laser focused at increasing depth along the nanotubes direction, going towards the sample substrate. Raman spectra were taken in the (500–3800) cm^{-1} wavenumber range.

2.5. Ion Bombardment on MWCNTs

Samples were measured at room temperature (RT) before annealing, to analyze the contamination content (O 1s core level signal at 6%). To remove the oxygen contamination [32] we annealed the pristine sample at 400 °C for 1 h reducing it to less than 1%. To avoid temperature healing effects, we avoided annealing the samples after ion bombardment. The ion bombardment was carried out with a sputtering ion source apparatus by Omicron Nanotechnology (ISE 10) on pristine MWCNTs samples coming from same batch of growth. Samples were bombarded with 3 keV Ar^+, Ne^+ and He^+ ions impinging from a direction parallel (TOP side) and perpendicular (LAT side) to the CNT axis. We used an ion current density of 10 μA/cm^2 and bombarded for 2400 s time, corresponding to a total number of 1.5×10^{17} bombarding ions.

3. Results and Discussion

3.1. SEM Analysis

The scanning electron microscopy images of the pristine MWCNTs displayed in Figure 1 (left), show the vertically aligned forests of MWCNTs grown on the Si substrate. At this high resolution spatial scale, a waviness in the orientation is visible that does not affect the average vertical alignment. On the top, a crust of twisted nanotubes of the order of few 10 nm is present, in agreement with previous observations on VA-CNTs [29]. All MWCNTs samples used in this experiment present heights of about 200 μm and diameters of about 10 nm, more details in the SI (Figure S1).

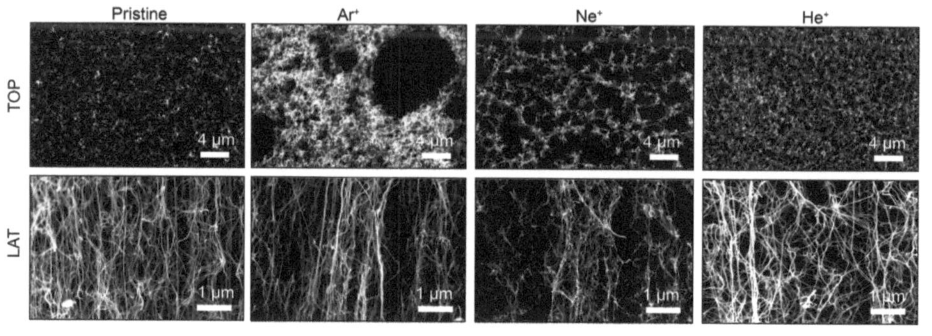

Figure 1. SEM image from the top side (TOP) and from the lateral side (LAT) of the CNTs; pristine (left), irradiated by 3 keV Ar^+ (center-left), Ne^+ (center-right) and He^+ (right) ion beams.

Ion bombardment with noble gas ions of different masses at the same energy (3 keV) affects both sides (top, [TOP] and lateral, [LAT]) of the samples. SEM images reveal that bombardment damages the surface of MWCNTs as shown in Figure 1, causing a rearrangement of the MWCNTs, with creation of holes across the entire surface due to coalescence in bundles, clusterization by combination of several CNTs and interlacing of external walls of individual CNTs with each other. From an analysis of the SEM images, we can extract the mean hole area by applying a grain analysis based on a watershed algorithm [33], implemented in the image-processing software Gwyddion (for more details see the SI Figure S2). Through this analysis we can find the area of the holes present in the TOP SEM image of each differently bombarded sample, that results to be roughly 0.07 ± 0.04 μm^2 (pristine), 0.09 ± 0.04 μm^2 (He^+), 2.8 ± 1.4 μm^2 (Ne^+) and 5.5 ± 2.6 μm^2 (Ar^+) (see SI Figure S3). We observe that the larger the ion mass, the larger the mean hole size induced on the sample. However, the He^+ bombardment does not induce a clear damage in the sample at least from a topographic perspective.

3.2. Core Level Analysis of the MWCNTs

An excellent technique to determine the density and the nature of defects present or produced in graphene-based materials, is core-level photoelectron spectroscopy. In fact, XPS is a very sensitive technique to finely determine the chemical bonding through the measured chemical shift in the C 1s core level components [34], like the sp^2 and sp^3 bonds, possible dangling bond states, etc.

The C 1s XPS spectrum of pristine MWCNTs taken after annealing to 400 °C is shown in Figure 2, along with the results of a fitting analysis. The experimental data have been deconvoluted using pseudo-Voigt line profiles (Lorentzian–Gaussian curves, with the Gaussian component taking into account the overall experimental uncertainty and the Lorentzian one the intrinsic excitation lifetime). We can single out five components: (i) the most intense associated to C−C sp^2 bonds [34] at 284.5 eV BE; (ii) the one due to C−C sp^3 distorted bonds at 285.1 eV (i.e., 0.6 eV higher BE than the sp^2 peak), where the presence of sp^3 bonds in pristine MWCNTs can be explained by the bending of the graphene surfaces in the nanotubes; (iii) an almost negligible component due to C-O_x bonds at 287.0 eV, related to residual oxygen contamination [19,35]; (iv) the expected π-plasmon (extended shake-up satellite) at 290.0 eV (i.e., 5.5 eV higher BE than the main peak); (v) a small fraction of dangling bonds (DB) component at low binding energy (283.9 eV) [36] that represents the tiny presence of vacancies in the MWCNTs.

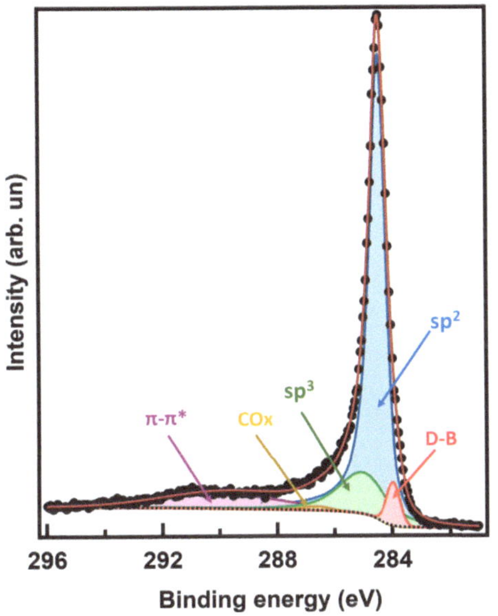

Figure 2. XPS C 1s core level of pristine CNT after annealing; experimental data (black dots), sp^2 fitting component (blue area), sp^3 component (green area), DB component (pink area), CO_x component (yellow area), π-plasmon component (violet area), Shirley background (dotted line) and fitting sum curve (red line).

The dominating sp^2 component is estimated to be 66% w.r.t the overall peak intensity. The sp^3 relative intensity ratio with respect to the sp^2 is ~19%, due to the bent nature of the cylindrical nanotubes [19,37–40]. We found very small concentration of oxygen, only 0.7% of overall intensity of the C 1s peak, in agreement with the measured intensity from the O 1s core level peak (see SI, Figure S4). The peak corresponding to vacancies in the honeycomb lattice, and associated to dangling bonds (DB), has a concentration of 3%. All fitting data of pristine MWCNTs are reported in the SI, Table S1.

Variation in the C 1s spectral lineshape of the bombarded MWCNTs with respect to the pristine sample is observed as an evident broadening, as shown in Figure 3b. The results of the fitting analysis are reported in the histogram of Figure 3c and in the SI Tables S2–S4.

After bombardment, both TOP and LAT, all the components do not shift in energy, while the sp^2 peak and the π plasmon [20] are reduced in intensity, and the sp^3 component representing the bond deformations in the lattice, shows a huge increase in intensity. The latter increase leads to a Θ ratio of about 48 $\pm$ 5%, where $\Theta = I(sp^3)/[I(sp^2) + I(sp^3)]$. Increase in dangling bond (from 3% to 8%) is also observed, due to an increase in vacancies in the honeycomb structure caused by bombardment. We do not observe any increase in the residual oxygen contamination onto the sample, thanks to the highly UHV controlled in situ experiment. XPS shows to be a very effective fingerprint of defect production in highly aligned MWCNTs after ion bombardment.

Finally, regarding the effect of different ion masses, we do not measure any significant difference in the C 1s spectral lineshapes, despite the different damage morphology observed by SEM. This can be explained by the local nature of defect production in the C mesh and on the very surface sensitivity of the XPS technique.

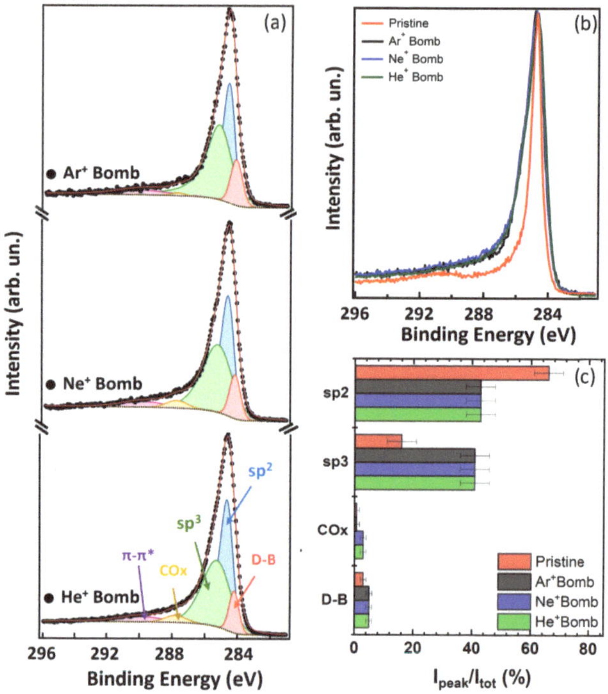

Figure 3. (**a**) XPS C 1s core level of the CNTs after ion bombardment with Ar^+, Ne^+ and He^+ (spectra stacked from top to bottom, respectively); experimental data (black dots), sp^2 fitting component (blue area), sp^3 component (green area), DB component (pink area), CO_x component (yellow area), Shirley background (dotted line) and fitting sum curve (red line). (**b**) Experimental C 1s data for the pristine and bombarded CNTs. (**c**) Histogram of the percentage of the C 1s core level components for pristine clean MWCNTs (red bars), Ar^+ bombarded CNTs (black bars), Ne^+ bombarded CNTs (blue bars) and He^+ bombarded CNTs (green bars); from the top to the bottom: sp^2, sp^3, CO_x and DB components.

We know from the SEM images that the morphology of CNTs is different from the TOP and LAT views, presenting circular voids from the top and elongated void parts among nanotubes on the side view. In order to obtain information about a possible spatial anisotropy in the defect production, we studied the XPS C 1s core level after bombardment of the two different sides of CNTs, TOP and LAT, as shown in Figure 4. To do these measurements we mounted two parts of the same sample oriented in the two sides (as shown in SI Figure S5), on the same sample holder, so to be able to bombard both sides at the same time in situ, with the same energy and dose.

We observe that there is not any lineshape change of the C 1s spectra of pristine CNTs taken for both orientations, as shown in Figure 4a. After bombarding with argon ions, we confirm the broadening of the lineshape, as discussed previously, but we do not observe any anisotropic effect depending on the nanotube orientation with respect to the impinging ion direction. This result confirms the local nature of the chemical damage induced by the ion beams, that does not depend on the spatial meso-scale morphology.

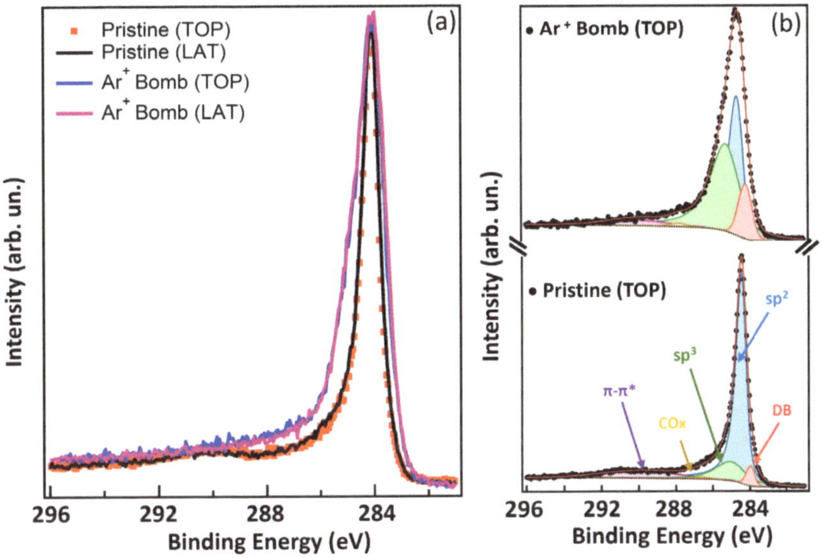

Figure 4. Comparative analysis of the XPS spectra of the CNTs after 3 keV Ar^+ bombardment on top (TOP) and perpendicular to the CNT axes (LAT). (**a**) superimposed C 1s XPS experimental data for the clean pristine (red dots and black line) and Ar^+ bombarded (blue and purple lines) CNTs. (**b**) fitting analysis of pristine TOP and Ar^+ bombarded CNTs TOP; experimental data (black dots) with fitting curve components, namely sp^2 (blue area), sp^3 (green area), DB (pink area), CO_x (yellow area), Shirley background (dotted line) and fitting sum curve (red line).

3.3. Raman Evidence of Defects

Raman spectroscopy is one of the most efficient and well-known techniques to determine defects and lattice deformations in graphene-based materials [41]. The Raman spectrum of pristine MWCNTs taken on the top of the sample after annealing to 400 °C, shows five most prominent peaks, the D, G, 2D, D+G and 2D′ bands, all expected for MWCNTs [20,42], as shown in Figure 5b. We fitted the Raman bands with Lorentzian curves (fitted values are reported in SI Table S9). The G band is representative of C−C stretching mode and appears at 1604 cm^{-1}, the D and D′ bands are associated with defects and distortions in the carbon hexagon rings and they appear at 1350 cm^{-1} and 1610 cm^{-1}, respectively, and the 2D band appears at 2690 cm^{-1}. Bumpy broad structures due to some residual amorphous carbon signal are considered below the main peaks by using Gaussian curves, in agreement with the literature on MWCNTs [19]. The integrated intensity ratios between the D and G band I(D)/I(G), and between the 2D and G bands I(2D)/I(G) are 0.90 and 0.47, respectively. The peak widths of the different peaks are 60 cm^{-1}, 81 cm^{-1}, 116 cm^{-1}, 145 cm^{-1}, 132 cm^{-1}, 198 cm^{-1} and 109 cm^{-1} for the G, D, D′, D+D″, 2D, D+G and 2D′ bands, respectively.

Raman data taken on the top of the He^+, Ne^+ and Ar^+ bombarded CNTs, as compared to the pristine one, are shown in Figure 5a. We fitted the Raman bands and fitting results are reported in Figure 5c. We observe a clear broadening of all main Raman bands, and also an important increase of the Gaussian component underlying the peaks, associated to the amorphous response [43,44]. In particular, broadening of the G and D bands after bombardment reaches 140 cm^{-1} and 253 cm^{-1}, respectively, (see the SI Tables S10–S12), due to the induction of defects by noble gas ions hitting the MWCNTS. A frequency shift is observed up to 1357 cm^{-1} for the D band and decrease down to 1565 cm^{-1} for the G band, probably associated to strain induced by defects introduced after bombardment. In the mid-range between these two peaks, there is a broad band at 1460 cm^{-1} related to increase in amorphous carbon [45]. Looking at the ion mass dependence of the produced lattice damage, it is evident that the lighter mass ions cause more diffused lattice damage than the heavier ones, as observable through the relative increasing broadening (values in the SI, Tables S10–S12) upon going from Ar^+, to Ne^+ eventually to He^+ ions. The results of Raman study correlate well with the SEM data.

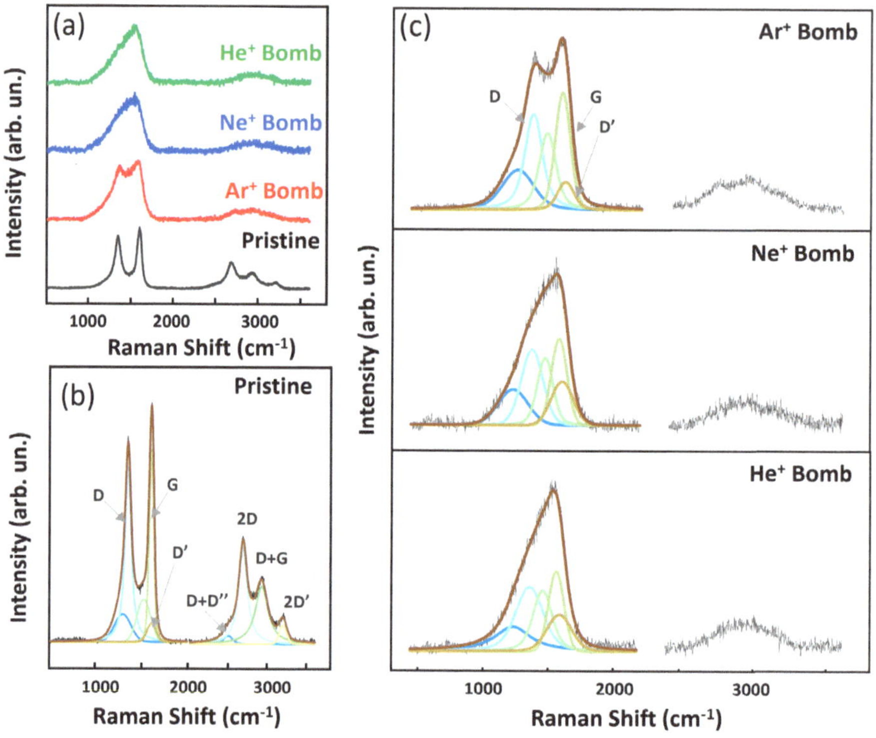

Figure 5. (**a**) Experimental data for Raman spectra (focused on the sample top) of Pristine and bombarded CNTs, (**b**) Fitted Raman spectra of pristine CNTs and Raman fitting components are reported with brown curves superimposed over the experimental data and (**c**) Fitted Raman spectra of the CNTs after ion bombardment with Ar^+, Ne^+ and He^+ (spectra stacked from top to bottom, respectively).

The Raman and SEM results suggest the following picture of produced damage: the energy loss of ions to CNT atomic electrons at velocity less than 0.01 c (where c is the speed of light in vacuum) can be described by Lindhard's model [46,47]. This model predicts an energy loss increasing with the ion velocity, as the data of CNT damage suggest: in fact, He ions are the fastest ($\sim 3.8 \times 10^5$ m/s), they easily reach a noticeable depth and consequently produce a more diffuse damage of the CNT lattice, while the Ar ions produce more localized defects in the lattice.

4. Conclusions

The effects of different masses of noble gas in ion irradiation on the chemical state, lattice morphology and local structure of MWCNTs were studied via XPS, Raman and SEM. It has been established that bombardment with ions on different geometries (TOP and LAT) of MWCNTs leads to defects and distortions in the MWCNT's structure. In particular, bundles and clustering of CNTs are clearly visible in the SEM images, where smaller voids on the top are produced by lighter ions. Raman analysis confirms the SEM results, showing lattice damage in the CNTs, as observed by broadening of Raman bands accompanied by partial amorphization.

The Raman spectroscopy data show more diffused lattice defects by bombarding with lighter ions (He^+) than heavier ones (Ar^+). The interaction of the heavier ions with the C lattice produces more localized defects; as a consequence, the damage depth reached by Ar is smaller than for He, the latter producing diffuse lattice defects.

The analysis of the C 1s core level spectra has shown an increase in the fraction of C atoms with sp^3 hybridized bonds from 19 $\pm$ 5% up to 48 $\pm$ 5%, and a reduction of the π-plasmon of delocalized electrons on the surface of MWCNTs. Furthermore, not any anisotropic effect is observed in the XPS study (TOP or LAT bombardment), due to very local, atomic and surface sensitive characteristics of XPS.

These ion-bombarded MWCNTs, where we demonstrated how to increase available sp^3 bonds thanks to clean and controlled UHV in situ ion bombardment, can constitute excellent scaffolds for further functionalization. For example, sp^3 bonds may strongly favor the uploading of hydrogen, offering highly useful and potential applications in new solid-state materials for energy storage.

Supplementary Materials: The following are available online at https://www.mdpi.com/article/10.3390/nano14010077/s1, Figure S1: Exemplary SEM images of MWCNTs from (a) (LAT) view and (b) (TOP) view, used for size estimation by the Gwyddion program.; Figure S2: mean hole size at the top of the MWCNT after ion bombardment; Figure S3: analysis of the SEM images to determine the mean hole size; Figure S4: survey XPS spectrum at pristine MWCNTs; Figure S5: MWCNTs mounted from TOP and LAT sides on same sample holder; Figure S6: FWHM evolution of Raman bands (D, G and D′) for Pristine and Bombarded MWCNTs (Ar^+, Ne^+ and He^+) after fitting analysis.; Table S1: C1s fitting parameters from the C1s core level of the pristine MWCNT sample; Tables S2–S4: C1s fitting parameters from the C1s core levels of the Ar-, Ne-, and He-bombarded MWCNT samples; Tables S5 and S6: C1s fitting parameters from the C1s core levels of the TOP and LAT pristine sample; Tables S7 and S8: C1s fitting parameters from the C1s core levels of the TOP and LAT Ar-bombarded sample; Table S9: Raman band fitting parameters for pristine MWCNT; Tables S10–S12: Raman band fitting parameters for Ar, Ne, He bombareded MWCNT. Reference [48] is cited in the Supplementary Materials.

Author Contributions: Conceptualization: M.G.B., G.C., C.M. and F.P.; Methodology: C.M.; Validation: C.M.; Formal Analysis: S.T., A.A., R.F., N.J.-A., C.M., A.P. and A.R.; Investigation: S.T. and C.M.; Data curation: S.T., E.B., C.M., I.R., M.S. and R.P.Y.; Writing—Original Draft: S.T. and C.M.; Writing—review and editing: All Authors; Visualization: S.T.; Supervision: C.M.; Project administration: C.M.; Funding acquistion: M.G.B., G.C., C.M. and F.P. All authors have read and agreed to the published version of the manuscript.

Funding: This research was funded by the European Union—Next Generation EU under the PRIN Grants 2D-FRONTIERS (20228879FT) and TUNES (2022NXLTYN) of Ministero dell'Università e della Ricerca (MUR), by ANDROMeDA (PRIN_2020Y2JMP5) of MUR, by PNRR-NEST project ENERGY SUSTAINABLE TRANSITION-NETWORK 4 Spoke 6 NextGenerationEU—PE0000021, by the INFN Ptolemy project, and by Sapienza University Ateneo and "Avvio alla Ricerca" funds. Part of this project received funding from the ATTRACT project funded by the EC under Grant Agreement 777222.

Data Availability Statement: Data are contained within the article.

Acknowledgments: The high-resolution XPS measurements were carried out at the SmartLab [31] departmental laboratory of the Department of Physics at Sapienza University of Rome. We acknowledge experimental assistance by F. Mura at CNIS for the SEM images. We thank Lara Benfatto for useful discussions.

Conflicts of Interest: The authors declare no conflicts of interest.

Abbreviations

The following abbreviations are used in this manuscript:

CNT	Carbon nanotube
MWCNT	Multi Wall Carbon nanotube
VA-MWCNT	Vertically Aligned Multi Wall Carbon nanotube
XPS	X ray Photoelectron Spectroscopy
UHV	Ultra High Vacuum
SEM	Scanning Electron microscopy
3D	Three Dimensional
2D	Two Dimensional
1D	One Dimensional

References

1. Iijima, S. Helical microtubules of graphitic carbon. *Nature* **1991**, *354*, 56. [CrossRef]
2. Andrews, R.; Jacques, D.; Rao, A.; Derbyshire, F.; Qian, D.; Fan, X.; Dickey, E.; Chen, J. Continuous production of aligned carbon nanotubes: A step closer to commercial realization. *Chem. Phys. Lett.* **1999**, *303*, 467–474. [CrossRef]
3. Ma, Y.; Foster, A.S.; Krasheninnikov, A.; Nieminen, R.M. Nitrogen in graphite and carbon nanotubes: Magnetism and mobility. *Phys. Rev. B* **2005**, *72*, 205416. [CrossRef]
4. Nakaharai, S.; Iwasaki, T.; Morita, Y.; Moriyama, S.; Ogawa, S. Electron transport tuning of graphene by helium ion irradiation. *Nano Express* **2022**, *3*, 024002. [CrossRef]
5. Ogawa, S. Anderson localization of graphene by helium ion irradiation. *Appl. Phys. Lett.* **2016**, *108*, 171605.
6. Pejcic, B.; Myers, M.; Ranwala, N.; Boyd, L.; Baker, M.; Ross, A. Modifying the response of a polymer-based quartz crystal microbalance hydrocarbon sensor with functionalized carbon nanotubes. *Talanta* **2011**, *85*, 1648–1657. [CrossRef] [PubMed]
7. Zhao, Q.; Yuan, Z.; Duan, Z.; Jiang, Y.; Li, X.; Li, Z.; Tai, H. An ingenious strategy for improving humidity sensing properties of multi-walled carbon nanotubes via poly-L-lysine modification. *Sens. Actuators B Chem.* **2019**, *289*, 182–185. [CrossRef]
8. Star, A.; Gabriel, J.C.P.; Bradley, K.; Grüner, G. Electronic detection of specific protein binding using nanotube FET devices. *Nano Lett.* **2003**, *3*, 459–463. [CrossRef]
9. Wong, M.; Paramsothy, M.; Xu, X.; Ren, Y.; Li, S.; Liao, K. Physical interactions at carbon nanotube-polymer interface. *Polymer* **2003**, *44*, 7757–7764. [CrossRef]
10. Rakhi, R.; Sethupathi, K.; Ramaprabhu, S. Field emission from carbon nanotubes on a graphitized carbon fabric. *Carbon* **2008**, *46*, 1656–1663. [CrossRef]
11. Lee, Y.H.; An, K.H.; Lim, S.C.; Kim, W.S.; Jeong, H.J.; Doh, C.H.; Moon, S.I. Applications of carbon nanotubes to energy storage devices. *New Diam. Front. Carbon Technol.* **2002**, *12*, 209–228.
12. Rivard, E.; Trudeau, M.; Zaghib, K. Hydrogen storage for mobility: A review. *Materials* **2019**, *12*, 1973. [CrossRef] [PubMed]
13. Wang, B.; Pang, B. Properties improvement of multiwall carbon nanotubes-reinforced cement-based composites. *J. Compos. Mater.* **2020**, *54*, 2379–2387. [CrossRef]
14. Tans, S.J.; Verschueren, A.R.M.; Dekker, C. Room-temperature transistor based on a single carbon nanotube. *Nature* **1998**, *393*, 49–52. [CrossRef]
15. Derakhshandeh, J.; Abdi, Y.; Mohajerzadeh, S.; Hosseinzadegan, H.; Soleimani, E.A.; Radamson, H. Fabrication of 100 nm gate length MOSFET's using a novel carbon nanotube-based nano-lithography. *Mater. Sci. Eng. B* **2005**, *124*, 354–358. [CrossRef]
16. Raffaelle, R.; Landi, B.; Harris, J.; Bailey, S.; Hepp, A. Carbon nanotubes for power applications. *Mater. Sci. Eng. B* **2005**, *116*, 233–243. [CrossRef]
17. An, K.H.; Kim, W.S.; Park, Y.S.; Choi, Y.C.; Lee, S.M.; Chung, D.C.; Bae, D.J.; Lim, S.C.; Lee, Y.H. Supercapacitors using single-walled carbon nanotube electrodes. *Adv. Mater.* **2001**, *13*, 497–500. [CrossRef]
18. Scardamaglia, M.; Amati, M.; Llorente, B.; Mudimela, P.; Colomer, J.F.; Ghijsen, J.; Ewels, C.; Snyders, R.; Gregoratti, L.; Bittencourt, C. Nitrogen ion casting on vertically aligned carbon nanotubes: Tip and sidewall chemical modification. *Carbon* **2014**, *77*, 319–328. [CrossRef]
19. D'Acunto, G.; Ripanti, F.; Postorino, P.; Betti, M.G.; Scardamaglia, M.; Bittencourt, C.; Mariani, C. Channelling and induced defects at ion-bombarded aligned multiwall carbon nanotubes. *Carbon* **2018**, *139*, 768–775. [CrossRef]
20. Lehtinen, O.; Nikitin, T.; Krasheninnikov, A.V.; Sun, L.; Banhart, F.; Khriachtchev, L.; Keinonen, J. Characterization of ion-irradiation-induced defects in multi-walled carbon nanotubes. *New J. Phys.* **2011**, *13*, 073004. [CrossRef]
21. Mathew, S.; Bhatta, U.; Joseph, B.; Dev, B. keV Ag ion irradiation induced damage on multiwalled carbon nanotubes. *Nucl. Instrum. Methods Phys. Res. Sect. B Beam Interact. Mater. Atoms* **2007**, *264*, 36–40. [CrossRef]
22. Krasheninnikov, A.; Nordlund, K. Channeling of heavy ions through multi-walled carbon nanotubes. *Nucl. Instrum. Methods Phys. Res. Sect. B Beam Interact. Mater. Atoms* **2005**, *228*, 21–25. [CrossRef]
23. Salonen, E.; Krasheninnikov, A.; Nordlund, K. Ion-irradiation-induced defects in bundles of carbon nanotubes. *Nucl. Instrum. Methods Phys. Res. Sect. B Beam Interact. Mater. Atoms* **2002**, *193*, 603–608. [CrossRef]

24. Krasheninnikov, A.; Nordlund, K. Irradiation effects in carbon nanotubes. *Nucl. Instrum. Methods Phys. Res. Sect. B Beam Interact. Mater. Atoms* **2004**, *216*, 355–366. [CrossRef]
25. Buchowicz, G.; Stone, P.R.; Robinson, J.T.; Cress, C.D.; Beeman, J.W.; Dubon, O.D. Correlation between structure and electrical transport in ion-irradiated graphene grown on Cu foils. *Appl. Phys. Lett.* **2011**, *98*, 032102. [CrossRef]
26. Betti, M.G.; Placidi, E.; Izzo, C.; Blundo, E.; Polimeni, A.; Sbroscia, M.; Avila, J.; Dudin, P.; Hu, K.; Ito, Y.; et al. Gap Opening in Double-Sided Highly Hydrogenated Free-Standing Graphene. *Nano Lett.* **2022**, *22*, 2971–2977. [CrossRef]
27. Junge, F.; Auge, M.; Zarkua, Z.; Hofsäss, H. Lateral Controlled Doping and Defect Engineering of Graphene by Ultra-Low-Energy Ion Implantation. *Nanomaterials* **2023**, *13*, 658. [CrossRef]
28. Marchiani, D.; Tonelli, A.; Mariani, C.; Frisenda, R.; Avila, J.; Dudin, P.; Jeong, S.; Ito, Y.; Magnani, F.S.; Biagi, R.; et al. Tuning the Electronic Response of Metallic Graphene by Potassium Doping. *Nano Lett.* **2023**, *23*, 170–176. [CrossRef]
29. Schifano, E.; Cavoto, G.; Pandolfi, F.; Pettinari, G.; Apponi, A.; Ruocco, A.; Uccelletti, D.; Rago, I. Plasma-Etched Vertically Aligned CNTs with Enhanced Antibacterial Power. *Nanomaterials* **2023**, *13*, 1081. [CrossRef]
30. Sarasini, F.; Tirillò, J.; Lilli, M.; Bracciale, M.P.; Vullum, P.E.; Berto, F.; De Bellis, G.; Tamburrano, A.; Cavoto, G.; Pandolfi, F.; et al. Highly aligned growth of carbon nanotube forests with in situ catalyst generation: A route to multifunctional basalt fibres. *Compos. Part B Eng.* **2022**, *243*, 110136. [CrossRef]
31. SmartLab. 2023. Available online: https://sites.google.com/uniroma1.it/smartlab (accessed on 5 December 2023).
32. Abdelnabi, M.M.S.; Izzo, C.; Blundo, E.; Betti, M.G.; Sbroscia, M.; Di Bella, G.; Cavoto, G.; Polimeni, A.; García-Cortés, I.; Rucandio, I.; et al. Deuterium Adsorption on Free-Standing Graphene. *Nanomaterials* **2021**, *11*, 130. [CrossRef] [PubMed]
33. Vincent, L.; Soille, P. Watersheds in digital spaces: An efficient algorithm based on immersion simulations. *IEEE Trans. Pattern Anal. Mach. Intell.* **1991**, *13*, 583–598. [CrossRef]
34. Lacovig, P.; Pozzo, M.; Alfe, D.; Vilmercati, P.; Baraldi, A.; Lizzit, S. Growth of dome-shaped carbon nanoislands on Ir (111): The intermediate between carbidic clusters and quasi-free-standing graphene. *Phys. Rev. Lett.* **2009**, *103*, 166101. [CrossRef] [PubMed]
35. Estrade-Szwarckopf, H. XPS photoemission in carbonaceous materials: A "defect" peak beside the graphitic asymmetric peak. *Carbon* **2004**, *42*, 1713–1721. [CrossRef]
36. Susi, T.; Kaukonen, M.; Havu, P.; Ljungberg, M.P.; Ayala, P.; Kauppinen, E.I. Core level binding energies of functionalized and defective graphene. *Beilstein J. Nanotechnol.* **2014**, *5*, 121–132. [CrossRef]
37. Díaz, J.; Paolicelli, G.; Ferrer, S.; Comin, F. Separation of the sp^3 and sp^2 components in the C1s photoemission spectra of amorphous carbon films. *Phys. Rev. B* **1996**, *54*, 8064–8069. [CrossRef]
38. Chu, P.K.; Li, L. Characterization of amorphous and nanocrystalline carbon films. *Mater. Chem. Phys.* **2006**, *96*, 253–277. [CrossRef]
39. Sun, C.Q.; Sun, Y.; Nie, Y.G.; Wang, Y.; Pan, J.S.; Ouyang, G. Coordination-Resolved C−C Bond Length and the C 1s Binding Energy of Carbon Allotropes and the Effective Atomic Coordination of the Few-Layer Graphene. *J. Phys. Chem. C* **2009**, *113*, 16464–16467. [CrossRef]
40. Al-Harthi, S.H.; Elzain, M.; Al-Barwani, M.; Kora'a, A.; Hysen, T.; Myint, M.T.Z.; Anantharaman, M.R. Unusual surface and edge morphologies, sp2 to sp3 hybridized transformation and electronic damage after Ar+ ion irradiation of few-layer graphene surfaces. *Nanoscale Res. Lett.* **2012**, *7*, 466. [CrossRef]
41. Malard, L.; Pimenta, M.; Dresselhaus, G.; Dresselhaus, M. Raman spectroscopy in graphene. *Phys. Rep.* **2009**, *473*, 51–87. [CrossRef]
42. Antunes, E.; Lobo, A.; Corat, E.; Trava-Airoldi, V. Influence of diameter in the Raman spectra of aligned multi-walled carbon nanotubes. *Carbon* **2007**, *45*, 913–921. [CrossRef]
43. Antunes, E.; Lobo, A.; Corat, E.; Trava-Airoldi, V.; Martin, A.; Veríssimo, C. Comparative study of first- and second-order Raman spectra of MWCNT at visible and infrared laser excitation. *Carbon* **2006**, *44*, 2202–2211. [CrossRef]
44. Jeet, K.; Jindal, V.K.; Bharadwaj, L.M.; Avasthi, D.K.; Dharamvir, K. Damaged carbon nanotubes get healed by ion irradiation. *J. Appl. Phys.* **2010**, *108*, 034302, [CrossRef]
45. Ferrari, A.C.; Robertson, J. Interpretation of Raman spectra of disordered and amorphous carbon. *Phys. Rev. B* **2000**, *61*, 14095. [CrossRef]
46. Lindhard, J. On the properties of a gas of charged particles. *Kgl. Danske Videnskab. Selskab Mat. Fys. Medd.* **1954**, *28*, 8.
47. Lindhard, J.; Scharff, M.; Schioett, H.E. Range Concepts and Heavy Ion Ranges (Notes on Atomic Collisions, II). *Kgl. Danske Videnskab. Selskab. Mat. Fys. Medd.* **1963**, *33*, 14.
48. Yeh, J.; Lindau, I. Atomic subshell photoionization cross sections and asymmetry parameters: $1 \leq Z \leq 103$. *At. Data Nucl. Data Tables* **1985**, *32*, 1–155. [CrossRef]

nanomaterials

MDPI

Article

Lateral Controlled Doping and Defect Engineering of Graphene by Ultra-Low-Energy Ion Implantation

Felix Junge [1], Manuel Auge [1], Zviadi Zarkua [2] and Hans Hofsäss [1,*]

[1] II. Institute of Physics, Georg-August-Universität Göttingen, 37077 Göttingen, Germany
[2] Quantum Solid State Physics, KU Leuven, 3001 Leuven, Belgium
* Correspondence: hans.hofsaess@phys.uni-goettingen.de

Abstract: In this paper, the effectiveness of ultra-low-energy ion implantation as a means of defect engineering in graphene was explored through the measurement of Scanning Kelvin Probe Microscopy (SKPM) and Raman spectroscopy, with boron (B) and helium (He) ions being implanted into monolayer graphene samples. We used electrostatic masks to create a doped and non-doped region in one single implantation step. For verification we measured the surface potential profile along the sample and proved the feasibility of lateral controllable doping. In another experiment, a voltage gradient was applied across the graphene layer in order to implant helium at different energies and thus perform an ion-energy-dependent investigation of the implantation damage of the graphene. For this purpose Raman measurements were performed, which show the different damage due to the various ion energies. Finally, ion implantation simulations were conducted to evaluate damage formation.

Keywords: graphene; ion implantation; ion implantation simulations; 2D materials; IMINTDYN

check for updates

Citation: Junge, F.; Auge, M.; Zarkua, Z.; Hofsäss, H. Lateral Controlled Doping and Defect Engineering of Graphene by Ultra-Low-Energy Ion Implantation. *Nanomaterials* **2023**, *13*, 658. https://doi.org/10.3390/nano13040658

Academic Editor: Muralidharan Paramsothy

Received: 19 January 2023
Revised: 2 February 2023
Accepted: 6 February 2023
Published: 8 February 2023

1. Introduction

As revolutionary materials with unique mechanical, thermal, and electrical properties, graphene and other 2D materials such as transition-metal dichalcogenides (TMDs) have been the subject of extensive research and development in various fields, from electronics to biomedicine. Some of the new applications include for example sensors for biomolecules and environmental contaminants [1–4], spintronic devices [5] or water cleaning [6]. To further modify the electrical, magnetic, structural or optical properties of 2D materials, foreign atoms can be introduced into the crystal lattice [7–9]. This can be achieved using various methods, such as adapting the growth process or diffusing atoms into the material [10,11]. A direct approach frequently used in the semiconductor industry is ion implantation. In this process, foreign atoms can be introduced into the materials in a targeted manner without altering the growth process, or reliance on chemistry or diffusion, which ensures a very atom specific incorporation. Conventional ion implantation (>1 keV), however, are insufficient for implanting in 2D materials, due to required implantation energies of only a few 10 eV [12]. Since the generation of an ion beam with such low energy and reasonable implantation currents (<10 nA on the sample) is a great challenge, the deceleration of the ions in front of the sample by applying an electrical potential to the sample a promising approach to reach the ultra-low energy regime (few 10 eV). This has the advantage that, in contrast to deceleration by a capping layer [13], no recoil atoms from the capping layer enter the sample and thus only the desired atomic species is implanted and no contamination occur. For graphene and TMDs it has already been shown that properties and damage of the sample can be successfully tuned by means of laterally uniform ULE ion implantation [14–20]. To further expand the capabilities of ULE ion implantation, we demonstrate how the surface of a graphene sample can be selectively modified by ion beams using laterally controlled implantation [21], paving the way for more complex implantation structures and thus the possibility of fabricating electrical 2D devices and the exploration of new scientific questions. For 2D target materials, electrostatic deflection has significant advantages compared to a conventional shadow mask. Indeed, when using physical masks, recoils can occur, which are also incorporated into the graphene. Another disadvantage of a shadow mask are the electrical field inhomogeneities, which occur at the edges of the mask due to the high electrostatic potential gradient required for the ion deceleration. Due to these inhomogeneities, sharp transition regions are prevented. This is amplified because a physical mask should also hover over the samples to avoid damage to the graphene. Electrostatic masking allows a contactless shielding of the sample and provides therefore a good possibility for the lateral selective implantation of one sample. Another application is the generation of an electrical potential gradient along the sample to obtain different implantation energies at different sample positions. This creates an opportunity to study the effects of implantations, such as damage formation, along a continuous ion energy gradient while all other experimental conditions, such as sample preparation, are the same. In this paper the electrostatic masks were used for boron (B) implantations of graphene to generate a transition between pristine and implanted regions on graphene. Additionally helium (He) was implanted in another graphene sample with an applied potential gradient to study the effects of different implantation energies.

Simulations of He implantation in graphene were conducted to provide a better understanding of the processes involved. Here, the damage formation can also be investigated depending on the implantation energy and the used fluence.

2. Materials and Methods

2.1. ULE Ion Implantation

The ion implanter ADONIS [22,23] was used for the implantations, which is currently the only accelerator in the world that can provide these ultra-low energies and can be used for direct implantation into 2D materials. In this setup, the ions are first accelerated with 30 kV, mass selected with the help of a 90°-sector magnet, with a mass resolution of $M/\Delta M = 150$ and finally slowed down again to achieve implantation energies of several 10 eV. This deceleration is carried out by placing the sample at the voltage of the ion source anode, minus the desired implantation voltage [24]. Thus, it can be ensured that the ions impinge the sample with a maximum energy, which is set beforehand in the range of 10 eV–600 eV. To exclude different energies due to fluctuations in different power supplies, the same power supply is used for acceleration and deceleration. Due to the high mass resolution only the desired ion species is implanted. Neutral particles are removed from the beam before entering the implantation chamber, so that all particles can be decelerated in front of the sample. All implantations were carried out at room temperature in ultra-high vacuum ($<10^{-8}$ mbar).

2.2. Introducing Lateral Inhomogeneities

2.2.1. Electrostatic Masking

To achieve electrostatic masking, the deceleration unit described in [21] is used. Here, an electrode is hovering at a distance of about 1 mm above the sample and is set to +100 V or +300 V with respect to the sample potential. This deflects the incoming ions, which have an energy of a few 10 eV at this point, and pushes them away from one side of the sample. A typical simulation for 100 V can be found in [21]. The sample itself is at −20 V in respect to the 30 kV accelerating voltage, which allows the ions to be implanted with a maximum energy of 20 eV.

Monolayer graphene on SiO_2 and Ni were used as samples, which were obtained from Graphene Supermarket (Graphene Laboratories Inc., Ronkonkoma, NY, USA, https://www.graphene-supermarket.com). B was implanted with fluences of 5×10^{14} at/cm^2 and 1×10^{15} at/cm^2. The samples were then analyzed by Scanning Kelvin Probe Microscopy (SKPM) measurements to determine the surface potential along the sample to analyze the sharpness of the transition between the implanted and un-implanted region.

2.2.2. Potential Gradient

To create an energy gradient for studying the damage of graphene when implanted with different energies, 20 nm copper contacts were first applied by sputter deposition on both sides of the graphene. Then, a constant voltage source was used to apply a potential difference of 100 V. The ground contact on the graphene was set to 0 V with respect to the 30 kV anode of the source, so that the incoming ions have an energy between 100 eV–0 eV. The current flow through the graphene layer results in a linear potential gradient along the surface. As sample, monolayer graphene on SiO_2 was used, so that the SiO_2 layer acts as an insulator and the voltage drops only through the graphene layer. The samples were also obtained from Graphene Supermarket. He ions were used for the implantation with a fluence of 1×10^{15} at/cm^2. The feasibility of laterally uniform He implantation into graphene has been shown in previous publications [25].

2.3. Sample Charaterization

For the measurement of the surface potential (SKPM measurements) of the sample after implantation using the electro static mask, measurements were performed with an Atomic Force Microscope, type MFP-3D Origin+ from Oxford Instruments Asylum Research (Oxford Instruments GmbH, Wiesbaden, Germany), in Scanning Kelvin Probe Microscopy (SKPM) non-contact constant height mode. Here the cantilever is held at a constant height above the specimen and the force on the cantilever is measured based on an applied voltage between the cantilever and the specimen. Subsequently, the value for each measurement was averaged over an area of 5 µm × 5 µm. Normal AFM measurements were not recorded separately and the selected resolution was sufficient for averaging of the surface potential but not for specific height analyses. The damage formation was determined using Raman spectrometry. Raman spectra were measured using a confocal Raman microscope (Monovista CRS+, S&I GmbH, Warstein, Germany) equipped with a 532 nm Nd:YAG laser. The laser was directed onto the sample surface through an objective (OLYMPUS, X43 100×, N.A. 0.7), with the maximum laser power remaining below 1 mW in order to avoid laser-induced modification. All the measurements were obtained in ambient conditions, at room temperatur.

3. Results and Discussion

3.1. Change of Surface Potential by B Implantation

The SKPM measurements, shown in Figure 1, reveal a change in the surface potential between the pristine and implanted region of the sample. It can be observed that the level of surface potential change correlates with the fluence used. Due to the stronger deflection of the ions at the 300 V mask (Figure 1b), a significant reduction of the transition zone from about 1 mm to below 500 µm has been achieved, compared to the 100 V mask. In addition, an increase in the surface potential, and thus the doping, at the interface can be seen at higher deflection voltages. This is due to the fact that the ions, which are deflected, are pushed to the sides and generate a higher fluence there. This only occurs on the implanted side, since the other side is completely shielded by the hovering electrode. An exact simulation with SIMION [26] of the ion trajectories can be found in [21]. Based on the magnitude of the surface potential, the high fluence of the region next to the transition can also be determined, since this potential corresponds to the surface potential of the 1×10^{15} at/cm^2 implanted sample. Thus, in the case of the 300 V deflection, the fluence directly at the interface is about twice as high as expected. This can also be used to dope various doping levels in different areas on the sample. By using different substrates (SiO_2 and Ni), and observing a similar change of the surface potential, both in values and in general behavior, it can be concluded that this effect is due to the doping of the graphene and not to the effects of the substrate.

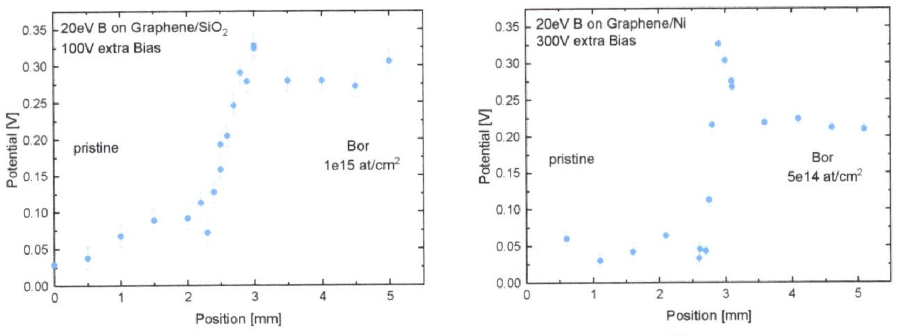

(**a**) (**b**)

Figure 1. (**a**) SKPM measurement of a graphene layer on SiO_2 previously implanted with 1×10^{15} at/cm² with B with an energy of 20 eV. In addition, only a part of the sample was implanted using an electrostatic mask. The voltage at the mask was set to +100 V corresponding to the sample bias. (**b**) SKPM measurement of a graphene layer on Nickel previously implanted with 5×10^{14} at/cm² with Boron at an energy of 20 eV. The voltage for masking was set to +300 V in respect to the sample bias. The comparison shows a sharper transition region ($\approx$1 mm to $\approx$0.5 mm) with increasing deflection voltage. The position indicates the relative position to the edge of the sample.

3.2. *Energy Dependent Defect Formation by He Implantations*

In Figure 2 different Raman spectra for different sample surface positions of the He implanted sample are shown. The positioning describes the distance from the one copper electrode connected to the 0 V. Therefore, lower positions corresponds to lower implantation energy. The gradient can be approximately assumed to be linear, since the resistance across the graphene layer should not change. Along the sample surface, a clear shift of the D-peak as well as a broadening can be seen. Another indication for a large damage is provided by the 2D-peak, which gets smaller with higher He energy until it almost disappears. For $x > 3$ mm the graphene becomes more and more amorphous, presumably due to the high fluence of 1×10^{15} at/cm². This is shown by the broadening of the G-peak and the merging with the D′-peak due to the broadening of the D band, which indicates the loss of the crystal structure is thus also a sign of higher damage to the graphene [27,28]. A typical Raman spectrum of graphene from Graphene Supermarket before and after an ion implantation can be found in [29]. Similar to our results, a decrease in the 2D-peak and an increase of the D-peak after the implantation was observed as damage increases. It should be noted that the first measuring point already shows a clear damage formation, since the 2D-peak is already smaller than the D-peak although at this position the implantation energy of around 1–5 eV is too small for vacancy formation. However, intercalation of He between the graphene and the substrate can still occur which leads to damage. Moreover, the graphene was probably slightly damaged by the sputtering process when applying the copper contacts. These measurements only serve as a proof of concept, since due to the strong initial damage caused by the copper contacts and the high fluence of the implanted ions, it is not possible to make a quantitative statement about the damage compared to the implantation energy. This is partly because the exact field gradient over the sample was assumed to be linear, but can deviate from this due to various disturbances in the graphene. Nevertheless, these measurements show a clear trend towards more damage at higher ion energies, similar to results of laterally uniform implantations [20].

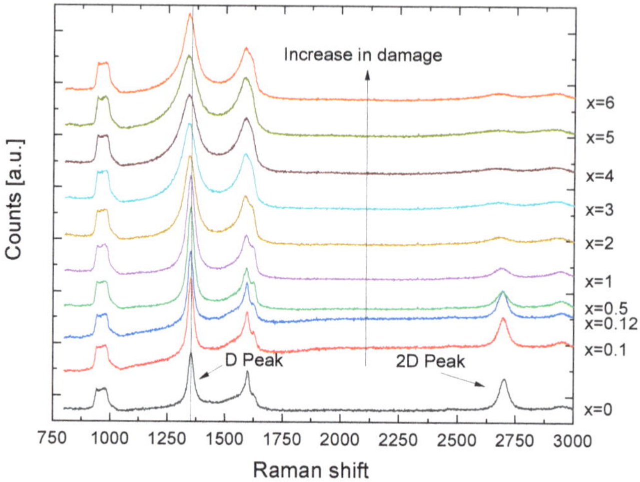

Figure 2. Raman spectra of the Helium implanted sample. The spectrum was measured on different positions on the sample. Here, x is in mm, where $x = 0$ describes the position directly at the 0 V contact. With higher x also the He energy increases from $E_{He} \approx 0$ eV at $x = 0$ mm to $E_{He} \approx 100$ eV at $x > 6$ mm. A clear shift of the D-peak and a broadening due to a higher implantation energy at larger x can be seen.

3.3. Simulations

To obtain a more profound understanding of the processes during implantation, simulations were performed using IMINTDYN [30], a binary collision approximation (BCA) Monte Carlo program for simulation of ion solid interactions based on SDTrimSP [31]. As most BCA simulation codes IMINTDYN also assumes an amorphous target structure. The interaction potential is the screened Kr-C potential and for electronic stopping the SRIM2013 database is used. IMINTDYN as well as SDTrimSP can perform dynamic simulations, which take stoichiometry changes, erosion and deposition during ion irradiation into account. The quality of different BCA simulation codes at lower ion energies in comparison with experimental data has been investigated by one of the authors in the past [32]. SDTrimSP and IMINTDYN simulations have been successfully applied to model surface pattern formation by low energy ion irradiation at grazing ion incidence [33,34]. We have also used BCA Monte Carlo simulations previously to model ultra low energy ion irradiation of graphen and other 2D materials to estimate the erosion effects and the ion retention as function of ion energy [29]. A special new feature of IMINTDYN is the possibility to insert vacancies as a target atom species. The vacancy does not influence the path direction of a moving particle but just increases its mean free path length. This offers the advantage that vacancies generated by recoil collisions and vacancies annihilated by stopped atoms dynamically change the composition of the sample and thus provides a more accurate representation of the final layer structure with increasing ion fluence. Furthermore, vacancies allow to model 2D materials much more realistic when the space in the center of six-ring atoms arrangements and the space between 2D layers is filled with vacancies. In other BCA simulations 2D materials are represented simply as homogeneous amorphous layers. The graphene monolayer with layer spacing of 3.35 Å, is simulated as a 1.1 Å thick carbon layer, followed by two layers of vacancies each 1.1 Å thick. The deeper layers, and thus the substrate, consist of SiO_2. The sublimation energy of 7.428 eV is used as the bulk binding energy (or surface binding energy) of carbon. This also corresponds to the vacancy formation energy of graphene [35]. For Si, O and other atoms we use the corresponding sublimation energies.

Since IMINITDYN does not simulate a lattice structure, but assumes the layers as amorphous, the graphene layer was adjusted slightly to take the hexagonal arrangement of the atoms into account. Because of the lattice structure there is the possibility that an ion flies through the middle of the carbon ring without colliding with an atom. Therefore, the first layer in the simulation consists of 2/3 C-atoms and 1/3 vacancies since two C-atoms and one empty center are to be assigned to one carbon ring. The density of the layer, with a layer spacing of 3.35 Å, was then increased to 0.4965 $\mathrm{at}/\text{Å}^3$, so that the atomic density of carbon atoms again corresponds to graphene. In the interpretation of the results, however, it must be taken into account that the first layer consists of only 66.7% carbon followed by two layers consisting solely of vacancies. In addition, not only collisions with the atoms that can hit directly but also collisions with nearest neighbor and next-nearest neighbor atoms (so-called weak collisions) were taken into account.

The He ion energies 20, 40, 60, 80 and 100 eV were simulated with fluences of both $5 \times 10^{14}\ \mathrm{at/cm^2}$ and $1 \times 10^{15}\ \mathrm{at/cm^2}$. The results in the composition of the first nanometer of the sample are shown exemplary in Figure 3. It is shown that the carbon content in the first layer decreases after implantation and is replaced by either vacancies, He or atoms from the substrate. In this extreme case, the carbon concentration drops by 10.9% from 66.7% to 59.4%. In addition, it can be observed that some of the carbon atoms can be incorporated into the substrate by recoil formation and that the uppermost layers of the substrate are also damaged, seen in a increase in vacancies and loss of Si and O. This shows that already at these low energies a certain amount of recoils are generated which are incorporated into the underlying layers, and therefore the application of a capping layer to slow down the ions can lead to further damage of 2D materials and to incorparation of unwanted foreign atoms. The trend of the carbon decrease with higher energy is shown in Figure 4. It can be noticed that with higher He energy and higher fluence, more carbon is removed from the first layer. The missing carbon content in the first layer in relation to the undamaged graphene can be comprehended as a measure of damage. Here the simulation confirms the higher defect density at higher implantation energies. In addition, it can be seen that in this energy regime, the fluence has a significantly greater influence on damage formation. Therefore, it is not surprising that the Raman spectrum shifts toward amorphous carbon for the fluence used in the experiment. The damage is caused on the one hand by the creation of vacancies and on the other hand by the incorporation of free atoms, either He or atoms of the substrate, which can be incorporated into the uppermost layers due to sputtering processes.

Figure 5 shows the fractions of introduced He, vacancies, Si, and O from the substrate into the top graphene layer after implantations versus implantation energy. For He (Figure 5a) it can be seen that a maximum is reached at 20 eV. This is due to the fact that at higher energies the He is implanted deeper into the sample and thus penetrates the graphene layer more and more. For the vacancies (Figure 5b) it can be seen that the vacancies first increase from the initial value of 33.3%. Subsequently, the concentration falls again, with higher decrease with higher fluence, so that initially at the fluence of $1 \times 10^{15}\ \mathrm{at/cm^2}$ and energies of up to just below 60 eV more vacancies are generated. At higher energies, more vacancies were generated in the top layer for the lower fluence of $5 \times 10^{14}\ \mathrm{at/cm^2}$. For Si (Figure 5c) and O (Figure 5d) it can be seen that with increasing energy the fraction of these substrate atoms in the graphene layer also increases, also more pronounced with higher fluence. The decrease in the vacancy concentration with a simultaneous decrease in carbon can be explained by the introduction of Si and O into the graphene layer, with O having a significantly greater effect. This effect is more pronounced at higher energies and can be explained by the scattering kinematics. When He hits Si or O, the resulting recoil can only go in the forward direction. In order to be incorporated into the graphene, a further scattering of the generated recoil with a substrate atom must subsequently take place. To have enough energy for the second scattering in the cascade, the first collision of He with Si or O must transfer sufficient high energy, thus this process becomes more likely with higher He energies. Another process is the backscattering of He in deeper substrate layers and the subsequent collision of He with Si or O on its way back. In this way, Si or O is also scattered towards the surface and can be incorporated there. The proportion to which these processes take place depends on the He energy. Thus, the process via the cascade has a fraction of 83% (17% via a direct collision with backscattered He) at 100 eV and of 53% (47% via a direct collision with backscattered He) at 20 eV. The energy transfer to C, O and Si can be calculated by the kinematic factor and has its maximum at 180° backscattering angle. Under these conditions, the maximum transferred energy is 75% for C, 64% for O and 43.75% for Si. Due to the $1/E^2$ dependency of the backscattering cross section, where E is the projectile energy, the high backscattering yield is reasonable in the ultra-low energy regime. The fact that more O than Si is incorporated into the graphene is due to the double O concentration in the substrate (SiO_2) as well as to the lower mass of O. Therefore the backscattering in the second collision of the cascade is more probable.

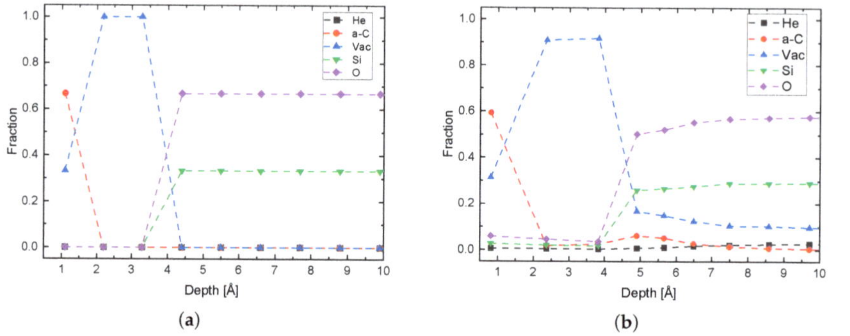

Figure 3. Start condition of the IMINTDYN simulation of graphene on SiO_2 (**a**) and after He implantation at 100 eV with a fluence of 1×10^{15} at/cm² (**b**). After the implantation, damage to the graphene due to He, vacancies and substrate atoms is clearly visible in the top atomic layer.

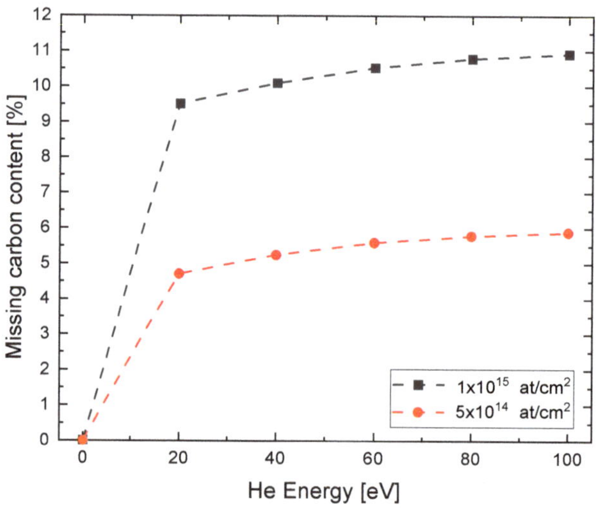

Figure 4. Values of missing carbon content in the top layer versus He implantation energy from the IMINTDYN simulations. A clear trend towards higher damage to the graphene at higher He energy is becoming apparent.

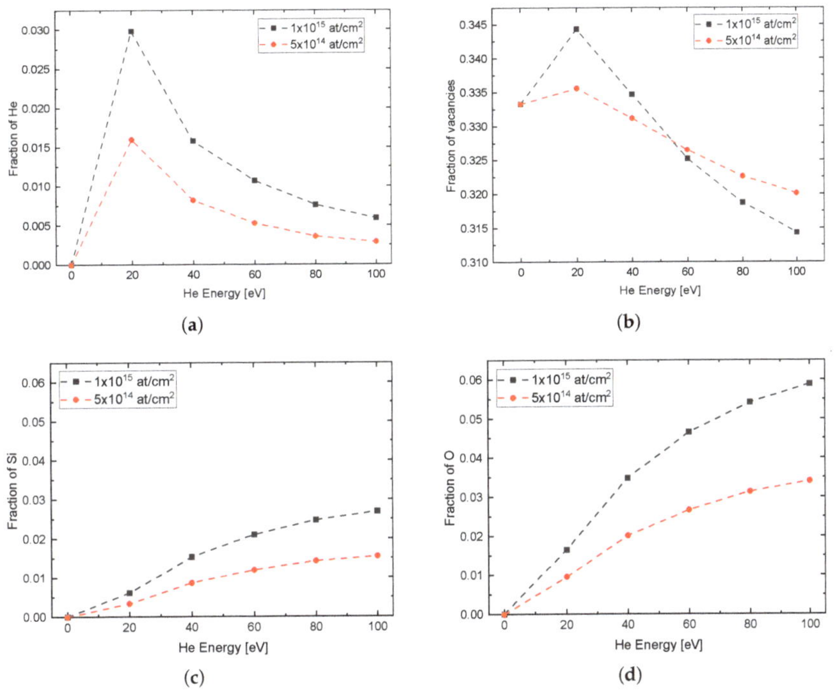

Figure 5. Concentrations of impurity atoms (**a**) He, (**b**) vacancies, (**c**) Si, and (**d**) O in the top graphene layer after implantation of He with different energies and fluences of 1×10^{15} at/cm² and 5×10^{14} at/cm².

4. Conclusions

We have shown that, by using electrostatic masks and potential differences, graphene can be doped with different fluences in only one implantation step. In addition, by adjusting the deflection voltage, the transition region between the doped and un-doped regions can be set. This is a great step towards the production of electronical graphene devices such as transistors using ion implantation. By being able to implant a sample at different energies at different surface positions, future measurements and investigations can be accelerated and eliminates unintended experimental variations in sample preparation as only one sample needs to be implanted instead of multiple.

These methods can be transferred to other 2D materials such as TMDs and therefore can be helpful for a better investigation and manipulation of these materials. Further experiments to sharpen the transition zone are planned. The simulations carried out can confirm the findings from the experiments and are therefore a valuable tool for predicting the processes during ultra-low energy ion implantation of 2D materials.

Author Contributions: Conceptualization, F.J., M.A. and H.H.; methodology, F.J.; software, F.J. and H.H.; validation, F.J., H.H. and Z.Z.; formal analysis, F.J.; investigation, F.J., M.A. and Z.Z.; resources, H.H.; data curation, F.J. and Z.Z.; writing—original draft preparation, F.J.; writing—review and editing, H.H.; visualization, F.J.; supervision, H.H.; project administration, H.H.; funding acquisition, H.H. All authors have read and agreed to the published version of the manuscript.

Funding: This research was funded by the DFG in the project DFG-G Ho 1125/29- and the Volkswagen Foundation. We acknowledge support by the Open Access Publication Funds of the Göttingen University.

Acknowledgments: We thank Marius Herold for his assistance in using the AFM for the SKPM measurements and Steven de Feyter for using his Raman setup.

Conflicts of Interest: The authors declare no conflict of interest.

References

1. Shangguan, Q.; Chen, Z.; Yang, H.; Cheng, S.; Yang, W.; Yi, Z.; Wu, X.; Wang, S.; Yi, Y.; Wu, P. Design of ultra-narrow band graphene refractive index sensor. *Sensors* **2022**, *22*, 6483. [CrossRef] [PubMed]
2. Shangguan, Q.; Zhao, Y.; Song, Z.; Wang, J.; Yang, H.; Chen, J.; Liu, C.; Cheng, S.; Yang, W.; Yi, Z. High sensitivity active adjustable graphene absorber for refractive index sensing applications. *Diam. Relat. Mater.* **2022**, *128*, 109273. [CrossRef]
3. Dastgeer, G.; Shahzad, Z.M.; Chae, H.; Kim, Y.H.; Ko, B.M.; Eom, J. Bipolar junction transistor exhibiting excellent output characteristics with a prompt response against the selective protein. *Adv. Funct. Mater.* **2022**, *32*, 2204781. [CrossRef]
4. Zhang, Z.; Cai, R.; Long, F.; Wang, J. Development and application of tetrabromobisphenol a imprinted electrochemical sensor based on graphene/carbon nanotubes three-dimensional nanocomposites modified carbon electrode. *Talanta* **2015**, *134*, 435–442. [CrossRef] [PubMed]
5. Dastgeer, G.; Afzal, A.M.; Jaffery, S.H.A.; Imran, M.; Assiri, M.A.; Nisar, S. Gate modulation of the spin current in graphene/wse2 van der waals heterostructure at room temperature. *J. Alloy. Compd.* **2022**, *919*, 165815. [CrossRef]
6. Cheng, Z.; Liao, J.; He, B.; Zhang, F.; Zhang, F.; Huang, X.; Zhou, L. One-step fabrication of graphene oxide enhanced magnetic composite gel for highly efficient dye adsorption and catalysis. *ACS Sustain. Chem. Eng.* **2015**, *3*, 1677–1685. [CrossRef]
7. Faccio, R.; Fernández-Werner, L.; Pardo, H.; Goyenola, C.; Ventura, O.N.; Mombrú, I.W. Electronic and structural distortions in graphene induced by carbon vacancies and boron doping. *J. Phys. Chem. C* **2010**, *114*, 18961–18971. [CrossRef]
8. Bhattacharya, B.; Sarkar, U. The effect of boron and nitrogen doping in electronic, magnetic, and optical properties of graphyne. *J. Phys. Chem. C* **2016**, *120*, 26793–26806. [CrossRef]
9. Bangert, U.; Pierce, W.; Kepaptsoglou, D.M.; Ramasse, Q.; Zan, R.; Gass, M.H.; Van den Berg, J.A.; Boothroyd, C.B.; Amani, J.; Hofsäss, H. Ion implantation of graphene—Toward ic compatible technologies. *Nano Lett.* **2013**, *13*, 4902–4907. [CrossRef]
10. Guo, B.; Liu, Q.; Chen, E.; Zhu, H.; Fang, L.; Gong, J.R. Controllable n-doping of graphene. *Nano Lett.* **2010**, *10*, 4975–4980. [CrossRef]
11. Wehling, T.O.; Novoselov, K.S.; Morozov, S.V.; Vdovin, E.E.; Katsnelson, M.I.; Geim, A.K.; Lichtenstein, A.I. Molecular doping of graphene. *Nano Lett.* **2008**, *8*, 173–177. [CrossRef]
12. Åhlgren, E.; Kotakoski, J.; Krasheninnikov, A. Atomistic simulations of the implantation of low energy boron and nitrogen ions into graphene. *Phys. B Condens. Matter Mater. Phys.* **2011**, *83*, 115424. [CrossRef]
13. Prucnal, S.; Hashemi, A.; Ghorbani-Asl, M.; Hübner, R.; Duan, J.; Wei, Y.; Sharma, D.; Zahn, D.R.T.; Ziegenrücker, R.; Kentsch, U.; et al. Chlorine doping of mose2 flakes by ion implantation. *Nanoscale* **2021**, *13*, 5834–5846. [CrossRef]
14. Petrović, M.; Rakić, I.Š.; Runte, S.E.A. The mechanism of caesium intercalation of graphene. *Nat. Commun.* **2013**, *4*, 2772. [CrossRef]
15. Kepaptsoglou, D.; Hardcastle, T.P.; Seabourne, C.R.; Bangert, U.; Zan, R.; Amani, J.A.; Hofsäss, H.; Nicholls, R.J.; Brydson, R.M.D.; Scott, A.J.; et al. Electronic structure modification of ion implanted graphene: The spectroscopic signatures of p- and n-type doping. *ACS Nano* **2015**, *9*, 11398–11407. [CrossRef]
16. Willke, P.; Amani, J.A.; Sinterhauf, A.; Thakur, S.; Kotzott, T.; Druga, T.; Weikert, S.; Maiti, K.; Hofsäss, H.; Wenderoth, M. Doping of graphene by low-energy ion beam implantation: Structural, electronic, and transport properties. *Nano Lett.* **2015**, *15*, 5110–5115. [CrossRef]
17. Lin, P.-C.; Villarreal, R.; Achilli, S.; Bana, H.; Nair, M.N.; Tejeda, A.; Verguts, K.; De Gendt, S.; Auge, M.; Hofsäss, H.; et al. Doping graphene with substitutional mn. *ACS Nano* **2021**, *15*, 5449–5458. [CrossRef]
18. Bui, M.N.; Rost, S.; Auge, M.; Tu, J.-S.; Zhou, L.; Aguilera, I.; Blügel, S.; Ghorbani-Asl, M.; Krasheninnikov, A.V.; Hashemi, A.; et al. Low-energy se ion implantation in mos_2 monolayers. *npj 2D Mater. Appl.* **2022**, *6*, 42. [CrossRef]
19. Lin, P.-C.; Villarreal, R.; Bana, H.; Zarkua, Z.; Hendriks, V.; Tsai, H.-C.; Auge, M.; Junge, F.; Hofsäss, H.; Tosi, E.; et al. Thermal annealing of graphene implanted with mn at ultralow energies: From disordered and contaminated to nearly pristine graphene. *J. Phys. Chem. C* **2022**, *126*, 10494–10505. [CrossRef]
20. Villarreal, R.; Lin, P.-C.; Zarkua, Z.; Bana, H.; Tsai, H.-C.; Auge, M.; Junge, F.; Hofsäss, H.; Tosi, E.; De Feyter, S.; et al. Bond defects in graphene created by ultralow energy ion implantation. *Carbon* **2023**, *203*, 590–600. [CrossRef]
21. Auge, M.; Junge, F.; Hofsäss, H. Laterally controlled ultra-low energy ion implantation using electrostatic masking. *Nucl. Instrum. Methods Phys. Res. Sect. B Beam Interact. Mater. Atoms* **2022**, *512*, 96–101. [CrossRef]
22. Uhrmacher, M.; Hofsäss, H. Ion accelerator facilities at the university of göttingen. *Nucl. Instrum. Methods Phys. Res. Sect. B Beam Interact. Mater. Atoms* **2005**, *240*, 48–54. [CrossRef]
23. Hofsäss, H.; Binder, H.; Klumpp, T.; Recknagel, E. Doping and growth of diamond-like carbon films by ion beam deposition. *Diam. Relat. Mater.* **1994**, *3*, 137–142. [CrossRef]
24. Junge, F.; Auge, M.; Hofsäss, H. Sputter hot filament hollow cathode ion source and its application to ultra-low energy ion implantation in 2d materials. *Nucl. Instrum. Methods Phys. Res. Sect. B Beam Interact. Mater. Atoms* **2022**, *510*, 63–68. [CrossRef]
25. Villarreal, R.; Lin, P.-C.; Faraji, F.; Hassani, N.; Bana, H.; Zarkua, Z.; Nair, M.N.; Tsai, H.-C.; Auge, M.; Junge, F.; et al. Breakdown of universal scaling for nanometer-sized bubbles in graphene. *Nano Lett.* **2021**, *21*, 8103–8110. [CrossRef]
26. Manura, D.; Dahl, D. Simion 8.1 Users Manual (Adaptas Solutions, LLC, Palmer, MA 01069). 2008. Available online: https://simion.com/info/manual.html (accessed on 18 January 2023).

27. Shahzadi, M.; Nisar, S.; Kim, D.-K.; Sarwar, N.; Rasheed, A.; Ahmad, W.; Afzal, A.M.; Imran, M.; Assiri, M.A.; Shahzad, Z.M.; et al. Highly efficient, non-covalent functionalization of cvd-graphene via novel pyrene-based supporter construct. *Chemosensors* **2023**, *11*, 83. [CrossRef]
28. Hang, S.; Moktadir, Z.; Mizuta, H. Raman study of damage extent in graphene nanostructures carved by high energy helium ion beam. *Carbon* **2014**, *72*, 233–241. [CrossRef]
29. Xu, Y.; Zhang, K.; Brüsewitz, C.; Wu, X.; Hofsäss, H.C. Investigation of the effect of low energy ion beam irradiation on mono-layer graphene. *AIP Adv.* **2013**, *3*, 072120. [CrossRef]
30. Hofsäss, H.; Stegmaier, A. Binary collision approximation simulations of ion solid interaction without the concept of surface binding energies. *Nucl. Instrum. Methods Phys. Res. Sect. B Beam Interact. Mater. Atoms* **2022**, *517*, 9–62. [CrossRef]
31. Mutzke, A.; Schneider, R.; Eckstein, W.; Dohmen, R.; Schmid, K.; von Toussaint, U.; Bandelow, G. *SDTrimSP version 6.00, ipp 2019-02*; Tech. rep.; MPI for Plasma Physics: Garching, Germany, 2019.
32. Hofsäss, H.; Zhang, K.; Mutzke, A. Simulation of ion beam sputtering with sdtrimsp, tridyn and srim. *Appl. Surf. Sci.* **2014**, *310*, 134–141. [CrossRef]
33. Hofsäss, H.; Bobes, O.; Zhang, K. Argon ion beam induced surface pattern formation on si. *J. Appl. Phys.* **2016**, *119*, 035302. [CrossRef]
34. Hofsäss, H.; Bradley, R.M. Tests of the modified sigmund model of ion sputtering using monte carlo simulations. *J. Appl. Phys.* **2015**, *117*, 174310. [CrossRef]
35. Özçelik, V. O.; Gurel, H.H.; Ciraci, S.; Self-healing of vacancy defects in single-layer graphene and silicene. *Phys. Rev. B* **2013** , *88*, 045440. [CrossRef]

MDPI AG
Grosspeteranlage 5
4052 Basel
Switzerland
Tel.: +41 61 683 77 34

Nanomaterials Editorial Office
E-mail: nanomaterials@mdpi.com
www.mdpi.com/journal/nanomaterials

www.ingramcontent.com/pod-product-compliance
Lightning Source LLC
Chambersburg PA
CBHW041351210926
51850CB00013B/3045